LAW IN THE LABORATORY

Law in the Laboratory

A Guide to the Ethics of Federally Funded Science Research

ROBERT P. CHARROW

THE UNIVERSITY OF CHICAGO PRESS Chicago and London

ROBERT P. CHARROW is a Washington, D.C., lawyer and former law professor who has served as principal deputy general counsel of the U.S. Department of Health and Human Services, counsel to a president's re-election campaign, a member of the Secretary's Advisory Committee on Organ Transplantation, and vice chair of the Clinical Research Interest Group of the Health Law Section of the American Bar Association. He currently serves on the Board of Advisors for the Institute of Human Virology at the University of Maryland.

The University of Chicago Press, Chicago 60637
The University of Chicago Press, Ltd., London

Printed in the United States of America

19 18 17 16 15 14 13 12 11 10   1 2 3 4 5

ISBN-13: 978-0-226-10164-4 (cloth)
ISBN-10: 0-226-10164-9 (cloth)
ISBN-13: 978-0-226-10165-1 (paper)
ISBN-10: 0-226-10165-7 (paper)

Library of Congress Cataloging-in-Publication Data

Charrow, Robert.
Law in the laboratory : a guide to the ethics of federally funded science research / Robert P. Charrow.
p. cm.
Includes index.
ISBN-13: 978-0-226-10164-4 (cloth: alk. paper)
ISBN-10: 0-226-10164-9 (cloth: alk. paper)
ISBN-13: 978-0-226-10165-1 (pbk.: alk. paper)
ISBN-10: 0-226-10165-7 (pbk.: alk. paper)
1. Research—Moral and ethical aspects. 2. Research—Law and legislation. 3. Science—Research grants—Handbooks, manuals, etc. 4. Bioethics. I. Title.
Q180.55.M67C47 2010
174′.95072—dc22

2009039094

♾ The paper used in this publication meets the minimum requirements of the American National Standard for Information Sciences—Permanence of Paper for Printed Library Materials, ANSI Z39.48-1992.

Contents

Preface

This book is about the legal and ethical norms that govern federally funded research. It is about quid pro quos—what the government expects in return for promising to fund your research, and the laws that are used to ensure that all parties keep their promises. This book is not aimed at lawyers, but rather at scientists—faculty and graduate students. In fact, for many graduate students, buying this book may be part of a quid pro quo. Graduate students who receive funding under a National Institutes of Health training grant are required by the government, as a condition of that funding, to take a course that covers certain topics related to the responsible conduct of research (e.g., research integrity, conflicts of interest).[1] This same type of requirement pertains to institutions seeking National Science Foundation funding on or after October 1, 2009.[2]

For those who are experienced in running a laboratory—getting grants, battling the administration for space, recruiting and mentoring graduate students and postdocs—this book can be used as a reference manual. For those who want to learn how federal legislation and regulations affect laboratory research or for those who are indifferent but are required to take a course on "research ethics," this book is designed to take you through each of the various areas of the law that affect research.

This book tries to bridge two disciplines—law and science. The first chapter gives a glimpse into the workings of the legal mind and our legal system. It examines, albeit briefly, how lawyers approach and analyze problems, how funding agencies issue rules, and how the courts operate to resolve disputes. At first, I was reluctant to include a chapter of this type because this is not a book for lawyers, nor is it a book about jurisprudence. I decided to include the chapter because I have found that many disputes between scientists and university administrators turn on misunderstandings about how the law functions and how one should approach an inherently legal problem.

The second chapter focuses on funding. It examines the legal nature of a grant as well as other funding mechanisms, and how agencies get and distribute their research dollars. The third chapter explores the seedier side of science—research misconduct. It presents a history of how and why the federal government got into the business of policing science, what constitutes research misconduct, and how allegations of misconduct are adjudicated at the institution and by the government. The fourth chapter provides an overview of the laws regulating research involving human subjects, including the privacy rules that pertain to human specimens. The fifth chapter deals with financial conflicts of interest, and the sixth chapter focuses on protecting your data and your obligations to share data and specimens with others. The seventh chapter provides an overview of the laws of intellectual property—patents and copyrights—and some practical considerations when licensing inventions. The final chapter examines the rules governing animals used in research. Each chapter, except the initial chapter, ends with one or more problems that you should reason through; many of the problems are taken from real cases or actual incidents. A guide to analyzing each of these problems is in appendix D.

At times, you will note that my own biases creep into the text. Whenever that occurs, I have tried to make it obvious that I am expressing my views as opposed to presenting legal gospel.

Whether you use this book as a manual or a text, I am always interested in receiving comments that can improve the book.

Many of my friends, colleagues, and acquaintances were kind enough to review portions or all of one or more chapters and provide comments, give criticisms, and spot errors, as follows (in alphabetical order): Reid Adler (Executive Vice-President and Chief Legal Officer, Correlogic Systems, Inc., and former director, Office of Technology Transfer, NIH), Dr. Paul Coates (Office of the Director, NIH), Hon. Arthur J. Gajarsa (Circuit Judge, United States Court of Appeals for the Federal Circuit), Dr. Daniel Kevles (Stanley Woodward Professor of History, Yale University), Dr. Robert Lederman (National Heart, Lung, and Blood Institute, NIH), Jack Kress (former Special Counsel for Ethics and Designated Agency Ethics Official, Department of Health and Human Services), Dr. Michael Saks (Regents Professor of Law and Psychology, Arizona State University), Dr. Robert L. Smith (Altarum/ERIM Russell D. O'Neal Professor of Engineering at the University of Michigan and currently Program Director for Operations Research, National Science Foundation), Mark Solomons (Shareholder, Greenberg Traurig, LLP), and Dr. Jon Strauss (President Emeritus, Harvey Mudd College and Member of the National Science Board).

Special thanks to the ultimate reviewer, my wife of forty years, Dr. Veda Charrow, who cut and edited with gusto.

This book would be incomplete without acknowledging the late C. P. Snow and the late Prof. Ted Waldman (Harvey Mudd College). C. P. Snow, in his Strangers and Brothers novels, first made me aware, as a college freshman, that scientists, like others, were bound by the ethical norms of their society. Ted Waldman, in his philosophy courses, helped me appreciate the complexities of the intersection of science and ethics.

Naturally, I take full responsibility for any errors, biases, and infelicities that remain. The views expressed in this book are mine and do not necessarily reflect the views of any reviewer, the law firm of which I am member, Greenberg Traurig, LLP, or any of its clients.

NOTES

1. *See* Ruth L. Kirschstein National Research Service Award (NRSA) Institutional Research Training Grants (T32), PA-08-226 (Aug. 1, 2008), *available at* http://grants.nih.gov/grants/guide/pa-files/PA-08-226.html; Required Education in the Protection of Human Research Participants, Notice NIH OD-00-038 (June 5, 2000) (revised Aug. 25, 2000), requiring all NIH principal investigators and other senior researchers to take courses on protecting human subjects if their research involves humans.

2. *See* 74 Fed. Reg. 8818 (Feb. 26, 2009) implementing section 7009 of the America Creating Opportunities to Meaningfully Promote Excellence in Technology, Education, and Science (COMPETES) Act, Pub. L. No. 110-69, 121 Stat. 572 (2007) (codified at 42 U.S.C. § 1862o-1), which provides as follows:

> The Director [of the National Science Foundation] shall require that each institution that applies for financial assistance from the Foundation for science and engineering research or education describe in its grant proposal a plan to provide appropriate training and oversight in the responsible and ethical conduct of research to undergraduate students, graduate students, and postdoctoral researchers participating in the proposed research project.

CHAPTER 1

Observations about the Law, the Legal Mind, and Our Legal System

THE NATURE OF THE LAW AND LEGAL THOUGHT IN A NANOSHELL

This book crosses two disciplines—science and law. The legal and the scientific minds have many things in common, and their approaches to problems are similar in many respects. Scientists and lawyers both strive to state the problem precisely, to gather whatever facts there might be, and to apply the appropriate scientific or legal principles to those facts to yield a solution. And both scientists and lawyers tend to be critical thinkers; they ask lots of questions to better understand what is happening.

Both scientists and lawyers attempt to synthesize a "rule" from observation. A scientist might call his "rule" a hypothesis and then set about testing it to see if it holds up. A lawyer might look to a group of cases dealing with similar situations and attempt to tease out of those court opinions a single legal principle that explains all of the courts' decisions in that group. Scientists and lawyers use their theories and principles, respectively, to predict the future. For example, a physicist might use the laws of motion to determine how long it will take for an apple to hit the ground when dropped by a fraternity pledge from the top of a university tower. A lawyer might use tort law and criminal law to tell her client, a college senior, what might happen to his fraternity if that apple were to hit a passerby.

There are, though, salient differences between the two disciplines. Some differences are obvious. For example, lawyers tend to be wordsmiths and are usually highly attuned to subtle differences in the meanings of words, phrases, or sentences. Other differences may not be so obvious. For example, the law is performative. If a court declares that someone is guilty or that someone is dead or that someone is divorced, those individuals are legally guilty, dead, and divorced, respectively. The external truth may not be relevant. Can you imagine a modern scientist declaring, based on his beliefs, that the human body has four humors and the key to good health is keeping them in balance? The law is also more concerned with setting up mechanisms for resolving disputes in socially acceptable ways than it is with establishing the truth. Finally, lawyers represent clients, and their job

Administrative Agencies Make Laws

How have regulations become such a prominent feature of our government? Some of this phenomenon can be traced to a curious crime committed more than 125 years ago where the National Gallery is now located in Washington, D.C. A former academic, who had developed one of the more ingenious proofs of the Pythagorean Theorem, was about to board a train for a reunion at Williams College. Before he could board the train, he was shot and killed. The crime would likely have engendered little public attention were it not for the fact that the victim, James Garfield, was president of the United States. The assailant, Charles J. Guiteau, was a disappointed and delusional office seeker who had failed to obtain a political appointment in the Garfield administration. Up until then, most government employees were political appointees, from the cabinet officers down to the lowly clerks. Garfield's assassination changed things. It led Congress to pass the Pendleton Civil Service Reform Act in 1883.

The Pendleton Act did two things. It created the first independent agency, the United States Civil Service Commission, and it created a career bureaucracy independent of who was in power. It is ironic that the first independent agency was one that was charged with generating the career civil servants who would later populate new agencies as they cropped up. Sure enough, agencies did crop up; slowly at first, and then with increasing speed. Five years after the creation of the Civil Service Commission, Congress created the Interstate Commerce Commission to regulate what railroads could charge, and thus began the proliferation of independent agencies. During the next fifty years, the number of federal agencies exploded. During the 1930s, it seemed that a week couldn't pass without some new agency and its acronym popping up.

But who controlled what the agencies did? Interestingly, these regulating agencies were themselves largely unregulated. Academics, including most notably Roscoe Pound, dean of the Harvard Law School and a botanist (not a lawyer), began questioning the wisdom of entrusting a significant slice of our government to unelected regulators who did their work in secret and were answerable to no one. With the end of both the Second World War and the Great Depression, Congress enacted the Administrative Procedure Act (APA) to regularize the operations of all administrative agencies, including cabinet departments; non-independent agencies, such as the National Institutes of Health (NIH), the Food and Drug Administration (FDA), and the Defense Advanced Research Projects Agency (DARPA); and independent agencies, for example, the National Science Foundation (NSF). The APA is the "constitution" of modern administrative law and has remained relatively unchanged over the past sixty years.

The APA divides the work of federal agencies into two broad categories—adjudications and regulations. An adjudication is a trial before an administrative agency aimed at resolving a past dispute. Many agencies have multitiered hearing processes, sometimes starting with a hearing before an administrative law judge (ALJ) and ending with a review by the head of the agency. For example, if a state feels that it has been shortchanged by the Medicaid program, it can seek a hearing before the Departmental Appeals Board. If someone believes that he or she was improperly denied Social Security benefits, that person can ask for a trial before an ALJ in the Social Security Administration. And if the holder of a television license believes that the Federal Communications Commission (FCC) should not revoke its license, it can ask for a hearing before an ALJ in the FCC. The hearing is usually between an individual or company and the agency itself.

Whereas adjudications are usually retrospective, the second category, creating regulations, or rulemaking, is prospective. Rulemaking is one of the most important prerogatives of an agency. The APA does a number of things, including defining what a regulation is and what an agency must do before it can issue one. Regulations come in two varieties—big ones and little ones. A big rule really makes law by filling in a vague congressional outline. The rules that define clean air and clean water or that give life to the Medicare prescription drug program are big rules, technically called legislative rules, as are the rules governing scientific misconduct, human subjects, and financial conflicts of interest. By contrast, a little rule, called an interpretive rule, does not really change much. For example, suppose that Congress enacts a detailed funding program for a certain type of research. The funding agency then issues legislative rules fleshing out the program and specifying that institutions that wish to apply for funding must do so within thirty calendar days after announcement appears in the *Federal Register*. However, the agency neglected to define whether the thirty calendar days would be satisfied if the proposal were mailed on the thirtieth day or received by the agency on the thirtieth day. The agency publishes a rule indicating that the agency must receive the proposal within thirty days. This latter issuance would be an interpretive rule because it clarifies an existing rule and addresses only a relatively small aspect of the overall program. Tomes have been written about the differences between a big rule and a little rule. Big rules make it into the Code of Federal Regulations; little rules rarely do. Big rules limit an agency's discretion by compelling agency personnel to act in a certain way; little rules do not. Big rules also have significant effects on those outside government; little rules usually do not.

But why all the concern? Should anyone care? After all, most practicing lawyers probably could not tell you the difference between a legislative rule and an interpretive one. You should care because most of your research is funded by federal agencies, and in many instances these agencies will attempt to regulate your research or your funding through rules or issuances, some of which may have been issued in the wrong way. An agency can issue a big rule only in certain limited ways; it can issue a little rule almost any way it wishes. If an agency misclassifies a legislative rule as an interpretive one and fails to adhere to the necessary formalities, a court is obligated to strike the rule down.

Most legislative rules are issued through what is called notice-and-comment rulemaking. The agency first publishes its rule as a proposed rule, which, among other things, sets out the agency's rationale for the rule and the options that it considered and rejected; the agency then solicits public comment. Some proposed rules are so technical and affect so few persons that they attract relatively few comments. Other proposed rules, though, may be considerably more controversial and can attract tens of thousands of comments.

Notice-and-comment rulemaking is time consuming and requires those in a bureaucracy to jump through many hoops. Many agencies, including NIH and FDA, do not even have independent authority to issue legislative rules; that authority rests with the Secretary of Health and Human Services, and the Secretary does not have authority to issue legislative rules without the approval of the Office of Management and Budget within the White House. Given the hassle associated with issuing big rules, there is a tendency within the bureaucracy to misclassify legislative rules as interpretive ones, thereby avoiding all the sign-offs, approvals, and analyses of comments. Misclassification is not a theoretical concern, especially within NIH. For example, the NIH rules governing research involving recombinant DNA appear to have been issued without notice-and-comment rulemaking even though they are legislative rules, but more about that later.

A recent case involving Yale University's hospital illustrates the significance of the distinction between legislative and interpretive rules. Physicians at Yale-New Haven Hospital had been conducting a clinical trial of a new implantable cardioverter defibrillator (ICD); the purpose of the trial was to gather data that the manufacturer could submit to FDA to gain approval of the device. Of the Yale-New Haven patients enrolled in the trial, forty-eight were Medicare beneficiaries (e.g., over sixty-five years of age). Medicare, however, refused to pay for their hospitalization and treatment because the ICD was "experimental" and had not been approved by FDA. Previously, Medicare had made these coverage decisions on a case-by-case basis, independently weighing the safety and effectiveness of the new unapproved device. However, before the clinical trial started at Yale, Medicare published a statement in one of its manuals that it would reimburse hospitals only if the medical device had been approved by FDA. It would no longer do a post hoc case-by-case assessment. Yale-New Haven Hospital sued, arguing that Medicare's new rule requiring FDA approval as a condition of coverage was a legislative rule and because it had not been issued through notice-and-comment rulemaking it was invalid. Medicare argued that it was an interpretive rule and could be issued informally. The trial court agreed with Yale-New Haven and vacated the manual's provision.[5] The appeals court affirmed, but on different grounds.[6]

Courts Make Laws

Our courts also make law. However, unlike Congress, which passes a statute, or an administrative agency, which issues a regulation, courts make law indirectly in the process of resolving a dispute between parties. This is what is referred to as the common law or case law. For example, suppose that you were involved in an automobile accident and were sued by the driver of the other car. He claims that you drove negligently (i.e., not as a reasonably prudent person would have driven under similar circumstances) because you exceeded the speed limit and your excessive speed caused the accident. You acknowledge exceeding the speed limit, but argue that you acted reasonably because everyone drives on that road at the speed you were driving; the posted speed limit is far below the safe speed. The plaintiff believes that he should win automatically because you admitted violating a traffic law. What weight should the violation be given? Should it be dispositive—if you violate a criminal statute or regulation or traffic law, then you are automatically negligent? Or should it merely be some evidence, but not dispositive, that you acted negligently? Or should it count for nothing? Suppose that your case goes up to the highest court in your state, and the court decides that violating a traffic law is merely some evidence of negligence. That court has effectively made law in the course of resolving a dispute between two drivers: Violation of a traffic ordinance can be some evidence of negligence, but it is not dispositive. Lower courts in that state will be obligated to follow the higher court's ruling. It becomes precedent.

When a court makes law, it usually does so to promote some underlying social or legal policy. For example, originally courts did not permit members of the same family to sue one another out of fear that it would undermine the structure of the family. This is called parental and spousal immunity. After the introduction of automobiles and automobile insurance, the policy against intrafamily litigation was strengthened out of concern that family members would collude and sue one another to collect the insurance proceeds. These two social policies—fear of undermining the family structure and fear that the family would be too harmonious and collude to defraud insurers—went in opposite directions, but supported the same rule.

Suppose that are you a trial judge in a state where your supreme court has precluded intrafamily litigation in tort. (Family members are still free to sue for divorce or battle over the proceeds of a will.) Also suppose that a fifteen-year-old girl is suing her father for having sexually molested her; there is no insurance involved. The father has already been convicted of rape and related crimes for having molested his daughter, and he is serving time in the state penitentiary. The father's attorney moves to dismiss the suit because one member of a family cannot sue another member in tort. What should you do? Does applying a rule designed to foster family harmony make sense in this setting? A court would normally look beyond the rule to the underlying social policy that the rule was designed to foster.[7] If applying the rule is inconsistent with that social policy, a court could well decline to apply the rule; by so doing, it would be carving out an exception to the general rule. For example, a court might decline to apply the intrafamily immunity rule where the defendant had committed a criminal act or where the act involved an intentional injury, as opposed to an accidental one. If, over time, the number of exceptions grows large, then eventually the jurisdiction's highest court might reconsider the rule's usefulness or propriety. This is one of the processes by which the common law evolves.

Let us look at another example of how the common law works. Since early times, courts have had little difficulty in dealing with cases involving physical injuries caused by an accident (e.g., broken bones, cracked skulls). Physical injuries were easy to see, document, and prove. Courts, though, have had considerably more difficulty dealing with emotional distress caused by an accident. Courts were concerned that it was too easy to fake emotional distress. As a result, most courts would not allow a plaintiff to recover for emotional distress unless he first suffered a direct physical injury that in turn led to the emotional distress. This was called the impact rule. The requirement of physical injury is designed to ensure that the emotional injury is legitimate.

What happens, though, when physical and emotional injuries take place without an impact? Take the case of Lillian Amaya, who observed her seventeen-month-old child run over by a truck.[8] At the time, Amaya was seven months pregnant. After witnessing her son killed, she became violently ill physically and suffered emotional distress as well. She sued the company that owned the truck, seeking to recover for both her physical and emotional injuries. Here, there was plenty of evidence indicating that the emotional injury was real. Not only was there a physical manifestation of the injury, but the circumstances themselves made fakery highly unlikely. This was the perfect factual setting for a court to reassess the impact rule. Amaya's case went all the way to the California Supreme Court. The California Supreme Court rejected the impact rule and held that a person could recover for emotional distress caused by witnessing a relative being injured, but only if the person were at risk of being physically injured, such as standing in the path of the truck. The court recognized that this narrow exception would not open the floodgates to fraudulent suits. This became known as the zone of danger rule. It did not help Mrs. Amaya because she was not in the zone of danger. Later courts expanded the zone of danger rule, replacing it with a more flexible standard.

The court's holding in *Amaya* would be binding on all state courts in California. This is called binding precedent: The lower courts are bound by the holdings of the higher courts in that same court system. Most precedent, though, is not binding but rather persuasive. Courts in one state may be influenced by how courts in other states have addressed a similar issue, but they are not required to follow those other decisions. Thus, for example, a New York court looking to California might be influenced by *Amaya*, but the New York court is not bound by *Amaya*. If the Supreme Court of the United States decides a given case, however, then all other courts are obligated to follow suit.

At the start of this section, I stated that courts make law, and we have seen some examples of this process. Nevertheless, many legislators and politicians claim that courts should not make law—that lawmaking is the job of Congress or the state legislatures and not the courts. Recently, the Supreme Court actually considered whether courts make law. The case involved a balloon catheter that been approved by FDA. Apparently, during an angioplasty procedure the physician overinflated the balloon and it burst inside the patient's coronary artery. The physician exceeded the limits set in the catheter's labeling. Nonetheless, the patient sued the manufacturer of the catheter, alleging that it been defectively designed and labeled. The catheter manufacturer argued that under federal law a court could not entertain this type of suit because, if it did, it could end up imposing a labeling or manufacturing requirement on the manufacturer that was different from the one imposed on the manufacturer by FDA. In fact, if a court finds that the product was defectively labeled, that means that it should have been labeled differently. But if were labeled differently, then that would conflict with the FDA-approved labeling. The Supreme Court had to evaluate whether court decisions based on state tort law were no different than legislation enacted by a state. Everyone agreed that state legislators could not enact medical device requirements that differed from those imposed by FDA. But what about a court decision? The Court reaffirmed the reality that court decisions based on state law or common law were no different than positive law enacted by a state legislature. The Court went on to hold that the federal law governing approved medical devices preempts most state tort laws, meaning that an injured party cannot, in most instances, sue the manufacturer of an approved device.[9]

Up to now we have been talking about state and federal courts, but really have not discussed how they interact. Usually, there is little interaction between the two systems; they are parallel. Federal courts decide federal issues (e.g., how to interpret a federal statute, whether a federal regulation was properly issued, whether a state acted constitutionally in prohibiting certain types of speech), disputes between citizens of different states (e.g., an automobile accident case between residents of New York and California when the amount in controversy exceeds $75,000), and federal criminal cases. If you sue the federal government, you have to do so in federal court. State courts hear local matters (e.g., state criminal cases, local contract disputes, personal injury cases), but can also hear many federal issues. For example, a person can sue in state court for violations of the federal civil rights laws.

The federal system has three tiers. The United States district court is the trial court; each state has at least one district (with many judges), and many larger states have many districts (e.g., California has four districts, as does New York). Decisions of a district court can be appealed as a matter of right to the United States court of appeals for the circuit in which the district court is located. The nation is divided into twelve geographic circuits with a court of appeals for each circuit. These geographic circuits are numbered 1 through 11; the extra circuit, called the Court of Appeals for the District of Columbia Circuit, sits in Washington, D.C. A disproportionate share of its caseload involves suits against the federal government because you can almost always sue the government in Washington, D.C. The lower numbered circuits, with the exception of the Eleventh Circuit (Atlanta), are in the eastern part of the country; the higher numbered circuits are in the West and Midwest. The United States Court of Appeals for the Second Circuit, for instance, hears appeals from district courts located in New York, Connecticut, and Vermont. There is one circuit with nongeographic jurisdiction: The Court of Appeals for the Federal Circuit located in Washington, D.C., hears patent cases from all over the country and also appeals from certain specialized agencies and courts.

What if you lose in a federal court of appeals? Normally that is the end of it. This is so because further review by the Supreme Court of the United States is discretionary and highly unlikely. The Supreme Court hears about seventy-five cases each year or about 1 percent of those in which review is sought.

While almost anyone can be sued, the federal government is protected from lawsuits by the doctrine of sovereign immunity. However, the government has waived its immunity in many areas. For example, the federal government has agreed to be subject to suit for breach of contract, for certain torts committed by its employees, and for issuing regulations that are inconsistent with law. It can also be sued for violating the Constitution.

LEGAL CITATIONS IN *THE BLUEBOOK*

In writing this book, I have used the legal citation system (called the *Bluebook* System), which is decidedly different from the system used in the sciences or any other discipline. I have done so out of habit, not to make the readers' lives more difficult. But since I have used a citation system foreign to most readers (indeed, foreign to most educated people), at the very least I am obligated to explain how the legal citation system works. I have set out a brief discussion of the legal citation system in appendix A.

NOTES

1. *See* BLACK'S LAW DICTIONARY 1046 (5th ed. 1979). The term "positive law" has many different meanings, including one that is extremely narrow and includes only laws that have been codified into the United States Code. *See* 1 U.S.C. § 204. This definition is so narrow that it excludes binding laws enacted by Congress that have not been compiled into the United States Code, as well as regulations issued by agencies.

2. *See* Pub. L. No. 107-368, 116 Stat. 3034 (2002).

3. *See* Pub. L. No. 108-173, 117 Stat. 2066 (2003).

4. *Compare* 117 Stat. 2072–2131 (2003) *with* 70 Fed. Reg. 4194–4585 (Jan. 28, 2005).

5. Yale-New Haven Hosp. v. Thompson, 162 F. Supp. 2d 54 (D. Conn. 2001).

6. Yale-New Haven Hosp. v. Leavitt, 470 F.3d 71 (2d Cir. 2006).

7. In fact, a case very much like this occurred about one hundred years ago in the state of Washington. The Washington Supreme Court, in one of the more absurd decisions of the century, actually ruled that the girl could not sue her family because it would undermine family harmony. *See* Roller v. Roller, 79 P. 788 (Wash. 1905). Most courts, though, would recognize that applying the rule of parental immunity in that case makes no sense—family harmony had already been destroyed, and there was no likelihood of collusive litigation to garner insurance proceeds. Justice Brachtenbach later observed that this decision "carries the doctrine of the sacredness of the family unit to the most absurd degree yet." Merrick v. Sutterlin, 610 P.2d 891, 893 (Wash. 1980).

8. *See* Amaya v. Home Ice, Fuel & Supply Co., 379 P.2d 513 (Cal. 1963).

9. *See* Riegel v. Medtronic, Inc., 128 S. Ct. 999 (2008).

CHAPTER 2

Government Funding of Research in the United States

HOW DID IT START?

Empirical research, whether in microbiology or high-energy physics, eats money, and the next tranche of funding is always a concern. That concern, though, is somewhat moderated because most funding for basic research comes courtesy of the federal government. The American Association for the Advancement of Science estimates that the government spent about $142.5 billion on research and development (R&D) in FY 2008, of which about $61 billion was allocated to nondefense R&D.[1] Of that amount, the National Institutes of Health (NIH) received about $29.5 billion. Overall, federal and private R&D spending totaled about 2.66 percent of the U.S. gross domestic product (GDP).[2] While this is a significant investment, much higher in terms of dollars than all other nations, the president hopes to increase this to about 3 percent of GDP.[3]

Almost from its inception, the U.S. government has been a major patron of basic and applied research. In 1803, Congress appropriated $2,500 for the Lewis and Clark expedition to explore and chart the western reaches of the American continent. Lewis and Clark actually spent $38,000, or about $488,888 in 2005 dollars—a relatively small investment per unit of knowledge gained. Notwithstanding the Lewis and Clark expedition, early legislators and even academics debated the wisdom and legal propriety of congressional funding for private scientific research. This debate was sparked in part by a clause in the Constitution that authorizes patents and copyrights, as follows: Congress shall have power "To promote the progress of science and useful arts, by securing for limited times to authors and inventors the exclusive right to their respective writings and discoveries[.]"[4] Many argued that this clause, called the patent clause (see chapter 7), was meant to be the exclusive means by which Congress was to promote the arts and sciences—by monopolies and not by money. Patents and copyrights were appropriate; grants were not. Others, less troubled with legal niceties, were concerned that if the government funded science, then the government would control science and that would not be healthy. As the importance of science to technology and industrialization became evident, starting in the 1830s, the philosophical and legal impediments to federal funding began to vanish, and the government started an "on again, off again" relationship with science and science funding. The evolution of federal funding is chronicled in two outstanding books: A. Hunter Dupree's *Science in the Federal Government: A History of Politics and Activities* and Daniel J. Kevles's *The Physicists: The History of a Scientific Community in Modern America.*[5] Suffice it to say that by the end of the Second World War, science funding had become an integral part of the appropriations process, and, following the Soviet launch of the Sputnik satellite in 1957, it took off.

One interesting political phenomenon is the disparity between NIH and the National Science Foundation (NSF) appropriations. In 1960, the NSF budget was about half the NIH budget ($1.2 versus $2.4 billion in 2008 dollars); in 2008 the NSF budget was about one-fifth the NIH budget ($6.1 versus $29.5 billion in 2008 dollars). During this forty-eight-year period, NSF's budget increased by a factor of 5.3 while NIH's increased by a factor of about 12.2, all in constant dollars. Why did the disparity grow? There are many reasons. First, every member of Congress gets sick and some even suffer rare and fatal diseases, the bailiwick of NIH; however, few members of Congress experience a psychotic sociology episode or a paralytic physics event; sociology and physics fall into NSF's area. And breakthroughs in physics, chemistry, or the social sciences do not have the same voter appeal as curing cancer or AIDS.

Second, NIH does a better job of fundraising than does NSF. Aside from the appeal of curing diseases, NIH has another advantage—numbered buildings just waiting to be named. Universities draw large donations by naming buildings after benefactors. NIH, with a large campus, is able to do the same thing. NIH and Congress name buildings after friendly chairs of appropriations committees or subcommittees while they are still alive. Just look at Buildings 33, 35, and 36 on the NIH campus. Building 33, a laboratory complex for the National Institute of Allergy and Infectious Diseases, is now called the C.W. "Bill" Young Center; Bill Young was chair of the House Appropriations Committee. Building 35, the Porter Neuroscience Center, is named after John Porter, former chair of the House Appropriations Subcommittee with jurisdiction over NIH. The NIH budget doubled during Porter's tenure as chair. Building 36, now known as the Lowell P. Weicker Building, is named after a former chair of the Senate Appropriations Subcommittee with jurisdiction over NIH. These are just a few examples of the naming game. By contrast, NSF is in a large office complex in northern Virginia. There is nothing much to name.

Finally, NIH has a more dependent constituency than does NSF. NIH researchers can receive 100 percent of their salaries from NIH grant funds; NSF will normally fund faculty salaries for no more than two months.[6]

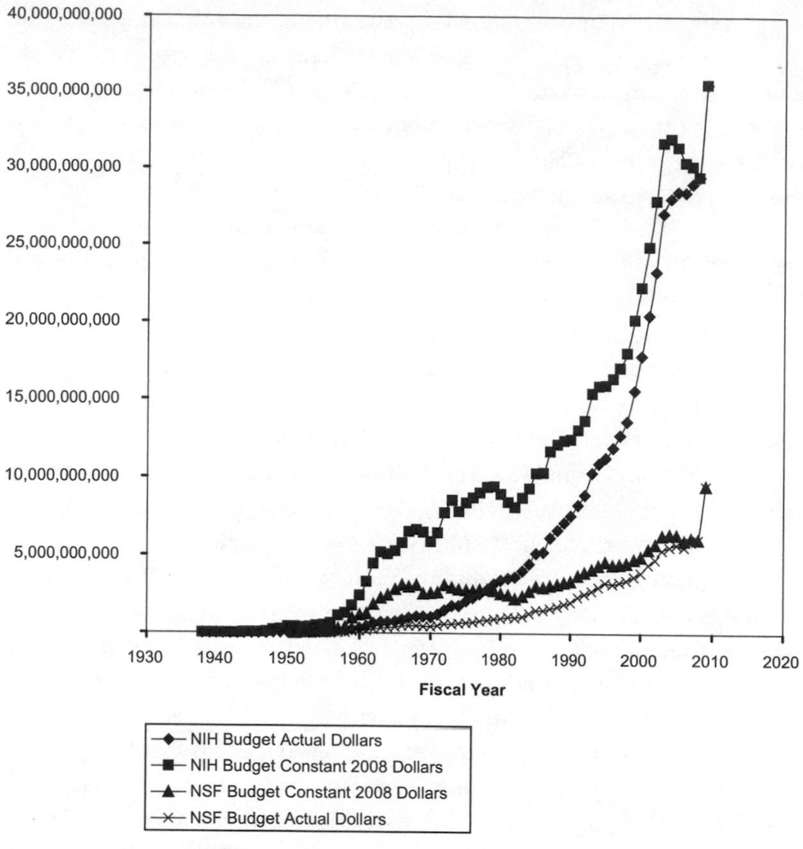

CHART 1. NIH and NSF Budgets, by Fiscal Year

Researchers who are more dependent tend to be more vocal and this is very helpful at appropriations time.

The growing disparity in budgets between the two major science funding agencies may prove counterproductive. Virtually all of the equipment that has fueled the revolution in the biological sciences was developed by physical scientists and engineers, many of whom received their funding from NSF. The budget imbalance could undermine the development of new equipment necessary for future biomedical research.

UNDERSTANDING THE GRANT AS A FUNDING MECHANISM

What Is a Grant?

The federal government spends money in three ways: grants, cooperative agreements, and procurement contracts.[7] All three vehicles are used to fund research. The grant is the mode of choice for funding research and is the closest thing there is to a government gift, but the very concept is a contradiction in terms. Government research grants are difficult to win, difficult to administer, and potentially difficult for a researcher to move from one institution to another. Grants come with terms and conditions that make them extremely unattractive to some, especially large profit-making corporations.

But precisely what is a grant? Is it a contract? Commercial contracts usually involve mutual promises: "I will deliver 100 pounds of top-grade raw cotton to your place of business on or before April 1, 2009, and you agree to pay me $250 cash, at the time of delivery." However, in the typical research grant, the grantee is not obligated to produce positive results; science is not like the ratchet or cotton business, and new knowledge cannot be produced on demand. In that regard, grants are the modern-day equivalent of the patronage system that prevailed throughout Europe in the Renaissance and later to promote science and the arts. So then, what is a grant?

As a historical fluke, one of the early influential Supreme Court cases involved precisely that issue. Our story begins in the 1760s with an event just as relevant to the twenty-first century—a college president in search of additional funding. The college was about to embark on a major fundraising campaign to pay for a new campus, new buildings, more faculty, and the like. It needed a large infusion of capital. The president, Dr. Eleazer Wheelock, decided that fundraising opportunities were far greater in England than in the Colonies, so he dispatched a friend, the Rev. Nathaniel Whitaker, to England to solicit contributions. Whitaker, a man ahead of the times, proved to be a master of the fundraising art. Within a year, Whitaker had surpassed the goals of the college's capital campaign. However, before he was permitted leave England with the cash, it was decided that the gifts and indeed the college itself should be memorialized and formally established, respectively. The largest single benefactor, one William Legge, prevailed upon King George III to establish the college and fund it through a grant. The grant, executed by the King in 1769, created a twelve-person board of trustees to govern the college. The college was named after Legge's title, the Earl of Dartmouth.

Fast forward about 35 years to 1815. The governors of New Hampshire and Vermont wanted to establish a state university for their two states, but did not particularly want to shell out the money to do so. The governor of New Hampshire hit upon a great idea: The two states could take over Dartmouth and transform it into a state university for both states. To accomplish this, New Hampshire enacted a series of laws in 1816 that expanded Dartmouth's board of trustees from twelve to twenty-one members and established a separate board of twenty-five to oversee the board of trustees.

This board of overseers was to be "packed" with state officials from Vermont and New Hampshire. Needless to say, Dartmouth College was not all that pleased with the governors' plans.

Dartmouth College sought to invalidate the attempted takeover by instituting suit against New Hampshire in a New Hampshire state court. It argued that the grant from the King that created the college was an ordinary contract and that the United States Constitution barred states from impairing contacts.[8] At issue was whether the grant was a contract. The New Hampshire Supreme Court held against Dartmouth College, but on appeal, the Supreme Court of the United States, speaking through Chief Justice Marshall, held that the grant was indeed a contract and that New Hampshire's efforts to wrest control from the appointed board of trustees "impaired the obligation of the contract" and were unconstitutional.[9]

Although a grant is a contract, it is no ordinary contract: The relationship between the government and the grantee "cannot be wholly captured by the term 'contract.'"[10] In an ordinary contract, there is some form of "meeting of the minds," usually some ability to negotiate, and some form of discrete deliverable (e.g., tanks, aircraft carriers, refrigerators, a report no one will read). That is not the case with a grant. A federal grant is a statute-based award designed to promote or accomplish some general public purpose. A grant, especially a research grant, usually has no discrete deliverable.

The rights and obligations created by a grant, especially a research grant, tend to be more amorphous than in a typical commercial contract. Thus, the grantee may be entitled to retain the grant funds even if the objectives of the grant are not met. This would not be the case in an ordinary contract. If a party fails to perform as required by the contract, then that party has breached the contract and can be required to return the funds or pay other damages. Research grants tend to have relatively few restrictions whereas other grants (e.g., Medicaid) are laden with extensive restrictions, requirements, reporting obligations, and penalties. Also, as we will see, the process of awarding a grant is considerably different from the process for awarding a procurement contract, sometimes referred to as a government contract. (Procurement contracts are discussed in more detail below.)

One final note on the legal character of grants. The contract nature of grants proved to be pivotal in early federal cases desegregating school systems, as well as in promoting certain environmental interests. For example, federal courts in some of the earlier school desegregation cases ordered desegregation of certain schools by enforcing a grant between the state and Department of Defense under which the Department of Defense agreed to supplement a county's education budget to accommodate students from nearby military bases. The grants required that the counties provide the same quality and type of education required by state law. In many southern states, the old post–Civil War statutes banning segregation were still on the books. They were just ignored. Those old state requirements became part of the grant. Federal courts promoted integration by enforcing those grant provisions.[11] On the environmental side, San Francisco received (and still receives) its water from Yosemite National Park under a grant from the federal government that precludes the city from diverting the water or any electrical power that the city generates with the water. When San Francisco tried to sell its excess electric power to Pacific Gas & Electric for resale, the federal government went to court and had the sale enjoined as inconsistent with the provisions of the grant.[12]

How Are Federal Grants Awarded?

Each federal funding agency has an assortment of different grant types and its own rules for awarding grants. Those rules may vary depending on the type of grant. NIH alone has more than 100 different types of grants, and when one includes cooperative agreements and procurement contracts, the number of funding mechanisms exceeds 190.[13] The grant processes for the Public Health Service agencies (e.g., NIH, Centers for Disease Control and Prevention [CDC]), NSF, Department of Energy (DOE), Defense Advance Research Projects Agency (DARPA), National Institute of Standards and Technology (NIST), the United States Department of Agriculture (USDA), and the National Aeronautics and Space Administration (NASA), differ in detail, but all rely to varying degrees on peer review. I focus on two agencies, NIH and NSF, primarily because those two agencies tend to award the greatest number of grants across the greatest number of disciplines. NASA grants tend to be highly specialized and largely limited to those in astronomy, astrophysics, and geophysics; DARPA and NIST grants are also highly specialized.

NATIONAL INSTITUTES OF HEALTH

NIH has a highly regularized system[14] mandated by statute that involves two steps: an initial review by a study section and then review by the advisory council for the institute that will do the funding. Although NIH has various types of grants, from large program project grants (P01) to single investigator basic research grants (R01), each denoted by an alphanumeric code, this two-step review applies to them all. In this section, I focus on R01 grants. Although there are some differences in the way the various types of grants are awarded, the general principles discussed here apply to all NIH grant types.

Applications for R01 grants, after being received and logged in by the Center for Scientific Review,[15] are assigned to the most appropriate study section,[16] based on subject area, where they are reviewed for scientific merit. There are about 180 regular standing study sections and an additional 230 special-emphasis study sections at NIH. Each study section panel consists of between one and two dozen scientists, usually drawn from academia, with expertise in the panel's area. Panelists usually have had experience as a principal investigator (PI) on a project comparable to those that they will be reviewing. Each study section is led by a Scientific Review Officer (SRO) who is a full-time employee of NIH. Most study sections meet three times a year. Prior to each meeting, the SRO will assign each worthy proposal to three members of the section who will act as the primary reviewers for that proposal. Meeting behind closed doors, the study section reviews and discusses each proposal for scientific merit, with the three primary reviewers taking the lead. Some have raised the concern that "while the three reviewers wrangle over a particular application, others [on the study section] are busy on their laptop computers."[17] Nevertheless, all members of the study section cast secret ballots assigning each proposal a priority score from 1 (outstanding) to 5 (acceptable). The votes are averaged, and the average is converted to a final priority score, ranging from 100 to 500, with the lowest being the best. Thus, an average score of 1.25 would be converted to a priority score of 125. NIH also assigns a percentile ranking to each priority score. Even priority scores of 500 are theoretically deemed "acceptable," although there is virtually no likelihood that a proposal with that score will be funded. About half the proposals received by a study section are deemed not good enough to warrant a full review. These proposals are not assigned a priority score and are not eligible for funding.[18]

After the meeting, the SRO for that study section prepares a "pink sheet," named in the days before photocopying, which summarizes the study section's discussion about that proposal. The pink sheet, priority score, and percentile ranking of proposals that score well enough to be candidates for funding are forwarded to the advisory council for that institute. Each advisory council consists of scientists and nonscientists and is intended to provide representation to all interested groups.

Under the Public Health Service Act, an NIH grant or cooperative agreement (which is treated legally like a grant) with direct costs greater than $50,000 "may be made only if such grant or cooperative agreement has been recommended after technical and scientific review [by a study section] and is recommended . . . by the advisory council for the national research institute involved."[19] Merely because a study section deems a proposal fundable does not mean it will be funded. The overwhelming majority of proposals actually reviewed are in fact deemed fundable, but relatively few, namely those with the lowest priority scores, will get funded. In fact, only those proposals with low priority scores will be forwarded to the council for review and a vote, also behind closed doors. All funding is technically done not by the institute and not by NIH, but rather by the Secretary of Health and Human Services. The Secretary cannot award a grant unless the proposal has been approved by both the study section and the institute's advisory council. The NIH system is very much a pyramid. About 50 to 70 proposals are assigned to each study section, of which about half are deemed sufficiently promising to warrant full study section review. Of those that are reviewed fully, only a fraction are forwarded to the council and even fewer are approved by the council for funding. The percentage of grant applications that are funded has been decreasing. In FY 1998, the overall success rate was about 32.4 percent.[20] In FY 2008, the overall rate had dropped to 21.8 percent across all institutes.[21]

There are two interesting points about the NIH system. First, it is technically unconstitutional because it vests the decision "not to fund" in individuals who are not officials of the government, but merely transient members of an advisory committee; advisory committees are only supposed to advise, not make binding decisions.[22] Even though the system is unconstitutional, it is unlikely that anyone would ever have legal standing to challenge the constitutionality of the system. Only a person (or, more accurately, a university) who submitted a grant application that was not fully reviewed and did not receive a priority score, but which the government wanted to fund, would technically have standing to sue. For example, suppose that Dr. Johann Cahnnot of Popper University submits a proposal to examine certain aspects of DNA replication. Cahnnot is a philosophy professor and has not been in a laboratory since high school, which really does not matter since he proposes to examine DNA replication through "thought experiments." The study section determines that the proposal is not worthy of review so no priority score is assigned. However, Dwayne Driver, the new Secretary of HHS and a former head of the Department of Motor Vehicles in the president's home state, likes the idea of thought experiments; they are so much cheaper than those conducted in a laboratory. By funding thought experiments, he can dramatically increase the number of R01 grants awarded without increasing the NIH budget. As Driver tells his staff, "it is a win win win." Driver, though, is prevented from funding Cahnnot's proposal because it was deemed unacceptable for funding by the study section. In this situation, Popper University could argue that it was harmed because it was denied funding solely because of an unconstitutional provision in the Public Health Service Act that precludes the Secretary from funding a grant if the study section or council deems the grant not fundable. Popper U would thus have legal standing to challenge the constitutionality of the NIH funding scheme. To my knowledge, this scenario has never occurred and is not likely to occur.

Second, it is debatable whether the study section and the advisory council meetings that consider proposals can legally be conducted behind closed doors. The Federal Advisory Committee Act and the Government in the Sunshine Act, discussed in greater detail in chapter 6, require collegial bodies to meet and vote in public unless they have a legitimate reason to meet behind closed doors. One can argue that because they will be discussing the competence of individuals to conduct research, this type of discussion is similar to personnel decisions and should be kept confidential and not publicly aired. But what happens if the scientist involved waives his or her privacy rights? In theory, the basis for closing the meeting when discussing that scientist's research evaporates. Even in such cases, though, NIH has refused to open the meeting.[23]

NATIONAL SCIENCE FOUNDATION

NSF also uses a peer review system to award research grants. However, the National Science Foundation Act of 1950,[24] which created the National Science Foundation, did not impose a rigid system for awarding grants. As a result, the NSF system is more flexible than the one mandated for NIH and the various Public Health Service agencies. While NIH has entire institutes devoted to particular areas, NSF is divided into twelve directorates, each directorate is further divided into divisions, and each division is further divided into programs. Each program is usually run or co-run by a few people, and, in many instances, those individuals are on two-year leaves from their universities to act as program directors. Thus, for example, the Directorate for Mathematical and Physical Sciences is divided into six divisions including one for chemistry and another for physics. The physics division consists of eleven programs, including one for theoretical physics (low direct costs, i.e., chalk and boards), and another for elementary particle physics (very high direct costs, i.e., particle accelerators). Funding decisions are made at the program level.

Rather than using advisory committees to review proposals for scientific merit, the NSF program director in charge of a given area will select at least three external peer reviewers for each proposal. Those reviewers can be ad hoc reviewers or part of a quasi-ad hoc group that that director appoints. At NSF, the external reviewers are not required to meet but rather may develop their reviews independently and submit them by mail—or now, electronic mail. Many program directors, however, convene meetings of their external reviewers to discuss the various proposals after the reviewers have submitted their electronic reviews. The program director makes her grant decisions based in part on the results or recommendations of her outside panel. The recommendations then move up the organizational chain to the head of a given directorate.

At NIH, the identities of those on each study section are publicly available. In contrast, NSF normally does not divulge the identities of those who peer-reviewed a given proposal. One disappointed applicant actually sued NSF in an attempt to learn the identities of the anonymous reviewers of his proposal. The court, in turning back the challenge, held that the federal Privacy Act provides peer reviewers with the expectation of privacy, and therefore, NSF properly declined to divulge the identities.[25] This case is discussed in greater detail in chapter 6.

Is There Any Recourse If a Grant Is Not Awarded?

Recourse is everywhere in American society. If, as a student, you are not happy with your grade in a class, you can complain to the dean, who might intervene depending on the circumstances. If you have been denied tenure at a university, there is normally some process available that permits you, under limited circumstances, to challenge that decision. And of course, you can always sue. If you fail to get a government contract and believe that you deserved it more than the shlub who walked off with it, you can sue or file a bid protest. If successful, you can end up with the contract. If the grant is a so-called entitlement grant, such as Medicaid, and you are denied benefits, there is usually an administrative process and then judicial review is available.[26]

Research grants, though, are different. The unsuccessful grantee has essentially no recourse.[27] Although many agencies permit unsuccessful PIs to file complaints, these processes are informal, have no legal effect, and are not a road to court. Research grants are so highly discretionary and so highly technical that courts have been extraordinarily reluctant to enter the fray; they have done so only in the rarest of circumstances. Correspondingly, the government has consistently argued that funding decisions on research grants are "committed to agency discretion," a term of art meaning that the agency decision is not reviewable in a court.

In only a handful of instances have researchers "successfully" sued a funding agency over a grant decision, and the facts of those cases were unusual. I use the term "successfully" to mean only that the researcher did not lose immediately. For example, in one case Dr. Julia Apter, a professor of surgery at Rush-Presbyterian-St. Luke's Medical Center, had applied for a five-year, $580,000 training grant for which she was to be the program director. While her application was pending, she testified before a Senate subcommittee concerning alleged conflicts of interest by members of NIH advisory committees, including the committee that was to be reviewing her grant application. One month after she testified, the advisory committee recommended against funding Apter's grant; she instituted suit against the Secretary, NIH, and others alleging, among other things, that the adverse recommendation was in retaliation for her testimony before the Senate subcommittee. The trial court dismissed the case, finding that only the putative grantee, namely Rush-Presbyterian, had standing to sue; the principal investigator did not. The court of appeals reversed this decision, finding that Apter suffered a sufficient injury to bestow standing on her. In returning the case to the trial court, the court of appeals noted that the Public Health Service Act confers broad discretion on NIH and that its decisions

> may be committed to the unreviewable discretion of the agency. However, that does not mean that NIH actions wholly escape judicial scrutiny. Where it is alleged that the agency has transgressed a constitutional guarantee or violated an express statutory or procedural directive, otherwise non-reviewable agency action should be examined to the extent necessary to determine the merits of the allegation.[28]

It is unclear whether Apter, a well-known ophthalmologist, was ever awarded the elusive training grant; she died in 1979.[29]

Most who opt to sue do not get as far as Dr. Apter did and instead have their cases dismissed at the first opportunity. One federal judge, in the course of dismissing a disappointed researcher's suit against NIH, candidly observed

> that as unfortunate as it might be, it is a fact of life that courts are simply not competent to step into the role of a medical research scientist faced with having to evaluate an applicant's technical expertise[,] the theoretical[,] chemical[,] and pharmacological underpinnings of his study methodology, the statistical validity of his test results and the conclusions drawn therefrom, or just about any other factor of importance to [the National Cancer Institute] in deciding who should get its research money. But in this court's opinion the preeminent consideration militating against general judicial review of research grant decisions is that such review would place a heavy burden of litigation on an agency with more important matters at hand, would delay the funding process to the detriment of potential grantees, and would perforce place in jeopardy a program designed to combat cancer.[30]

In short, unless your grant application was denied for reasons having nothing to do with its scientific merit or germaneness to the funding agency's research agenda, the prospect of successfully challenging an adverse funding decision is bleak at best.

Who Are the Parties to a Grant, Who "Owns" the Grant, and How Long Is Funding Assured?

A research grant usually involves two essential parties and one nominal party: the grantor (i.e., the funding agency), the grantee (i.e., the recipient of the funds), and the principal investigator (PI) (e.g., the scientist who will oversee the research). The essential parties are the grantor and the grantee; the PI is listed in the grant application, but is legally not a necessary party to the grant. This means that if you're a scientist, the $475,000 NIH grant is not really your grant, but rather is owned by your university. A PI's minimal legal status is evident from the grant application itself. In PHS 398, the Public Health Service Grant Application form, there is not even a place for the proposed PI to sign. The only signature required is that of the designated institutional official. That official certifies, on behalf of the organization,

> that the statements herein are true, complete and accurate to the best of my knowledge, and [I] accept the obligation to comply with the Public Health Service terms and conditions, if a grant is awarded as a result of this application. I am aware that any false, fictitious, or fraudulent statements or claims may subject me to criminal, civil, or administrative penalties.

The corresponding Notice of Grant Award (PHS), now known as a Notice of Award Letter, the official document from the federal government confirming that the grant you sought has in fact been awarded, is directed to the institutional official who signed the grant application.

The National Science Foundation electronic grant application form can be submitted only by the institution's authorized representative, which usually means the university's vice president for research. DARPA, the folks who brought us the Internet, among other things, uses the NSF application forms.

Although I have used the term "principal investigator" in the singular, some agencies, such as NSF, have had a history of co-principal investigators. NIH, which traditionally has had a single-PI policy for most awards, is in the process of modifying that policy to permit multiple PIs and multiple awardee institutions. How this policy will be implemented remains to be seen.[31]

Because the institution, and not the PI, is the legal "owner" of the grant, a grant can be transferred only if both the grantor (e.g., NIH, NSF), the grantee (e.g., university that previously employed the PI), and the new grantee (e.g., university that just hired the PI) agree to the transfer. As a practical matter, most institutions permit their faculty to move a grant to another institution. Sometimes (and this occurs rarely), an institution may decline to sign off on the transfer, a situation that can occur if the institution and the PI have had a serious falling out. This is a no-win situation for everyone. While the grantor (e.g., NIH, NSF, NASA, DARPA) will rarely insinuate itself into the fight, it could advise the jilted university that if it does not approve the transfer, the grantor will not renew the grant (if that is an option) or, in fact, may terminate the grant and then reaward it the following cycle to the university that hired the PI away. The unwillingness of a university to permit a PI to move "his" or "her" grant to another institution can have many short- and long-term negative ramifications. If the university, as a grantee, ever needs a favor from the grantor-funding agency, that favor may be a long time in coming. And the university that was blocked from taking over the grant may decide to increase its efforts to recruit more faculty away from the jilted university.

If the university or other grantee actually owns the grant, does that mean that it also owns the data, the specimens, and anything else purchased with the grant funds? This is an important issue without a simple answer and has been the subject of intense litigation. We examine the issue in more detail in the chapter on intellectual property (see chapter 7).

Although the PI does not own the grant, he or she has an abiding interest in making certain that it continues to be renewed. Grants are normally awarded for a fixed term, usually three or five years. This does not mean that you are assured of funding for the term of the grant. The federal government, like the owner of an apartment building who hopes to convert to condominiums, operates on a year-to-year basis. Funding is not automatic but comes through an annual or almost annual appropriations bill; therefore, agencies have to go back to the trough for funding for the upcoming federal fiscal year (October 1 to September 30). In this regard, NSF differs from other agencies. The program directors at NSF have the discretion of awarding a grant either as a "standard grant" or a "continuing grant." In a standard grant, the grantee receives the full, multiyear award at the beginning of the grant. Thus, if you were awarded a three-year, $100,000 per year standard grant from NSF, NSF would transfer $300,000 to your institution. A continuing grant, in contrast, is paid out year to year, much like at NIH. Grants that are funded yearly depend on Congress enacting its yearly appropriations act for the funding agency.

Sometimes Congress and the White House play chicken over appropriations bills. When neither one blinks, sometimes the government or a portion of it shuts down. Usually, these shutdowns are short-lived, lasting for a few hours or perhaps two or three days.[32] During the Reagan administration, the Department of Interior's appropriation bill was hung up. The Department of the Interior is responsible for running most of the operations at the White House, including cooking for the president. The White House staff and their guests, on the other hand, eat in the White House mess, which, as its name implies, is run by the Navy. One lunch day, staffers and their guests were surprised to see the president stroll into the mess for lunch; his Department of Interior–funded staff had been sent home for want of an appropriation, and the Navy mess was the only food establishment open for business in the White House compound.

Even when things run smoothly, grants, other than NSF standard grants, remain a year-to-year affair with no guarantee that the second year of your supposed three-year grant will be funded. Most agencies that award multiple-year grants require PIs to submit yearly a noncompetitive renewal application, which is actually a progress report. Only rarely will an agency decline to fund a noncompetitive renewal application, but if it does, it does not have to give a reason for declining to renew the grant. But can NIH, NSF, or other funding agencies actually terminate a grant in midyear? Most funding agencies have only limited authority to terminate (or "suspend") a grant. For example, NIH "may terminate any grant in whole or in part before the date of expiration, whenever it determines that the grantee has materially failed to comply with the terms of the grant."[33] It is normally easier for the funding agency to wait until the end of the grant year and decline to renew the grant, assuming that this option is available.[34] For PHS-funded grants, the decision to terminate or suspend is subject to an administrative appeal (discussed below).[35]

What Are the Terms and Conditions of a Grant?

A federal grant is an odd blend of extreme freedom and extreme regulation. On the one hand, most federal grants give the researcher wide latitude when it comes to conducting or even changing the research without prior approval. This flexibility obviously varies from agency to agency, with NIH being extremely flexible. On the other hand, by accepting a grant, the institution and the researcher agree to a host of terms and conditions (i.e., quid pro quos, including adopting and implementing policies on scientific misconduct, human subjects, financial conflicts of interest, patient privacy, recombinant DNA, and vertebrate animals in research). These conditions are discussed in subsequent chapters. The institutions also agree to handle the money they receive in accordance with the federal government's financial management principles and to incur only allowable costs. (I will discuss the financial aspects of the grant in the following section of this chapter.)

Does this mean that a funding agency can impose whatever terms and conditions it wishes in a grant award? One might think that because a grant is a voluntary commitment by the government to fund a particular activity, it can impose whatever conditions it wishes. If the putative grantee is dissatisfied with those conditions, it has an obvious choice—it can forgo funding. Coupled with this notion of free will is a legal principle, consistently affirmed by courts, that individuals and institutions can waive various constitutional rights and that those waivers are enforceable. Thus, a person can, as a condition of employment or receipt of a federal grant, waive his or her First Amendment rights. For example, in one case, the Supreme Court enforced the provisions of the Central Intelligence Agency's employment agreement that prohibited an employee or a former employee from writing and publishing any book without submitting the manuscript to the CIA for review and approval.[36]

This principle applies outside the world of spooks. For example, in *Rust v. Sullivan*,[37] the Supreme Court upheld a family planning grant program that prohibited grantees (family planning clinics) from counseling about abortion. The Court ruled that clinic "employees' freedom of expression is limited during the time that they actually work for the project; but this limitation is a consequence of their decision to accept employment in a project, the scope of which is permissibly restricted by the funding authority." The Court, though, went on to note that the government's ability to restrict speech as a condition of receiving federal funds is not unrestricted and in certain areas may in fact be impermissible. Significantly, the Court noted that "the university is a traditional sphere of free expression so fundamental to the functioning of our society that the Government's ability to control speech within that sphere by means of conditions attached to the expenditure of Government funds is restricted. . . ."[38] *Rust*'s broad carve-out for universities was put to the test a few months later.

In 1989, the National Heart, Lung, and Blood Institute (NHLBI) announced that it would fund two academic centers to engage in research about artificial hearts. Each recipient would receive a $1.5 million award over the five-year length of the project. The announcement indicated that the contract might include a clause requiring researchers to obtain government approval before publishing or otherwise disseminating preliminary research results. Stanford University responded to the request for proposal (RFP), but objected to the confidentiality clause. Ultimately, Stanford was awarded one of the two contracts, but refused to execute it, arguing that its charter prevented it from doing so and that the confidentiality clause constituted an unconstitutional prior restraint. After the parties were unable resolve their differences through negotiations, Claude Lenfant, the NHLBI director, withdrew the offer and awarded the contract to St. Louis University, which apparently had no qualms about the clause. Stanford filed suit shortly thereafter.

Four months after the Supreme Court decided *Rust*, the district court hearing the Stanford case held that the confidentiality clause was an unconstitutional attempt to regulate scientific speech on a university campus and ordered NHLBI to award the contract to Stanford. The court held that

> [t]he subject of this lawsuit is the very free expression that the *Rust* Court held to be so important for the functioning of American society that it may be curtailed through conditions attached to grants or contracts only if these conditions are not vague or overbroad. Yet, the conditions imposed by the defendants are plainly in that category [because the contracting officer has unlimited discretion to decide whether to approve a publication].[39]

Even though the Stanford case involved a procurement contract, which we will discuss later, and not a grant, the same principles apply, namely a government grantor may not seek to impose unconstitutional restrictions on the grant award.

There is another interesting class of grant conditions that I consider improper, namely those terms and conditions issued informally as binding agency policy. If an agency such as NIH requires that of all its grants incorporate a specific restriction, that restriction is actually a legislative rule (i.e., a big rule) and can be issued only through notice-and-comment rulemaking.[40] If a condition in a grant looks unusual and potentially onerous, ask about its provenance; it may not be kosher.

One set of restrictions that likely falls into the category of "non-kosher" was the severe limitations on funding research using human stem cells that was enforced by NIH for nearly eight years. Since these restrictions precluded those within NIH from funding research that they might otherwise have funded, the restrictions should have been issued through notice-and-comment rulemaking, but they were not. As a result, the restrictions were likely improper procedurally.

Can Work under a Grant Be Contracted Out?

A grant can provide funds to more than one institution; this can occur in one of two ways. Either the grant can be jointly awarded to two institutions, each with its own PI, or it can be awarded to a single institution, which is authorized to provide funding downstream to recipients that participate either as subgrantees or contractors to the main grantee. The term "subgrantee" is normally used when the subawardee is performing a portion of the research in much the same way that it would have done had it received the grant. The term "contractor" is used when the downstream institution is performing more rudimentary tasks that would normally be funded through a procurement contract. For example, if you will need glass equipment made to certain specifications or DNA sequencing performed on certain specimens, your grant funds could be used to pay a contractor for these services.

As a general rule, if the grantee intends to undertake the research jointly with another institution, then that ought to be built into the budget. However the arrangement is structured, it should be memorialized in a detailed written agreement between the institutions in which the subgrantee or contractor agrees to abide by all of the terms and conditions that pertain to the primary recipient. This is important because the grantee is responsible for ensuring that the terms and conditions of the grant are met irrespective of who is performing the work. The primary grantee will be held responsible if a subgrantee or contractor fails to comply with the terms of the grant.

Developing the budget when a subgrantee or contractor is involved can be challenging because the cost principles that apply to a grantee may not apply to a subgrantee. For example, if the grantee is an institution of higher education and the subgrantee is a hospital, then each is governed by different cost principles and different indirect cost rates, as discussed in the next section. Also, if the "downstream" recipient is a contractor, there may be special procurement requirements, including competitive bidding, imposed on the grantee to ensure that the ultimate downstream recipient was selected openly and fairly.[41]

How Are the Economics of a Grant Administered?

DIRECT VERSUS INDIRECT COSTS

Researchers get their first glimpse of grant economics when they complete their first grant application. The application invariably requires a detailed budget broken down by direct costs (e.g., salaries and benefits, equipment, supplies, travel) and indirect costs, now known as facilities and administrative costs, or F&A. Indirect costs are those costs incurred for common or joint purposes that cannot be readily identified with a specific project or program. For example, the cost to heat, light, air-condition, and maintain academic buildings on a university campus, the cost of university administrators' salaries, and the cost of maintaining libraries and the like make up the indirect costs. Each research institution has an indirect cost rate that is used to determine the indirect costs. For example, suppose you submit a grant application for $200,000 in direct costs consisting of $170,000 for salaries and benefits, $15,000 for equipment, $10,000 for supplies, and $5,000 for travel. And further suppose that your university's indirect cost rate is 60 percent. The indirect costs of the grant would be $120,000 (i.e., 60% × $200,000), and the total grant (direct and indirect), assuming it were awarded, would be $320,000.

The general rules governing direct and indirect costs, referred to as cost principles, are issued by the Office of Management and Budget (OMB) within the White House and other agencies and vary by the type of organization that is to receive the grant. There are three general sets of OMB costs principles, as follows: (1) OMB Circular A-21 for Institutions of Higher Education;[42] (2) OMB Circular A-87 for State and Local Governments;[43] and (3) OMB Circular A-122 for Nonprofit Organizations.[44] There are also cost principles governing hospitals and for-profit organizations.[45]

There is nothing simple about cost principles. It is not even obvious which OMB Circular or set of cost principles applies when an organization is potentially subject to two or three OMB Circulars. For example, is a state university hospital governed by A-21, A-87, or 45 C.F.R. pt. 74? In this instance, higher education predominates over the institution's hospital character, which predominates over its state ownership.

ALLOWABLE DIRECT COSTS VERSUS UNALLOWABLE DIRECT COSTS

Government grants cover only "allowable costs." A cost is allowable if it is reasonable and necessary for your research, relatively consistent with the budget, and not otherwise unallowable. If you have budgeted $100,000 for the salaries of your fellows and others and instead spend the money on a cruise and a Mercedes, you do not need to read this book to appreciate that the expenditure is not only an unallowable cost, but one that is likely to provide a one-way ticket to "Club Fed" (i.e., federal prison for white collar offenders). Another route to federal prison is to use federal grant funds to lobby Congress for additional grant funds and then certify that you did not.

At a more subtle level, the cost principles help differentiate between allowable and unallowable costs. For example, if, as part of your grant, you travel to London to attend a conference, you are obligated to fly on an American-flag carrier, unless no American carrier services that route. You cannot fly British Air or Virgin Air; you have to use United, American, Delta, USAir, and the like.[46] If you opt to fly British Air, your grant is not supposed to reimburse you for the travel. You are also obligated to fly at the lowest reasonable commercial airfare—steerage class is appropriate, business or first class is not. The general travel rules that apply to federal employees usually pertain to those on federal grants.

There are a few pitfalls worth noting. First, the salaries of individuals employed on grants are usually capped; the cap varies from agency to agency. Those receiving NIH grants are not permitted to receive a salary that exceeds the salary of a member of the cabinet.[47] Thus, if you are normally paid $250,000 per twelve-month year and anticipate devoting 10 percent of your time to an NIH grant, you would expect to receive $25,000 from the grant. That would not be the case here, though. A member of the president's cabinet received, in 2009, $196,700 per annum; since that is the cap on your salary, as far the federal government is concerned, your 10 percent effort generates $19,670 rather than $25,000, plus benefits based on $19,670.

Second, beware of the time clock. Federal grants bring to academia that same sort of shop mentality more appropriate to the assembly line—clock in, clock out, put in your thirty years, and retire. If you have a federal grant, you will be required to keep sufficient records of how you spent your days to enable you to complete the required "effort reports." Those who are on more than one grant or who are on a grant and have nongrant responsibilities, such as teaching or administering, have to take care to make sure that they properly account for their "effort." Effort reporting has become a major headache and a major source of fines and penalties. In the past few years, the government has extracted significant settlements from Northwestern University ($5.5 million), Johns Hopkins University ($2.6 million), University of Alabama ($3.4 million), The Mayo Foundation ($6.5 million), and Cornell University ($4.4 million) over allegations that time and effort were not properly reported.[48] Recently, Cornell settled a second case in which it had been alleged that one of its principal investigators failed to list all of the grants on which she was working and, as a result, overcommitted her time; Cornell settled the case for $2.6 million.[49]

Some of the major problems arise when researchers on multiple grants report spending more than 100 percent of their time on various grants (e.g., spending 70 percent of their time on one grant and 60 percent of their time on another grant). Under federal law, you get only 100 percent. Suppose that you spend forty hours per week on NIH grant 1 and ten hours per week teaching. Is your grant-related effort 100 percent because 40 hours is the standard work week, or is it 80 percent because that is actual grant effort, or is it some variant? When it comes time to complete your university's effort report, which is required if you are on a federal grant, be careful to avoid software that automatically uses a forty-hour week. The best way to allocate time is to take into account your complete university effort. While an eighty-hour week may raise red flags for auditors, trying to adjust it down to forty hours will introduce improper distortions.[50]

Third, beware of reprogramming budgets. Most granting agencies, including NIH and NSF, provide the grantee with flexibility to reallocate budget items, but this flexibility is not unbounded.[51] Before rebudgeting, check with your institution's financial management folks, and if there is the slightest doubt, speak with your funding agency. If preapproval is necessary, many agencies require that the request be submitted at least thirty days before the expenditures are made.

Finally, when clinical research is being conducted on subjects who would ordinarily be covered by Medicare, the cost rules and reimbursement principles become extremely complex, well beyond the scope of this text. If you are planning this type of research, you should meet with your hospital's chief financial officer or comparable person to discuss the reimbursement implications of the research. You do not want to admit patients who would ordinarily be covered by insurance only to learn that such coverage vanishes if any portion of the hospitalization and attendant procedures were part of a research protocol.

ESTABLISHING INDIRECT COSTS

Your institution's indirect cost rate will affect the size of your grant.[52] The majority of indirect cost rates for institutions of higher education are between 40 and 70 percent, with an average of slightly above 50 percent. These rates are negotiated between the institution and its so-called cognizant agency. The cognizant agency is the one that provided the most funding to the institution in the year in which the institution's rate was first set. For most research universities, the cognizant agency is Health and Human Services (HHS). There are notable exceptions. Stanford's cognizant agency is the Office of Naval Research (ONR) because when Stanford's indirect cost rate was first established, it received more funding from ONR than from any other agency.

The negotiated indirect cost rate, in theory, ought to apply across the board to all grants irrespective of the funding agency. That, however, is not the case. Some grants and some agencies dramatically limit indirect costs that can be recovered. Thus, for example, NIH limits indirect costs for training grants to 8 percent, and USDA limits indirect cost recovery to only 14 percent on all of its grants.[53]

What goes into determining the indirect cost rate can be highly debatable. In 1991, Stanford was involved in a highly publicized Washington-esque scandal revolving around its indirect cost rate. In calculating its indirect cost rate, Stanford was accused by an ONR auditor of including in its indirect cost rate the costs of various luxury items for the president's house, including a $12,000 pair of urns, an expensive shower curtain, and a fruitwood commode. Congressman John Dingell held caustic hearings, *60 Minutes* did its usual exposé, and the Navy auditor filed a False Claims Act suit against Stanford University seeking to recover a substantial reward. In the end, Stanford was largely vindicated, and the Navy auditor lost his suit.[54]

Can Post-award Decisions Be Challenged?

Not infrequently, a grantee will have a disagreement with the funding agency. It could be over whether certain costs are allowable or it could more serious, such as the refusal of the agency to continue a grant scheduled for renewal because the agency believed that the grantee or its PI had failed to comply with the terms and conditions of the grant. Many granting agencies have mechanisms for attempting to resolve these disputes short of court. In fact, if the agency has such a system, you can get into court only if you work your way through all of the administrative steps first.

If your dispute is with NIH and certain other Public Health Service agencies, the first step in the appeals process is the PHS Grant Appeals Board.[55] The PHS Grant Appeals Board can hear cases in which the dispute is over (1) whether costs are allowable, (2) whether a funding agency's decision to terminate a grant for not adhering to the grant's terms and conditions was proper, or (3) whether an agency's decision to deny a noncompeting continuation award for failure to comply with the terms and conditions of a prior award was justified. The Grant Appeals Board, which consists of at least three members, is largely ad hoc, with members being selected when needed, and the process is relatively informal. If the grantee is dissatisfied with the decision of the PHS Grant Appeals Board or if the granting agency does not offer this level of appeal, the next stop is at what is currently called the HHS Departmental Appeals Board (DAB), formerly the Grant Appeals Board.[56]

The DAB, which sits in panels of three, has jurisdiction to hear an appeal from the PHS Grant Appeals Board or from the decision of an agency that does not permit appeals to the Grant Appeals Board. Indeed, you are precluded from seeking DAB review unless you have first gone through the PHS Grant Appeals Board, where available. The DAB was established voluntarily for a political reason. Before the DAB, decisions concerning grants including Medicaid were made by the Secretary of HHS. When large disputes loomed, members of the congressional delegation of the affected state would bombard the Secretary with letters and attempt to exert other forms of pressure. To depoliticize and regularize grant appeals, the DAB was created by regulation, not by statute. Hearings that are mandated by statute are governed by the Administrative Procedure Act (APA), which provides litigants with a full array of protections. Because the DAB hearings are not mandated by statute, the protections of the APA do not apply and hearings may be somewhat less formal than APA-type hearings.

NSF also has a dispute resolution process for cost-related issues, but that system is not as well defined as the one used by PHS and involves individual decision makers as opposed to panels.[57] The final arbiter at NSF over cost-related disputes is the deputy director or his or her designee.[58] Most agencies have some form of internal appeals process, and those processes can differ dramatically from agency to agency. These processes are usually outlined on an agency's Web site and set out in the Code of Federal Regulations.

If you remain dissatisfied after bringing your dispute before the DAB or the NSF deputy director, your next stop is federal court. But be forewarned: Your battle in federal court will be uphill. This is because courts are required to affirm the decisions of federal agencies unless those agencies acted either in contravention of the law or "arbitrarily or capriciously."[59] This is a very difficult standard for most plaintiffs to meet, and, as a result, plaintiffs rarely prevail.

WHAT IS A PROCUREMENT CONTRACT?

A procurement contract, also known as a government contract, resembles a normal commercial agreement, except that the purchaser is the government. A procurement contract is supposed to be used when an agency wants to purchase specific goods or services.[60] The fact that the government is the purchaser affects everything from the way the contract is awarded, negotiated, and implemented to the way that disputes are resolved. The procurement process is detailed, time-consuming, and subject to highly objective criteria.

Most researchers will never need to worry about procurement contracts because these contracts normally are ill suited for funding small, basic research projects. Procurements are expensive to administer and, unlike grants, are subject to challenge and litigation at very nearly every stage. They are used primarily when purchasing meta-research, namely purchasing the services of an entity that will undertake or manage all basic and applied research in a given area at a given facility. For example, Lawrence Livermore National Laboratory, which is owned by the Department of Energy, is operated by a government contractor (a consortium that includes the University of California); the Jet Propulsion Laboratory, which is owned by NASA, is also operated by a government contractor, Caltech; the Stanford Linear Accelerator, which (except for the land) is owned by the Department of Energy, is operated by Stanford University; and the National Cancer Institute's Frederick, Maryland, facility (located in the U.S. Army's Fort Detrick) is owned by NCI (except for the land) and operated by a set of government contractors. These big science operations are all called GOCOs—government owned, contractor operated.

How is it that a procurement contract is an appropriate vehicle for purchasing large amounts of research, but inappropriate when purchasing small amounts of research? The best way to look at the difference is that a procurement contract is used when the government wants an entity to perform functions that would otherwise have to be performed by the government, such as overseeing research. It is akin to contracting out the NIH intramural program. The procurement does not specify the scientific results that need to be achieved to warrant payment; instead, the procurement specifies the management functions that the awardee agrees to undertake. The contract is more a management contract than it is a "research contract." When you get down to the specific types of research, management is no longer the critical element, but rather science and innovation become the key elements. The latter are much more consistent with the goals of a grant than with a procurement contract.

Aside from the obvious difference (i.e., discrete deliverables), procurement contracts differ from grants in the way they are awarded, the way in which the amount of the government payment is determined, the applicable cost principles, the remedies available to a disappointed applicant, and the way in which post-award disputes are resolved. The rules governing procurement contracts are set out in the Federal Acquisition Regulations, which vary slightly from department to department.[61]

Procurement contracts are awarded through two primary methods—sealed bidding and competitive proposals—although certain circumstances allow agencies to dispense with bidding or proposals.[62] In sealed bidding, an invitation for bid (IFB) spelling out the metes and bounds of the items or services that the government seeks to purchase is published in the *Commerce Business Daily* (CBD). The CBD is as scintillating as the *Federal Register*. Sealed bidding is used when the agency knows what it wants to purchase and is more interested in price and the ability of a contractor to deliver the item or service in a timely fashion than with other factors. In fact, in awarding a contract under a sealed bid, the agency is required to make the award based on price and other price-related factors. John Glenn, the first American to orbit the earth, once purportedly responded to a question about how it felt to go into space by noting that "I felt about as good as anybody would, sitting in a capsule on top of a rocket that were both built by the lowest bidder."[63] In fact, most of the rocket and capsule that Glenn rode into space was likely not procured using the sealed-bid method, but rather pursuant to competitive proposals where factors other than price, such as design and the like, are key elements in the selection process. In times of emergency, agency heads have discretion to award a sole source contract, meaning that only one company is permitted to negotiate for the contract.

Whatever the process by which a contract is awarded, disappointed bidders or submitters can challenge the award by filing what is called a bid protest. The bid protest is adjudicated by the Government Accountability Office (GAO), an arm of Congress. GAO issues its decision within one hundred days of the date that the protest is filed.[64] Bid protests can also be filed in the United States Court of Federal Claims or the district courts.[65]

Once you have been awarded a procurement contract, you must agree to the terms and conditions of the award. Those terms and conditions differ, in many respects, from the ones that pertain to grants; even the cost principles are somewhat different. A government procurement contract looks more like something out of the old joke club than a typical commercial contract. Members of the joke club, as you might remember from your youth, knew the jokes so well that they found it unnecessary to relate the story and punch line; instead, they just yelled out the number by which the joke was known and everyone laughed. In the government contract, numbers replace text; those numbers refer to sections of the Federal Acquisition Regulations (FAR) that govern the particular contract.[66] Thus, a procurement contract will frequently list the provisions of the FAR that govern the particular contract.

Finally, if a dispute arises in the course of the contract, it is handled differently than disputes arising under a grant. For example, if there is a dispute over whether certain costs in an HHS contract are allowable, the matter can be resolved in one of two ways. The contractor can seek to have the matter resolved by the Armed Services Board of Contract Appeals or, alternatively, the contractor can file an administrative claim with HHS under the Contract Disputes Act. If the matter is not satisfactorily resolved, the contractor can file suit in the Court of Federal Claims.

Although there are many more formalities and protections at the front of a contract than there are for a grant, the reverse is the case at the back end should the government wish to terminate the arrangement. When the government wishes to terminate a grant, it can do so only "for cause." However, the government can terminate a procurement contract for the "convenience of the government" without giving a reason.[67]

WHAT IS A COOPERATIVE AGREEMENT?

A cooperative agreement is a strange duck. Legally, it is treated as a grant, except "substantial involvement is expected between the [grantor agency] . . . [and the] recipient when carrying out the activity contemplated in the agreement."[68] A cooperative agreement is normally used for large research projects where the government expects to be an active participant. For example, certain AIDS-related clinical trials are funded by the National Institute of Allergy and Infectious Diseases pursuant to a cooperative agreement. The NCI funds the National Surgical Adjuvant Breast and Bowel Project (NSABP), which it has done for more than fifty years. NSABP, run out of the University of Pittsburgh, conducts clinical trials of various cancer treatments; its multicenter research has transformed the treatment of breast cancer.

Cooperative agreements have certain advantages for the grantee and certain disadvantages, both arising out of the cooperative nature of the research.

In the advantage column, bureaucrats would prefer to continue working with a known entity. This means that once you are awarded a cooperative agreement, if you do not screw it up, you are far more likely to continue receiving funding than if you were seeking a competitive renewal of an R01 grant. Large cooperative agreements, like Congress, favor incumbency. In the disadvantage column, bureaucrats are not particularly adept at handling scandals. If one arises out your research, a bureaucrat's first inclination is to foist the blame elsewhere, and the PI is the usual target of choice. This happened to the NSABP.

In the summer of 1990, NSABP personnel audited records of one of the institutions, St. Luc Hospital in Montreal, that had been participating in a large clinical trial comparing various breast cancer surgical interventions (mastectomy and lumpectomy) and the drug tamoxifen. During the course of that audit, NSABP personnel discovered anomalies in the St. Luc's data, including instances where patients' birthdates and tumor sizes were falsified so that these patients would qualify for the study when they otherwise would not have. In February 1991, shortly after the audit was completed, Dr. Roger Poisson, the principal investigator at St. Luc, confessed to Dr. Bernard Fisher, the head of NSABP, that he had altered the patient records. Fisher promptly reported this to all of the relevant federal agencies, including NCI, FDA, and the Office of Research Integrity. Poisson's falsification represented a minuscule percentage of the patients participating in the study and had no impact on the study's conclusions. However, a debate raged within NCI and between NCI and NSABP on how to handle Poisson's data. Because of the study design, an equal number of ineligible patients had been randomly assigned to each of three study arms, further minimizing the likelihood that Poisson's transgressions would have any impact on the study's results even at St. Luc. Nonetheless, some within NCI wanted NSABP to discard all data from St. Luc's while others wanted to discard only the problematic data. NSABP statisticians, however, argued that discarding any of St. Luc's data would be inconsistent with the protocol, which was designed to accommodate a large error rate, and Poisson's rate of fabrication was lower than the maximum error rate. In the end, NSABP preliminary publications included all of the data.

However, the press soon learned of the problems, and it was alleged that NSABP's results were tainted. One reporter, with little understanding of science, wrote a lengthy article that got the attention of Congressman John Dingell, who held hearings on NCI's management of NSABP and on Fisher's management of NSABP. In the end, the head of NCI, Sam Broder, sought to preserve his own job by pressuring Pittsburgh to relieve Fisher of his position as head of NSABP. Pittsburgh capitulated; Fisher sued Pittsburgh, NIH, and others; shortly thereafter, the head of NCI quietly resigned, the contract for the president of the University of Pittsburgh was not renewed, and Pittsburgh briefly lost a portion of the cooperative agreement to the Allegheny Health, Education and Research Foundation, which had the good grace to go into a scandal-ridden bankruptcy.[69] Broder also tried to annotate NASPB studies that had been published in PubMed, the NIH database, and that too resulted in litigation (see chapter 6). In 1997, Fisher settled his lawsuit for a substantial sum, well into the millions according to newspaper reports. An interesting side note: The United States sued St. Luc's Hospital for return of the funds paid to it as subgrantee. To my knowledge, that litigation in a Quebec court has not been resolved. The entire NSABP affair was more suited to *People* magazine than to the journal *Science*. Nevertheless, the affair highlights some of the hidden perils of large-scale cooperative agreements.

THE FALSE CLAIMS ACT AND OTHER SANCTIONS

No matter how (e.g., grant, procurement, cooperative agreement, subgrant) or from whom (e.g., NIH, NSF, NASA, USDA) you receive your federal funding, there are special laws designed to ensure that you "toe the line," namely that you comply with the terms and conditions of the grant, cooperative agreement, or procurement contract. At the benign end are technical disputes over allowable costs and things of that nature that are usually resolved informally or through a hearing at a grant appeals board, if the funding agency has such a mechanism. At the sinister end is overt criminal behavior, such as spending your grant funding on a vacation with your mistress or absconding with the funding to one of the many nations without extradition treaties with the United States. In between these two extremes are two sets of laws that every recipient of federal funding ought to know: (1) the False Claims Act, and (2) debarment rules.

What Do Army Mules and Research Grants Have in Common?

It all started with a mule or, more precisely, a thousand mules. It was the summer of 1861, and the Union Army was preparing for a series of major forays into territory held by the Confederacy. Major J. McKinstry, a man so obscure that even his first name has been lost to history, was ordered to procure the necessary beasts of burden. From outward appearances, McKinstry was ideally suited to the task, or so his superiors thought. He was a career procurement officer with an uncanny knack for filling requisition orders, no matter how difficult. McKinstry, though, had a personality flaw—he was a crook. To fill the requisition, he purchased at $119 apiece (the going price for a healthy animal) one thousand blind, lame, or diseased mules. A significant portion of the payment found its way into McKinstry's private account. McKinstry, though, was not the only Army officer engaged in creative procurements. As a result of this rampant fraud, Congress enacted the False Claims Act (FCA) criminalizing McKinstry's procurement practices.

Over the years Congress broadened the FCA, adding a civil remedy, broadening its scope, making it easier for the government to prevail, and enabling private citizens to sue on behalf of the United States and share in the proceeds.[70] A law that once dealt exclusively with out-and-out fraud in the purchase of war implements now reaches virtually all conduct involving federal funds, from the construction of high-tech weaponry to basic research underwritten by the Public Health Service. The Department of Justice reported that it recovered more than $3 billion under the FCA in FY 2006 and $2 billion in FY 2007. In both years, more than 70 percent came from the healthcare sector.[71]

The FCA penalizes anyone who, among other things, knowingly presents or causes to be presented a false or fraudulent claim for payment or approval. The penalties can be staggering. The government can collect not only treble damages, but also up to $11,000 for each false claim. For example, take the case of a physician who, over the course of a year, falsely bills Medicare a total of $4,000 for one hundred office consultations that never occurred. The government could conceivably collect $12,000 (three times $4,000) plus an additional penalty of $1.1 million (100 times $11,000). At first blush, you might wonder if there is any rational connection between a fraudulent physician bilking Medicare and an honest researcher who has little to do with the financial arrangements between his or her institution and the funding agency. The FCA draws no distinction between the types of activities involved and more significantly, between overt dishonesty and ignorance.

As its name implies, the FCA deals with "false" claims, as opposed to fraudulent claims, the latter being a subset of the former. Fraud is normally difficult to prove because it requires the government to establish (1) that the statement was false, (2) that the defendant actually knew it was false, (3) that the defendant made the false statement with the intent to deceive, and (4) that the government relied to its detriment on the false statement (i.e., the government was deceived). By contrast, under the FCA the government merely has to show that a false statement was "knowingly" made and that it resulted in a false claim being presented to the government. The person making the false statement does not have to present the claim to the government, if someone else does, nor does the government even have to pay the claim. For example, suppose that Johnson, a salesman for WidgetCo, tells Smith, WidgetCo's billing manager, that he has just sold 100 Major Widgets to the Navy at $10,000 per widget and Smith should bill the Navy $1 million. If, in fact, Johnson has made no such sale and has told a fib to Smith, Johnson has violated the FCA. He has caused Smith and WidgetCo to submit a false claim to the government. Suppose that a government auditor discovers that there has been no sale of widgets to the Navy and disapproves the $1 million payment. Johnson has still violated the FCA since he caused a false claim to be presented to the government. The fact that the claim was not paid is not relevant. WidgetCo may also have violated the FCA. How can WidgetCo be culpable? Doesn't the word "knowingly" differentiate between the real miscreants (e.g., Johnson) and those acting out of ignorance of the facts (e.g., WidgetCo, Smith)? In the strange world of legal linguistics, where words can be redefined in ways that would surprise even Lewis Carroll, "knowingly" does not really mean "knowingly." If you *should* have known that something was false, but did not out of significant ignorance, that could suffice under the FCA. The FCA penalizes those who bury their heads in the sand. For instance, suppose a researcher is conducting several different research projects within a university hospital and is also supervising clinical work outside the scope of his grants. As noted in an earlier section, the institution is obligated to keep time records of its employees reflecting the relative percentages of their time devoted not only to each grant, but also to non-grant-related clinical work. If the PI misallocates an employee's time so that a grant ends up being charged for non-grant-related work, then the PI, depending on what he or she knows or should have known, could face liability under the FCA. But is the government really likely to prosecute a PI under the FCA for misallocating costs? Normally, the answer would be no. However, the FCA is no ordinary statute because it comes equipped with a provision that makes whistleblowing not only fun, but also potentially profitable. The provision transforms every disgruntled employee or former employee into a private attorney general.

Under the so-called *qui tam* provisions of the FCA, anyone can file suit in the name of the United States against anyone else. The term *qui tam* is a truncated version of the phrase "*qui tam pro domino rege quam pro se ipso in hoc parte sequitur*," meaning roughly "he who sues for the king as well as himself." The person bringing the suit is called the relator. The suit is filed under "seal," meaning that the defendant is not aware that it is a defendant in an FCA suit. The case remains under seal until the government decides whether to intervene in the case. If the government intervenes (which it does in about 25 percent of the cases), then it takes over the case and the relator is entitled to receive a reward of up to 15 percent of the recovery. If the government declines to intervene, then it is up to the relator to prosecute the case on his or her own; should the relator prevail, he or she is entitled to receive up to 30 percent of the recovery.

Take the case of Dr. Janet Chandler and the Cook County Hospital. The National Institute of Drug Abuse awarded a $5 million grant to the hospital and later to a research institute affiliated with the hospital. Chandler ran the study from September 1993 to January 1995, when the institute fired her. In 1997, Chandler filed a *qui tam* action, claiming that Cook County and the institute had submitted false statements to obtain grant funds, "had violated the grant's express conditions, had failed to comply with the regulations on human-subject research, and had submitted false reports of what she called 'ghost' research subjects."[72] She further alleged that she was fired for reporting the fraud to doctors at the hospital and to the granting agency, rendering her dismissal a violation of both state law and the whistleblower provision of the FCA.[73] The government declined to intervene in the action.

In 2000, while Chandler was prosecuting her FCA case, the Supreme Court held in another case that the FCA does not apply to states.[74] Cook County sought to extend this ruling by arguing that because states cannot be sued under the FCA neither can political subdivisions of a state government, such as counties. The case went all the way to the Supreme Court. A unanimous Court held that although states are immune from prosecution under the FCA, counties and other political subdivisions are not. Ten years after Chandler filed her suit, the parties settled after the trial court denied defendants' motion to dismiss. Most research-related relators have not been as successful as Chandler, and their cases have usually been dismissed well before trial.[75]

The government has consistently argued that the FCA champions an infinite domino effect, meaning that if you can cause someone to cause someone to cause someone to submit a false claim, then you are just as liable as if you had submitted the claim directly. In 2008, the Supreme Court held that the FCA does not countenance an infinite daisy chain.[76] Rather, a sub-sub-subcontractor who makes a false statement to a sub-subcontractor may not be liable, even if the information is passed on by the contractor to the government and the government relies on that information to pay the contractor. Congress was not pleased with the way in which the Supreme Court had interpreted the FCA, and in 2009, Congress responded by amending the FCA to make it easier to hold subcontractors liable under the act no matter how far removed they may be from the contractor and the government.[77]

What Are Debarment and Suspension?

The FCA requires the government to file suit in federal court, but before this occurs, the funding agency has to get approval from the Department of Justice, which is responsible for prosecuting the case. If the case involves "small" bucks (e.g., less than $20 million) or complex scientific issues, it is unlikely that Justice will take up the case. A funding agency, though, is not without remedies of its own. When it believes that a recipient of federal funds has acted improperly or irresponsibly, it can disallow certain costs, it can withhold funds, it can seek to terminate funding or refuse to renew, and it can "debar" the investigator or the institution or both. In the context of research grants, as noted above, midyear termination poses logistical problems for an agency such as NIH; nonrenewal is the safer and easier course. However, when the conduct is truly egregious, an agency may seek to debar.

If you are suspended or debarred, that means that you cannot participate in any grant, cooperative agreement, or procurement contract (or any subcontract or subgrant) funded by any federal agency. If you are an individual, that means that you cannot work as a bottlewasher on an NIH-funded grant. Once a person is suspended or debarred by one federal agency, all federal agencies are required to honor that action; and once a person is suspended or debarred from receiving grant funding, that person is also suspended or debarred from receiving procurement funding.[78] In this section, I discuss both suspension and debarment, starting with debarment.

DEBARMENT

As its name implies, debarment is a sanction that precludes an individual from participating in contracts, grants, or cooperative agreements for a specified time, usually three years, but it can be longer. An agency may seek to debar a person for a variety of reasons, including the following, among others: (1) criminal or civil judgment involving fraud or certain other crimes, antitrust laws, embezzlement, theft, forgery, false statements, or similar offenses indicating a lack of integrity; (2) failure to perform in accordance with one or more public agreements; (3) a history of unsatisfactory performance of one or more contracts, grants, or cooperative agreements; (4) willful violation of a legal requirement applicable to a funding agreement; (5) violation of specific regulations, including knowingly doing business with a debarred person; and (6) "any other cause of so serious or compelling a nature that it affects the present responsibility of a person."[79] Debarment is not viewed as a penalty but rather as a way of protecting the government from doing business with those who lack the integrity necessary to fulfill their contractual obligations. Since debarment can be triggered if an agency believes that a contractor or grantee is not "presently responsible," it can seek to debar a person on the basis of an indictment alone before there has been a conviction or before the person has exhausted his or her appellate remedies.

One of the more famous debarments involved the now-defunct accounting firm Arthur Andersen, LLP, which was indefinitely debarred on April 10, 2002, following its March 14, 2002, indictment for obstruction of justice in connection with its destruction of documents relating to its auditing work for Enron Corporation. On June 15, 2002, a jury convicted Andersen of obstruction of justice, and less than two months later, on August 31, 2002, it surrendered its licenses and its right to practice before the Securities and Exchange Commission (SEC), effectively ending the company's operations. Arthur Andersen had nothing left to salvage when a unanimous Supreme Court overturned the company's conviction in 2005.[80]

Once an agency has decided to debar a person, it must notify that person and provide that person with an opportunity to respond to the allegations. In the event of a dispute over material facts, an agency may convene a fact-finding hearing.[81] In any debarment proceeding, the agency must prove its case by a preponderance of the evidence. As will be discussed later, a hearing is required in cases involving allegations of scientific misconduct, where one of the penalties is invariably debarment. A debarment takes effect when the debarring official enters his or her debarment order. As with any agency action, a debarment can be challenged in court, although the prospects of prevailing are usually meager. Once you are debarred, your name is entered on the master list maintained by the General Services Administration.[82]

SUSPENSION

Debarment has a twin—suspension. Suspension of a person under the suspension and debarment rules is not the same as the suspension of a grant. Suspension of a grant is a temporary event and does not necessarily indicate that the grantee has engaged in improper or illegal conduct. In contrast, suspension of a person is a prelude to debarment and occurs only when matters are so serious that corrective action cannot wait until formal debarment proceedings have been completed. Suspension is appropriate when (1) there is sufficient evidence, including an indictment, to indicate that a person has committed a crime to support debarment or there is sufficient evidence to indicate that a person has committed another transgression that would warrant debarment, and (2) immediate action is necessary to protect the public interest.[83] In such a setting, the agency may suspend the person pending a further investigation and an opportunity to contest the suspension.

SUMMARY

Federal funding has been the driving force behind most basic research in the United States for well over half a century. The results of that research have proven to be critical to scientific and biomedical advances as well as to economic development. As funding agencies have matured and grown, so have their requirements. Thirty years ago, federal oversight of research was scant. That is no longer the case. Regulation of federally funded research, though, is by mutual consent, since the money flows through contracts—grants, cooperative agreements, and procurement contracts. Contracts are bilateral agreements in which the government promises to provide a specified sum of money to accomplish certain purposes, subject to certain restrictions, and the recipient agrees to abide by those restrictions as a condition of receiving the money. About two hundred years ago, when Congress was debating the wisdom of funding research, some of the opponents argued that government funding of research would lead to government control of research. Those concerns are very much alive today, especially when the research appears to conflict with the moral beliefs of some in the body politic. As long as science is federally funded, it will reflect to some degree, rightly or wrongly, the political bent of those in power. Scientists have to be ever vigilant to ensure that government intrusion in the scientific enterprise is held to a minimum, especially because government regulation is less efficient and more pernicious when the activities subject to regulation are beyond the understanding of most in Congress and many of the putative regulators.

CASE STUDIES AND PROBLEMS

Case 1: The Case of the Sporting Scientist

Marie Defarge was the principal investigator on an NIH grant. A portion of the grant was allocated toward travel permitting her to meet with her peers or attend relevant conferences. The line item for travel permitted her to make between six and ten trips per year depending on when and where she traveled. Defarge's son, Hector, was the first-string quarterback on Robespierre University's football team. Defarge arranged three of her trips to correspond to cities where Hector would be playing that weekend; she would fly out either Thursday night or Friday morning, hold meetings all day Friday, and then attend the football games on Saturday, returning home on Sunday. She would bill the grant for her Saturday stay, but usually she saved an equal amount in airfare by returning on Sunday rather than Friday. Using her son's football schedule, she also arranged to have her friends invite her to deliver papers at conferences or symposia or small groups of graduate studies on Fridays in cities where her son was to play football the next day.

After the football season ended, Defarge arranged to attend a three-day conference in February in Acapulco, Mexico; the six-hour conference was spread out over the three days. Defarge spent much of the time on the beach with her boyfriend Peter Lavoisier, a stock broker. Defarge billed the entire trip to the grant, including most of Lavoisier's meals.

Defarge's grant spending is audited by her university. Does she have problems? If so, describe and discuss.

Case 2: The Case of the Dictatorial Director

For as long as anyone could remember, Richie Lieu had been the director of a major science funding agency. Lieu ran a very tight ship. He was marvelous at dealing with the press and Congress, explaining complicated science in ways that the media and congressional members could understand. Each year he was able to extract more money from Congress than had been recommended in the president's budget. Indeed, while other science funding agencies were facing significant cuts, his agency's budget was growing in real dollars. Lieu was charming, but charm alone did not explain his very high success rate.

A significant portion of Lieu's budget was devoted to cookie-cutter science: applied science involving little real research. For example, one project was aimed at counting the number of emperor penguins in a certain area of Antarctica between June 1 and June 15 each year. Another was aimed at collecting and sequencing DNA from polar bears living in certain areas above the Arctic Circle and grizzly bears living outside national parks. All told, Lieu had about twenty projects like this to give out every three years. Early on, Lieu awarded these projects by cooperative agreements. Universities that had given him honorary degrees invariably received these cooperative agreements.

Each recipient was also secretly required to return about 10 percent of the amount awarded to Lieu's agency. Lieu used that money to host parties for members of Congress all over the United States and Europe.

Mary Stewart is the vice president for research at Etonbridge University. Etonbridge originally had the bear grant. However, the cooperative agreement was not renewed shortly after she refused to fork over the 10 percent. Stewart visits her general counsel to discuss the university's options. What are the issues? Does Lieu have problems?

NOTES

1. *See* http://www.aaas.org/spp/rd/09ptbii1.pdf (table II-1) (Mar. 21, 2008).

2. *See* Jeannie Baumann, *Research Funding: Obama Boosts Medical Research Funding But Questions on Long-Term Support Linger*, 8 MED. RES. & L. RPTR. 319 (2009).

3. *See id.*

4. U.S. CONST. art. I, § 8, cl. 8.

5. A. HUNTER DUPREE, SCIENCE IN THE FEDERAL GOVERNMENT: A HISTORY OF POLITICS AND ACTIVITIES (1986); DANIEL J. KEVLES, THE PHYSICISTS: THE HISTORY OF A SCIENTIFIC COMMUNITY IN MODERN AMERICA (1995).

6. "NSF regards research as one of the normal functions of faculty members at institutions of higher education. Compensation for time normally spent on research within the term of appointment is deemed to be included within the faculty member's regular organizational salary.

As a general policy, NSF limits salary compensation for senior project personnel to no more than two months of their regular salary in any one year." NATIONAL SCIENCE FOUNDATION, GRANT PROPOSAL GUIDE ch. 2 (Apr. 2009), *available at* http://www.nsf.gov/pubs/policydocs/pappguide/nsf09_29/gpg_2.jsp#IIC2gi.

7. The federal government also spends money on its direct operations to pay employees' salaries and overhead.

8. *See* U.S. CONST. art. I, § 10 ("no state shall pass any . . . law impairing the obligation of contracts.").

9. *See* Dartmouth Coll. v. Woodward, 17 U.S. 518 (1819).

10. Am. Hosp. Ass'n v. Schweiker, 721 F.2d 170, 182–83 (7th Cir. 1983).

11. *See* United States v. County Sch. Bd., Prince George County, Va., 221 F. Supp. 93 (E.D. Va. 1963).

12. *See* United States v. City & County of San Francisco, 310 U.S. 16 (1940).

13. National Institutes of Health, Office of Extramural Research, Activity Codes, http://grants.nih.gov/grants/funding/ac_search_results.htm (last visited July 26, 2009).

14. The rules governing how NIH grants are to be awarded do not necessarily apply to the other Public Health Service agencies (e.g., the Health Resources and Services Administration, the Indian Health Service, the Substance Abuse and Mental Health Services Administration, the Agency for Healthcare Research and Quality, the Agency for Toxic Substances and Disease Registry, the Centers for Disease Control and Prevention). However, in most instances the process used by the other PHS agencies to award grants is similar, especially in its reliance on peer review, although not identical to the one used by NIH.

15. The Center for Scientific Review was previously known as the Division of Research Grants, or DRG.

16. Study sections were previously called Initial Review Groups, or IRGs; however, the term "IRG" is now used to describe the Integrated Review Group. According to the Center for Scientific Review, review activities "are organized into Integrated Review Groups (IRGs). Each IRG represents a cluster of study sections around a general scientific area. Applications generally are assigned first to an IRG, and then to a specific study section within that IRG for evaluation of scientific merit." National Institutes of Health, Center for Scientific Review, http://www.csr.nih.gov/review/irgdesc.htm/ (last visited July 26, 2009).

17. Michele Pagano, *American Idol and NIH Grant Review*, 126 CELL 637 (2006).

18. Many proposals are not even subject to a full review. About a week before the study section meets, the SRO solicits from all members a list of those applications believed not to rank in the top half for scientific merit, and from those lists, the SRO compiles a list of proposals that will not be scored. These applications are considered to have been "streamlined." *See* National Institutes of Health, Center for Scientific Review, http://www.csr.nih.gov/review/peerrev.htm (last visited July 26, 2009). While NIH states that "streamlining is not equivalent to disapproval" (*id.*), as a purely legal matter streamlining and disapproval are synonymous.

19. Public Health Service Act § 405(b)(2)(B), 42 U.S.C. § 284(b)(2)(B).

20. *See* National Institutes of Health, Research Project Success Rates by NIH Institute for 2008, http://report.nih.gov/award/success/Success_ByIC.cfm (last visited July 26, 2009).

21. *See id.*

22. *See* Robert Charrow, *Motherhood, Apple Pie, and the NIH Grant-Review System: Are They Unconstitutional*? 5 J. NIH RES. 86 (1993).

23. *See* Robert Charrow, *A Fly on the Wall: Opening the Peer Review Process to the Public*, 2 J. NIH RES. 79 (1990).

24. Pub. L. No. 81-507, 42 U.S.C. § 1861 *et seq.*

25. *See* Henke v. Dep't of Commerce, 83 F.3d 1445 (D.C. Cir. 1996).

26. There are also "mandatory grants" that require the federal agency to make the award so long as the statutory criteria have been met. Once these criteria have been met, the agency has no choice but to make the grant. *See* City of Los Angeles v. Coleman, 397 F. Supp. 547 (D.D.C. 1975). Nor may the agency award less than mandated by the statute. *See* Train v. City of New York, 420 U.S. 35 (1975).

27. There is limited review where a grantee's noncompetitive renewal request is denied because of its failure to comply with the terms and conditions of a previous award. *See* 42 C.F.R. § 50.404(a)(4).

28. Apter v. Richardson, 510 F.2d 351, 355 (7th Cir. 1975).

29. *See also* Finley v. Nat'l Endowment for the Arts, 795 F. Supp. 1457 (C.D. Cal. 1992) (finding that the NEA's refusal to award certain grants was improper because the decisions were based on considerations improper under the First Amendment).

30. Grassetti v. Weinberger, 408 F. Supp. 142, 150 (N.D. Cal. 1976); *see also* Tsau v. Nat'l Sci. Found., No. 00-cv-0006 (N.D. Ill. 2000) (dismissing a suit by a disappointed grant applicant against NSF).

31. *See* National Institutes of Health, http://grants.nih.gov/grants/multi_pi/Mult_PI_FAQ_6_Feb_2006.doc/.

32. In December 1995, a Republican Congress and Democrat president could not reach a budget accord, and various components of the government were idled for up to four months.

33. 45 C.F.R. § 74.115.

34. This option would not be available for a multiyear standard grant from NSF.

35. *See* 42 C.F.R. § 50.404(a)(4).

36. *See* Snepp v. United States, 444 U.S. 507 (1980).

37. 500 U.S. 173, 199 (1991).

38. *Id.* at 200.

39. Bd. of Trustees of Leland Stanford Jr. Univ. v. Sullivan, 773 F. Supp. 472, 477 (D.D.C. 1991).

40. The same rule does not apply to NSF. Normally, restrictions on grants can be issued informally without going through rulemaking. However, a number of agencies, including HHS, but not NSF, have voluntarily agreed to comply with rulemaking requirements in the case of grants.

41. *See* 45 C.F.R. § 74.43.

42. *See* 2 C.F.R. pt. 220.

43. *See id.* pt. 225.

44. *See id.* pt. 230.

45. *See* 45 C.F.R. pt. 74, app. E; 48 C.F.R. pt. 31.

46. *See* Fly America Act, 49 U.S.C. § 40118.

47. Every year since 1990 Congress has legislatively mandated a provision limiting the direct salary that an individual may receive under an NIH grant and other PHS-funded grants. For FY 2009, the Omnibus Appropriations Act, 2009, Pub. L. No. 111-8, 123 Stat. 524 (2009), restricts the direct salary to the amount paid to individuals serving at Executive Level I of the Federal Executive Pay scale (i.e., members of the president's cabinet). Effective January 1, 2009, the Executive Level I salary level is $196,700.

48. *See* Mark Barnes, Anne Claiborne & Clint Hermes, *Compliance Focus on Time and Effort Reporting*, Ass'n Healthcare Internal Auditors, Fall 2005, at 18, 21.

49. *See* United States *ex rel.* Resnick v. Weill Med. Coll. of Cornell Univ., 04-cv-3088 (S.D.N.Y. Order Mar. 5, 2009).

50. *See* OMB Circular A-133.

51. *See* 45 C.F.R. § 75.25 (NIH).

52. There are two fundamental indirect cost schemes, with variations on each: (1) the Modified Total Direct Cost (MTDC) method, and (2) the Salary and Wage (S&W) method, which is used primarily by smaller institutions. Under the MTDC method, one calculates the indirect cost rate of the institution by dividing the institution's total indirect costs by the MTDC (direct costs less equipment costs, tuition credits, and certain other items). This MTDC indirect cost rate is then applied to the modified total direct costs for each grant, yielding the indirect costs for that grant. Under the S&W method, one calculates the indirect cost rate of the institution by dividing the institution's total indirect costs by the total salary and wages (either with or without benefits). The S&W indirect cost rate will be different from (usually higher than) the MTDC rate. However, as long as one compares apples to apples and oranges to oranges, the indirect costs should be about the same.

53. *See* Office of Science & Technology Policy, Analysis of Facilities and Administrative Costs at Universities (July 2000), *available at* http://clinton4.nara.gov/WH/EOP/OSTP/html/analysis_univ.html/.

54. *See* United States *ex rel.* Biddle v. Bd. of Trustees of Leland Stanford Jr. Univ., 147 F.3d 821, *as amended*, 161 F.3d 533 (9th Cir. 1998).

55. *See* 42 C.F.R. pt. 50, subpt. D. The PHS Grant Appeals Board applies to grant- or cooperative-agreement-related disputes between grantees and the following Public Health Service agencies: (1) National Institutes of Health; (2) Centers for Disease Control and Prevention; (3) Agency for Toxic Substances and Disease Registry; (4) Food and Drug Administration; and (5) the Office of Public Health and Science. The PHS Grant Appeals Board does not have jurisdiction over grants from the other Public Health Service agencies, such as the (1) Substance Abuse and Mental Health Services Administration, (2) Agency for Healthcare Research and Quality, and (3) Health Resources and Services Administration. *See* 42 C.F.R. § 50.402.

56. *See* 45 C.F.R. pt. 16.

57. *See* NSF Grant Policy Manual §§ 922–23 (2005), *available at* http://www.nsf.gov/publications/pub_summ.jsp?ods_key=gpm.

58. *See id.* § 923.

59. *See* 5 U.S.C. § 706(2)(A).

60. *See* 31 U.S.C. § 6303.

61. *See* 48 C.F.R. pts. 1–9999.

62. *See* Competition in Contracting Act of 1984, Pub. L. No. 98–369, 98 Stat. 494.

63. Press Conference, John Glenn, Colonel USMC (Ret.), Kennedy Space Center, Orlando, Fla. (1998).

64. *See* 4 C.F.R. pt. 21.

65. *See* 28 U.S.C. § 1491(b).

66. *See* 48 C.F.R. pts. 1–99 (all agencies); *id.* pts. 300–99 (HHS); *id.* pts. 2500–99 (NSF).

67. *See* Lockheed Martin Corp. v. England, 424 F.3d 1199 (Fed. Cir. 2005).

68. 31 U.S.C. § 6305.

69. *See* Stephen Burd, *Head of U.S. Cancer Research Takes Corporate Job*, Chron. Higher Educ., Jan. 6, 1995.

70. *See* 31 U.S.C. § 3729 *et seq.*

71. *See* Press Release, U.S. Dep't of Justice, Justice Department Recovers Record $3.1 Billion

in Fraud and False Claims in Fiscal Year 2006 (Nov. 21, 2006), *available at* http://www.usdoj.gov/opa/pr/2006/November/06_civ_783.html; *id.*, Justice Department Recovers $2 Billion for Fraud Against the Government in FY 2007; More Than $20 Billion Since 1986 (Nov. 1, 2007), *available at* http://www.usdoj.gov/opa/pr/2007/November/07_civ_873.html.

72. Cook County v. United States *ex rel.* Chandler, 538 U.S. 119, 124 (2003).

73. *See* 31 U.S.C. § 3730(h); *Chandler*, 538 U.S. at 124.

74. *See* Vt. Agency of Natural Res. v. United States *ex rel.* Stevens, 529 U.S. 765 (2000). Even though the Court held that states could not be sued under the FCA, whether directly by the federal government or through a relator, the Court said nothing about the employees of a state institution. It is still very much open to question whether employees of a state university could be sued under the FCA, even though the university could not be.

75. *See, e.g.*, United States *ex rel.* Prevenslik v. Univ. of Wash., No. 03-2180 (4th Cir. 2004) (relator was an engineer; research was funded by NSF); United States *ex rel.* Berge v. Bd. of Trustees of the Univ. of Ala., 104 F.3d 1453 (4th Cir. 1997) (relator was a doctoral candidate; research was funded by NIH); United States *ex rel.* Milam v. Regents of the Univ. of Cal., 912 F. Supp. 868 (D. Md. 1995) (relator was a postdoc; research was funded by NIH).

76. *See* Allison Engine Co. v. United States ex rel. Thacker, 128 S. Ct. 2123 (2008).

77. *See* Fraud Enforcement and Recovery Act of 2009, § 4, Pub. L. No. 111-21, 123 Stat. 1617, 1621-1623.

78. *See* 2 C.F.R. pts. 180 & 215 (Guidance for Governmentwide Debarment and Suspension (Nonprocurement)); 48 C.F.R. §§ 9.400–9.409 (procurement). The HHS rules were previously at 45 C.F.R. pt. 76, but were modified and moved to 2 C.F.R. pt. 376 on March 1, 2007. *See* 72 Fed. Reg. 9233 (Mar. 1, 2007).

79. 2 C.F.R. § 180.800.

80. *See* Arthur Andersen, LLP v. United States, 544 U.S. 696 (2005).

81. *See* 2 C.F.R. §§ 800.835–800.845 (nonprocurement); 48 C.F.R. § 9.406-3 (procurement).

82. *See* Excluded Parties List System, http://www.epls.gov (last visited July 26, 2009).

83. *See* 2 C.F.R. § 180.700.

CHAPTER 3

How Is the "Integrity of Research" Regulated: What Is Research Misconduct?

Whether the falsification or fabrication or plagiarism of data is called scientific misconduct, research misconduct, scientific fraud, or culinary science (i.e., cooking data), it is the antithesis of good science. Such misconduct is also relatively rare, far rarer than even allegations of misconduct themselves, namely charges that someone has falsified or fabricated data or taken inappropriate liberties with the words or ideas of others. For every five scientists accused of misconduct, less than one is found culpable. According to statistics from the Office of Research Integrity, which are summarized in table 1, it would appear that about 17 percent of new cases of misconduct end in a finding that the accused committed misconduct.[1] This chapter examines what constitutes scientific misconduct and the process that is supposed to be used to separate the guilty from the innocent.[2]

The process will be examined from four perspectives—those of the person bringing the allegation, the person against whom the allegation is brought, the university officials who received the complaint and invoke the university's process for dealing with allegations of misconduct, and the faculty committee that is appointed to take evidence and resolve the allegations. As is discussed in greater detail below, federal law requires universities and other grantees to deal with their own dirty laundry. As a condition of receiving National Institutes of Health (NIH) or National Science Foundation (NSF) funding, the grantee promises to take the first cut at resolving allegations of misconduct involving its faculty or other employees. The government becomes involved only when it is time to mete out punishment and publicize the moral lapse, or in that unusual circumstance in which it believes that a crime has been committed.

If you come away from this chapter with only one bit of useful information, I hope it is that the process for resolving allegations of scientific misconduct is not a scientific process but a legal one. Institutions with little experience in dealing with allegations of misconduct sometimes naively be-

TABLE 1. Research Misconduct Cases, by Year

Year	New Cases Opened by Institutions[a]	Cases with ORI Finding of Misconduct
2007	no data	10[b]
2006	86	15[c]
2005	92	8[d]
2004	81	8[e]
2003	105	12[f]
2002	83	13[g]
2001	72	14[h]
2000	62	6[i]
1999	63	13[j]
1998	54	9[k]
1997	64	14[l]
1996	70	17[m]
1995	81	24[n]
1994	64	11[o]
1993	77	no data
Average per year	75.29	11.6
Fraction of opened cases with misconduct findings	0.154	

a. Data in this column were taken from OFFICE OF RESEARCH INTEGRITY, ORI ANNUAL REPORT 2007, at 40 (table 9) (June 2008).
b. *Id.* at 7 (table 5).
c. ORI ANNUAL REPORT 2006, at 7 (table 5) (May 2007).
d. ORI ANNUAL REPORT 2005, at 6 (table 5) (May 2006).
e. ORI ANNUAL REPORT 2004, at 7 (table 5) (May 2005).
f. ORI ANNUAL REPORT 2003, at 6 (table 5) (Aug. 2004).
g. ORI ANNUAL REPORT 2002, at 13 (table 5) (Nov. 2003).
h. ORI ANNUAL REPORT 2001, at 7 (table 5) (July 2002).
i. ORI ANNUAL REPORT 2000, at 4 (table 5) (2001).
j. ORI ANNUAL REPORT 1999, at 7 (2000).
k. ORI ANNUAL REPORT 1998, at i (1999).
l. ORI ANNUAL REPORT 1997, at 2 (1998).
m. ORI ANNUAL REPORT 1996, at 1 (1997) (the Annual Report indicates that there were seventeen findings of misconduct, and one was overturned on appeal; it is unclear whether the seventeen includes the overturned case).
n. ORI ANNUAL REPORT 1995, at 12, 21 (table 3) (July 1996).
o. ORI ANNUAL REPORT 1994, at 30 (table 4) (Apr. 1995).

lieve that the scientific method can somehow be harnessed to magically resolve the allegation and case. I have been involved in more than one hundred misconduct cases, and whenever someone tries to replicate the experiments at issue or bring other forms of complex analysis to bear, the case only spirals out of control. I have never seen a scientific misconduct case resolved using science; I have only seen resolution of that case needlessly delayed to the detriment of all concerned. After all, scientific misconduct is a creature of the law; as such, the tools of the law ought to be used to resolve the case. Legal tools and processes are designed to protect the accused from capricious officials and to ensure that all concerned have had an opportunity to present their case in an orderly fashion. The legal process, rather than slowing things down, actually speeds things up. The history of the federal rules governing scientific misconduct is important because it illustrates these points.

This chapter is divided into four sections. The first recounts the history of the regulation of research misconduct; the second discusses the federal definition of misconduct and differentiates between conduct that is misconduct and conduct that may be antisocial but is not misconduct. The third section discusses the actual procedures at the university level for resolving allegations of misconduct from the four perspectives noted above—complainant, target, institution, and faculty committee. And the fourth section spells out the procedures at the federal level and discusses the hearing process. The government's misconduct rules impose various paperwork obligations on universities and research institutions; detailed discussion of those obligations is beyond the scope of this guide.

A BRIEF HISTORY—FROM THE PILTDOWN MAN TO THE CONGRESSMAN

The Early Years

On the front page of the May 25, 1974, *New York Times*, nestled between a story on the death of Duke Ellington and one on Watergate, was an article that would have as significant an impact on the scientific community as Watergate was having on the political community. The item, headlined "Inquiry at Cancer Center Finds Fraud in Research," detailed the controversy surrounding William T. Summerlin, then chief of transplant immunology at Sloan-Kettering Institute for Cancer Research.[3]

According to the *Times*, Summerlin had admitted to an internal investigating committee at Sloan-Kettering "that [on March 26, 1974] he darkened the skin of two white mice with a pen to make it appear that the mice had accepted skin grafts from genetically different animals" without the use of immunosuppressant drugs.[4] In addition, Summerlin also admitted that on four occasions he had misrepresented other data. The committee recommended that Summerlin be terminated.[5] The Summerlin case marked the first publicly disclosed instance in contemporary times of what would later be termed "scientific misconduct."[6] The case raised eyebrows in the scientific community and led some to question whether the Sloan-Kettering researcher was an anomaly. The matter, though, quickly passed, and all returned to normal—at least on the surface.[7]

In 1980, however, four misconduct cases burst onto the scene: Vijay Soman (Yale University), Marc J. Straus (Boston University), John Long (Harvard University), and Elias Alsabti (various universities).[8] The public revelations about the four cases, coming as they did in quick succession, gave the impression that misconduct was rampant in the scientific community. More critically, the cases revealed that neither universities nor NIH had procedures in place for promptly and fairly resolving allegations of scientific fraud. Coupled with the Summerlin affair six years earlier, the four cases proved to be the seed of the misconduct crisis that was to grip the scientific community for the next decade and a half.

In 1980, Vijay Soman, an assistant professor at Yale, published an article with his laboratory chief, Philip Felig. The article suffered from two fatal flaws. First, much of the article had been plagiarized from a manuscript that Soman had been asked to review for a journal. And second, those data that were original to Soman had been fabricated. Although the matter was discovered prior to publication by the authors of the original article that Soman had reviewed, neither Yale nor Felig aggressively pursued the allegations. Yale lacked procedures for dealing with such allegations, and Felig could not bring himself to believe that his coauthor and student would fabricate and plagiarize. As a result, nothing was done for more than a year. The case finally went public in late summer 1980. Soman was fired, and Felig, who just taken a new position at Columbia, was asked to resign because of his "ethical insensitivity."[9]

Marc J. Straus was a clinical investigator at Boston University who "devised chemotherapy regimes which he claimed produced remissions in 93 percent of patients with small-cell lung cancer."[10] Straus daydreamed aloud about curing cancer and winning the Nobel Prize. In 1978, nurses and other researchers who worked with him charged that he had ordered them to falsify research records of clinical drug trials funded in part by NIH. In an unusual twist, NIH and the Food and Drug Administration (FDA) both asserted jurisdiction and conducted a joint investigation into the allegations, although it is difficult to ascertain from the official record when NIH became actively involved.[11] The matter became public shortly before FDA formally charged Straus with, among other things, falsifying data.[12]

On May 17, 1982, Straus acknowledged that false reports were submitted, that Institutional Review Board (IRB) rules had been violated, that some patients had received doses not in accord with the approved protocol, and that some ineligible patients were enrolled in studies. Throughout the affair, though, Straus maintained his innocence and never acknowledged that he falsified data. He continued to maintain that "false data had been created by disgruntled subordinates."[13] Straus was debarred from receiving government grant funds and declared ineligible to receive investigational drugs for four years.[14]

On the same day that the *Boston Globe* reported on the travails of Boston University's Straus, it carried a story about Harvard's John Long, a thirty-six-year old immunologist at the medical school. According to the press account, Long resigned his position in January 1980, "when he was confronted . . . with evidence that he had faked a portion of the data for a team research project on Hodgkin's disease."[15]

The Elias Alsabti case is different in most respects from those discussed above. Soman, Straus, and Long were legitimate scientists who, for one reason or another, appear to have gone astray. Alsabti was not a scientist as much as a peripatetic miscreant and charlatan who happened to have graduated from a medical school. Over a two-year period, starting in 1978, Alsabti moved from Temple University to Thomas Jefferson University to M.D. Anderson Cancer Center, never lasting more than a few months at any institution. At each place, he managed to commit some form of misconduct; sometimes it was data fabrication and other times it was plagiarism. In fact, while at M.D. Anderson he managed to purloin a manuscript that had been sent for peer review by a journal to a professor at M.D. Anderson. Unbeknownst to the journal, though, the professor had died. Alsabti took the manuscript, substituted his name for that of actual author, and submitted it for publication to a Japanese journal. In 1980, he was publicly exposed and, as a result of his significant academic transgressions, was relieved of his clinical privileges at the Roanoke Veterans Administration Hospital, where he had just started his residency in internal medicine.

Congress Enters the Fray

At least two of these cases—Long and Soman—caught the attention of then congressman Albert Gore Jr. (D-TN), chair of the Oversight and Investigation Subcommittee of the House Committee on Science and Technology. On March 31 and April 1, 1981, the subcommittee held hearings "(1) to ascertain whether the reported cases are anomalies or instead the tip of the iceberg, and (2) to determine whether universities and federal funding agencies were doing an adequate job of detecting and resolving cases of misconduct."[16]

The hearings were remarkably predictable. John Long testified about his misdeeds; Rep. Gore responded by "compliment[ing] [him] on [his] courage in being willing to talk about these very painful events. . . ."[17] Correspondingly, government and university representatives testified that scientific fraud was extremely rare, that the funding agencies and research institutions were capable of handling any allegations, and that science itself is a self-correcting discipline.[18] Accordingly, government regulation to ensure the integrity of science not only was unnecessary but would also set a dangerous precedent. Based on these representations, the hearings ended with a whimper: No report was issued and no legislation recommended. Indeed, the hometown paper, the *Washington Post*, did not even mention the hearings. Timing is everything. Less than two months later, the Darsee case began to unfold.

The Darsee Affair and Its Fallout

In the spring of 1981, two postdoctoral research fellows in Eugene Braunwald's Cardiac Research Laboratory at Harvard Medical School harbored suspicions that John Darsee, another postdoc in the lab, was fabricating data. In mid-May 1981, the two postdocs and a technician observed Darsee fabricating data out in the open. According to Braunwald, "[Darsee] was observed to be labeling recordings that he was making on an instrumented dog, '24 hours,' '72 hours,' 'one week,' and 'two weeks,' with only seconds or minutes between obtaining these tracings."[19] Darsee acknowledged a single transgression, but denied any other wrongdoing. Although Braunwald learned of the strong allegations in May, Harvard failed to undertake a full-scale investigation until November 21, 1981, when the dean appointed an ad hoc committee to investigate the matter. On January 25, 1982, the ad hoc committee concluded that Darsee had fabricated data in three instances.[20]

In fact, as we were later to learn, Darsee had been fabricating data since his undergraduate days at Notre Dame. While at Emory (1974–79), "Darsee's fabrications compromised at least eight published papers," and he appears to have fabricated data for at least thirty-two abstracts.[21] At Harvard, Darsee's dishonesty undermined the integrity of at least twenty-one abstracts and nine published papers.[22] The Darsee case struck a chord—the level of dishonesty was shocking, it was flagrant, and it occurred at one of the most prestigious institutions in the country. If it was happening with such abandon at Harvard, what about the other institutions?

After the Darsee affair, other cases of misconduct began surfacing, and the pressure on both the government and the universities to do something concrete began to mount. On March 22, 1982, two months after the Darsee affair was made public, Congressman Henry Waxman (D-CA), a member of the House Energy and Commerce Committee, introduced legislation to require the director of NIH to respond promptly to allegations of misconduct.[23] The bill died in committee.

Soon thereafter, William Broad and Nicholas Wade, two respected science reporters, authored *Betrayers of the Truth*, a well-written account of selected episodes of cheating in science, including among others Long, Soman, Darsee, and Alsabti. The book was at its strongest when it recounted, fact by fact, the specific tales of dishonesty. The book was at its weakest, though, when it attempted to extrapolate from a few examples the real incidence of misconduct in science.[24] According to the authors, "we would expect that for every case of major fraud that comes to light, a hundred or so go undetected. For each major fraud, perhaps a thousand minor fakeries are perpetrated."[25]

Although Broad and Wade had virtually no data to support their extrapolations, developing events appeared to be consistent with their thesis. Misconduct appeared to be occurring everywhere and at a greater rate than ever before. Between 1981 and 1985, seven new cases of confirmed or admitted scientific misconduct were reported in the press—Mark Spector (Cornell Medical School), Zoltan Lucas (Stanford), Joseph Cort (Mt. Sinai Hospital), Wilbert Aronow (UC Irvine and Creighton), Kenneth Melmon (Stanford), and Robert Slutsky (UC San Diego).

Congress Legislates, but Only a Little

By 1985, statements from the scientific community that misconduct was an extremely rare phenomenon and that intrusive prophylactic measures were unwarranted had little impact on a Congress that appeared to be fast losing patience. Responding to evidence that the procedures used by universities and NIH for resolving allegations of misconduct were ad hoc, less than prompt, and frequently inadequate, Congress, as part of the Health Research Extension Act of 1985, Pub. L. No. 99-158, amended the Public Health Service Act to require that the Secretary of Health and Human Services issue a regulation requiring awardee institutions to establish an administrative process to review allegations of "scientific fraud."[26] The Extension Act provision applied to biomedical or behavioral research funded by grant, contract, or cooperative agreement under the PHS Act. In enacting the legislation, the House Energy and Commerce Committee noted that although "instances of scientific misconduct are relatively infrequent, they are destructive of the public trust. . . ."[27] The committee went on to note that legislation was necessary to help formalize what had been a relatively ad hoc system of responding to allegations of misconduct.[28]

Soon after the Extension Act was signed into law, NIH amended its *Guide for Grants and Contracts* to include "Policies and Procedures for Dealing with Possible Misconduct in Science."[29] The *Guide* was intended to apply to all research activities conducted, funded, or regulated by the Public Health Service. The *Guide* defined "misconduct" as

> (1) serious deviation, such as fabrication, falsification, or plagiarism, from accepted practices in carrying out research or in reporting the results of research; or (2) material failure to comply with Federal requirements affecting specific aspects of the conduct of research—e.g., the protection of human subjects and the welfare of laboratory animals.[30]

In accordance with the Extension Act, the *Guide* placed primary responsibility for resolving allegations of misconduct with the legal recipient of grant funds, the awardee institutions (e.g., universities, research institutions). However, the Department of Health and Human Services (HHS) did not develop rules to implement the Extension Act, as the law expressly required.

In addition, to fulfill its obligations under the Extension Act to establish a process for responding to allegations of misconduct, NIH established a one-person unit within its Office of Extramural Affairs. Within two years, the unit would find itself at the center of congressional hearings. However, in 1986, it appeared that all was well and the likelihood of either significant regulation or further congressional action was minimal. To help emphasize that further government intervention would be inappropriate, the late Daniel Koshland, then editor of *Science*, wrote in what was to become a famous editorial that the "procedures established by the National Institutes of Health and various universities to deal with fraud seem admirable and appropriate." He went on to state that "99.9999 percent of reports are accurate and truthful. . . . There is no evidence that the small number of cases that have surfaced require a fundamental change in procedures that have produced so much good science."[31]

A Confluence of Unfortunate Events and Personalities

The early history of scientific misconduct in the United States involved a rare confluence of unfortunate events—a ferociously zealous congressman, a tenacious but untrained bureaucrat, and a system for resolving allegations of misconduct that was remarkably devoid of basic due process.[32]

THE CONGRESSMAN

Perhaps the most enduring aspect of the history of the regulation of misconduct is the role played by Congressman John Dingell. Dingell chaired the House Committee on Energy and Commerce and also its Subcommittee on Oversight and Investigations. During a four-year period from 1988 to 1992, Dingell relentlessly investigated scientific misconduct, or what he believed was scientific misconduct. In the course of his investigations and hearings, he publicly vilified Nobel Laureate David Baltimore and his colleague, Thereza Imanishi-Kari;[33] Robert Gallo and Mikulas Popovic, two of the discoverers of the AIDS virus; and the famed breast cancer researcher and early lumpectomy advocate, Bernard Fisher (see chapter 2 for the details of the Fisher case). As a result, Baltimore was forced to resign the presidency of Rockefeller University, Imanishi-Kari was suspended from her position as a researcher at Tufts, Popovic found himself unemployable in the United States, and Fisher was forced to resign as head of the prestigious National Surgical Adjuvant Breast and Bowel Project (NSABP) at the University of Pittsburgh.

Dingell's first set of hearings on scientific misconduct on April 12, 1988, focused on Baltimore and Imanishi-Kari.[34] The impact of the hearings transcended the Baltimore case; the hearings stimulated HHS to move quickly to issue regulations that would formally define scientific misconduct and set out procedures to be used by universities and the government for dealing with allegations of misconduct in federally funded research.[35]

THE PROCESS

HHS was concerned that either Dingell or Waxman, also a member of the Energy and Commerce Committee, would seek to enact legislation to deal with misconduct and thereby eliminate the department's flexibility. However, the political appointees within the department were sharply divided on the types of processes that ought to be put in place. Those at NIH and the Office of the Assistant Secretary for Health[36] believed that the process ought to be collegial and nonadversarial. They believed that scientific-based disputes ought to be resolved as scientifically as possible. In contrast, the senior lawyers within HHS believed that the process was inherently legal and that when a person's career is at stake, labeling the process as "collegial" and "nonadversarial" makes little sense. A peephole into this debate was the department's September 19, 1988, Advance Notice and Notice of Proposed Rulemaking, which asked the research community to comment on the type of process that ought to be used and the entity or entities that ought to oversee the process.[37]

In the end, the NIH view prevailed and a nonadversarial system was established, a decision that most scientists would later regret. Specifically, on March 16, 1989, HHS established within NIH an Office of Scientific Integrity (OSI); this new office was charged with investigating and resolving allegations of misconduct.[38] Shortly thereafter, HHS issued its final misconduct rule, which set up a system for dealing with misconduct.[39] Universities were given the primary responsibility for resolving misconduct allegations against their faculty and employees, and they were supposed to exercise this responsibility using a two-step process—the inquiry and the investigation. The inquiry was viewed as the preliminary fact-finding process and could be conducted by a committee or single individual. Its purpose was to determine in sixty days whether there was sufficient reason to believe that misconduct had been committed. If the inquiry determined that there was sufficient evidence that a given researcher had committed misconduct, the university had to report this to OSI and then had to convene a formal investigation (e.g., hearing) to determine whether the researcher had in fact committed misconduct. Although the universities had the primary responsibility for resolving allegations of misconduct in PHS-funded research, OSI could, at any time, swoop in and take the case away from the university.

THE BUREAUCRAT

Creating the OSI proved to be far easier than finding someone willing to direct the new office. Few scientists were eager to put their careers on hold to play data cop, and as cases were piling up and congressional pressure building, the new office was unable to find anyone who was both qualified and willing to act as its first director. However, the leadership void was quickly filled by the deputy director of OSI, Suzanne Hadley. Hadley, a psychologist from the National Institute of Mental Health (NIMH), had been a misconduct officer at NIMH and had cut her teeth on the Stephen Breuning case in the mid-1980s. Breuning, a psychologist at the University of Illinois and later at the University of Pittsburgh, had conducted a series of apparently seminal experiments questioning the use of certain drugs on the mentally retarded; his research had had a direct influence on treatment policy. It turned out, though, that much of his work, largely funded by NIMH, was fabricated. Following a lengthy investigation by NIMH and later the Department of Justice, Breuning ultimately pled guilty in 1988 to two counts of making false statements in federal grant applications. On November 10, 1988, he was sentenced to serve sixty days in a halfway house and five years of probation, and was ordered to pay back $11,352, serve 250 hours of community service, and abstain from psychological research for at least the period of his probation.[40]

Under Hadley, OSI moved quickly to target those cases that it deemed significant and take them away from the university or research institute. Not surprisingly, OSI took jurisdiction over the Imanishi-Kari case and the Gallo-Popovic cases. While universities were required to comply with relatively strict timelines for resolving allegations of misconduct, OSI did not labor under those constraints. Its investigations dragged on for years, and its policies kept changing. Hadley eventually had a nominal boss (Dr. Jules Hallum, a retired microbiologist from the University of Oregon);[41] however, she appeared to be operating on her own and largely under the protection of and in cooperation with Congressman Dingell and his staff. Scientists with cases pending before Hadley's OSI found it a surreal experience, as basic indicators of due process were overtly lacking. First, accused scientists were denied access to both the precise allegations and the evidence against them. Second, OSI's procedures, to the extent one could call them procedures, were more akin to random variables than to a set of definitive rules. The situation was proving untenable for the research community.

The Tide Begins to Turn

The tide began turning against OSI far from Washington, D.C.—in Madison, Wisconsin. James Abbs, a professor of neurology and neurophysiology at the University of Wisconsin, had been charged with misconduct in that he was alleged to have "published certain curves in the journal *Neurology* that were traced from curves he had published previously, rather than being from two different patients as [Abbs] represented."[42] A university committee, after conducting an inquiry into the allegations, concluded that they were "unsubstantiated and [did] not justify or require a more formal investigation."[43]

The university in 1987 forwarded its conclusion to NIH's Office of External Affairs, OSI's predecessor office. Although the university resolved the Abbs case in about one month, it banged around NIH for nearly three years before Hadley concluded that the university's findings were incorrect and an investigation was warranted. Her team of investigators sought to interview Abbs, but he terminated the interview when the OSI team refused to permit his attorney to be present during the interview.

Thereafter, Abbs and the University of Wisconsin instituted suit in federal court against HHS, NIH, and OSI's director Jules Hallum, alleging that Abbs was being denied due process and that OSI's procedures, which changed every few months, had not been published, as required by the Administrative Procedure Act (APA). During oral argument, Chief Judge Barbara Crabb stated that "she was 'shocked' by the procedures used by OSI, that those procedures 'were the work of amateurs,' and that she would find it 'embarrassing . . . to defend them.'"[44] On December 28, 1990, Crabb concluded that OSI's procedures were invalid because they had violated the APA.[45]

Shortly after Crabb's decision, Hallum began asserting himself and attempting to rein in Hadley. This led to Hadley's departure from OSI, although she retained responsibility for completing the Gallo–Popovic and Imanishi-Kari cases. Following Hadley's departure, Bernadine Healy, former head of the Research Institute of the Cleveland Clinic Foundation, became the director of NIH; she realized the OSI lacked real procedures, and that Hadley, who was still overseeing some cases, lacked the analytical skills, legal acumen, and scientific background necessary to bring cases to a reasoned end.[46] She soon relieved Hadley of all responsibility for any OSI case, including the Gallo–Popovic and Imanishi-Kari cases.

Hadley had also been involved in another rather innocuous case, that of Dr. Rameshwar K. Sharma. Sharma had been accused of misstating the status of research in an NIH grant proposal by claiming that he had completed experiments with certain proteins when, according to OSI, he had not. The case probably would have garnered little attention but for the fact that Sharma happened to be on the faculty of the Cleveland Clinic Foundation while Healy had been its director. Dingell hoped to show that the Cleveland Clinic under Healy had bungled the Sharma investigation. Also, during this time, someone at NIH had been feeding Dingell's staff information about the various cases, including the Sharma case. Suddenly, a minor case involving a minor player took center stage. Following congressional prodding, OSI concluded that Sharma had engaged in misconduct.

By this time, no one trusted anyone. Dingell and his staff believed that an OSI under the thumb of NIH, and hence Healy, was not apt to dispense justice—that is, not willing to find scientists culpable.[47] The scientific and legal communities had had their fill of OSI decision making, and they wanted a system that assured those accused of misconduct due process. Lack of mutual trust led to two changes. First, in June 1992, OSI was disbanded and a new entity called the Office of Research Integrity (ORI) was established—not at NIH, but rather downtown in the Office of the Assistant Secretary for Health.[48] Not only did this permit the Secretary of HHS to respond to Dingell's concerns but also to re-retire Hallum, who was viewed by most as in way over his head. Second, on November 6, 1992, HHS announced that scientists accused of misconduct by ORI would be provided with an opportunity for a trial-like hearing before three administrative law judges at the Departmental Appeals Board (DAB).[49] This minor change led to the eventual undoing of ORI's broad jurisdiction.

The first scientist to benefit from this new procedure was Sharma. OSI, now ORI, had charged that Sharma had misstated in an NIH grant application how advanced his work was. Specifically, ORI alleged that Sharma had intentionally claimed in the grant application to have performed experiments with a specific protein when in fact he had not. Sharma agreed that he had not performed the experiments, but he had performed them with another protein and had mistyped the protein at issue. Now, ORI had to prove to three independent judges—two lawyers and one scientist—that Sharma's "error" was intentional. The Sharma trial set the tone for those that were to follow. The government's case against Sharma was weak to begin with. To make matters even worse, the government lawyers, most with little real trial experience, were up against seasoned trial veterans from large Washington law firms, many providing their time essentially pro bono. Following a one-week trial, Sharma was acquitted.[50] The DAB's conclusion in *Sharma* is most telling:

> The sole error—the core of this dispute which has consumed so much time and attention in PHS and elsewhere—is a subscript error in a single sentence. Subscripts are important in distinguishing one protein from another in the same family. The subscript error in question, however, is inconspicuously located and forms no part of any obvious web of deceit. The application contains none of the data or explanations which reviewers of the applications would have expected to see if the erroneous statement were true. Consequently, the error was unlikely to mislead reviewers—and, in fact, the evidence does not establish that it did. On the other hand, the record does show, virtually without challenge, that a typographical error could have occurred as a result of one wrong keystroke using word-processing macros. (The two subscripts which were interchanged in the error in question appeared about 128 times in that one application.) The record also shows that the same subscript transposition by Dr. Sharma and others occurred many times in circumstances where it was much more obviously an inadvertent error.

The next case was one of the big ones—the one against Mikulas Popovic, a co-discoverer of the AIDS virus and co-inventor of the immortalized cell line that allowed scientists for the first time to grow and study the virus; Popovic also was one of the co-inventors of the AIDS antibody test kit. The case against Popovic centered on the meaning of a few words in one of his seminal *Science* articles about the AIDS virus. In acquitting Popovic, the board concluded that "one might anticipate that from all this evidence, after all the sound and fury, there would be at least a residue of palpable wrongdoing. That is not the case."[51]

The Sharma and Popovic cases revealed that ORI had little real evidence against either scientist, notwithstanding years of investigation. The lawyers and bureaucrats at ORI, for whatever reason, failed to analyze the relative merits of either case critically before proceeding to trial. As a result, not only did the government lose both *Sharma* and *Popovic*, but they lost both cases badly. These were cases that should never have been brought; they had been pushed along more by political pressure than by reason.

The political pressure behind both *Sharma* and *Popovic* was nothing when compared to the pressure that had moved an obscure article in *Cell* to the forefront of the lay press and congressional hearings.[52] The case against

Thereza Imanishi-Kari would have had all the indicia of a made-for-TV drama, were it not for the fact that the science itself was beyond the ken of even the average educated American and virtually all in Congress. Nevertheless, the *Cell* article was the focus of three sets of congressional hearings; the Secret Service investigated Imanishi-Kari's data printout to ascertain whether it been altered or forged, and the case itself, once it was tried, featured two Nobel laureates as witnesses for opposing sides—David Baltimore, Imanishi-Kari's defender and coauthor on the *Cell* paper, and Walter Gilbert from Harvard. The DAB

> held a six-week hearing beginning in June 1995. [The DAB] amassed voluminous exhibits, including more than 70 original laboratory notebooks, and a 6500-page hearing transcript. The parties submitted lengthy factual and legal arguments, and ORI proposed thousands of findings of fact and conclusions of law. The record was completed in April 1996.[53]

In acquitting Imanishi-Kari, the DAB "found that much of what ORI presented was irrelevant, had limited probative value, was internally inconsistent, lacked reliability or foundation, was not credible or not corroborated, or was based on unwarranted assumptions."[54] The Imanishi-Kari case had dragged on for about a decade. When it ended, it further confirmed that ORI was simply not up to the task of objectively evaluating evidence. Its standing in the scientific community had become a joke, and many researchers, rightly or wrongly, regarded those at ORI to be failed scientists who were seeking retribution against those who had been more successful. The system had simply broken down.

On May 12, 2000, HHS quietly declawed ORI by stripping it of its authority to investigate and prosecute cases of scientific misconduct and transferring that authority to the Office of Inspector General, as has always been the case at NSF.[55] According to the HHS notice, the "role and structure of ORI will be changed to focus more on preventing misconduct and promoting research integrity through expanded education programs."[56]

WHAT IS RESEARCH MISCONDUCT?

Some General Legal Background

In discussing hard-core pornography, Justice Potter Stewart once wrote that "I know it when I see it, and the motion picture involved in this case is not that."[57] Many may believe that they know research misconduct when they see it, but that is beside point. Unlike hard-core pornography, which is more a function of societal norms and morés than anything else, research misconduct is a creature purely of the law.

Research misconduct is defined in much the same way as one would define a criminal offense (e.g., burglary, murder) or an intentional tort (e.g., assault, battery), namely through a set of elements, some of which involve conduct and some of which involve intent. Usually, the law does not punish conduct alone, but rather requires a marriage of conduct and intent. The prohibited conduct is called the *actus reus* of the crime (or civil wrong), and the associated evil intent is called *mens rea*, literally "evil mind."[58] Thus, for example, the common law defines burglary as the "breaking and entering of a dwelling house of another in the nighttime with the intent to commit a felony therein."[59] Notice how the law separates the elements of the act—(1) breaking and entering, (2) the dwelling house of another, and (3) at night—from the intent—to commit a felony therein. Furthermore, all of the elements must be present to constitute the crime or offense. Thus, if one were to break and enter a barn at night, that would not be sufficient because a barn is not a dwelling house. Correspondingly, suppose someone broke into and entered another person's house at night merely to get in out of the rain. After the rain abates, our would-be felon notices the homeowner's expensive laptop computer. He picks up the laptop on his way out and, shortly thereafter, is arrested. Has he committed burglary? Under the old common law definition, he has not because he did not develop the intent to commit a felony (e.g., stealing the laptop computer) until some time after he broke and entered the house. To commit common law burglary, the intent and the breaking and entering must coexist temporally. While our laptop admirer may not have committed common law burglary, he did commit common law larceny.

Many civil wrongs (called torts) also require this marriage of conduct and intent. For example, a person commits a battery if he or she acts intending to cause a harmful or offensive contact with another person, or an imminent apprehension of such a contact, and harmful contact with a person occurs.[60] Notice how precisely the law defines the term "battery." Suppose, for example, you throw a rock toward your friend John hoping to scare him, but the rock accidentally hits Harry. Have you committed a battery? Yes, because the person you intended to place in imminent apprehension does not have to be the same person who actually gets hit. The law calls this "transferred intent."

It might appear at first blush that the law is unrealistic. How can anyone ever know a person's intent? Does the law really require a judge or jury to ordain a person's intent independently of a person's acts? In fact, the law imposes no requirement of independence. To the contrary, the law permits one to infer a person's intent from that person's acts. A set of trivial errors, for instance, that has little bearing on the vitality of a scientific paper's results would tend to indicate that the errors were inadvertent; that same set of errors all going in the same direction and likely to enhance the paper's chances of being published may suggest something else.

The law also draws fine distinctions between various types of intent. Actual intent to commit a crime (usually called specific intent) differs from intent to commit the specific acts that may be a crime (usually called general intent). Specific intent focuses on the crime itself while general intent focuses on the conduct.[61] For example, burglary is a specific intent crime—one has to have a specified intent (i.e., an intent to commit a felony). Murder is also a specific intent crime—one has to intend to take the life of another without justification. However, statutory rape and manslaughter are general intent crimes. In the case of statutory rape, one merely has to intend to have sexual intercourse; one does not have to intend to have intercourse with someone under a given age. Thus, a person can be guilty of statutory rape even if the person believed that the girl was in fact above the age of consent. Correspondingly, voluntary manslaughter is usually a general intent crime. The person merely has to intend to perform the acts that led to the death, as opposed to murder, where the person has to intend to cause death. As we will see, research misconduct is now more akin to a general intent crime than it is to a specific intent crime.

What Is Fabrication and Falsification?

With this background in mind, let's examine how the law defines "research misconduct." Bear in mind that while the HHS definition applies only to research funded under the Public Health Service Act (e.g., NIH, Human Resources and Services Administration [HRSA], Substance Abuse and Mental Health Services Administration [SAMHSA]), other funding agencies (e.g., NSF) use a similar definition, as do most universities and research institutions for all research conducted at their institution irrespective of the funding source.[62] The HHS definition is set out below, and a link to the complete rule, as well as to the NSF rule, are set out in appendix C:

> **42 C.F.R. § 93.103 Research misconduct.**
> Research misconduct means fabrication, falsification, or plagiarism in proposing, performing, or reviewing research, or in reporting research results.
> (a) Fabrication is making up data or results and recording or reporting them.
> (b) Falsification is manipulating research materials, equipment, or processes, or changing or omitting data or results such that the research is not accurately represented in the research record.

(c) Plagiarism is the appropriation of another person's ideas, processes, results, or words without giving appropriate credit.
(d) Research misconduct does not include honest error or differences of opinion.

42 C.F.R. § 93.104 Requirements for findings of research misconduct.
A finding of research misconduct made under this part requires that—
(a) There be a significant departure from accepted practices of the relevant research community; and
(b) The misconduct be committed intentionally, knowingly, or recklessly; and
(c) The allegation be proven by a preponderance of the evidence.

As you will note, section 93.103 focuses on "conduct" and section 93.104 focuses on "intent." Each section sets out elements in much the same way that the law defines burglary.

CONDUCT: WHAT ACTS CONSTITUTE FABRICATION OR FALSIFICATION?

Scientific or research misconduct is defined in terms of (1) fabrication, falsification, or plagiarism (referred to as FFP) in (2) proposing, performing, or reviewing research, or in reporting research results. The terms "fabrication," "falsification," and "plagiarism" all connote some form of evil intent, and therefore it is difficult to tease out the conduct from the intent. The idea of unintentionally plagiarizing or unintentionally fabricating is difficult to fathom. Let's put aside, for the moment, the intent and focus exclusively on the nature of the conduct itself.

Fabrication and falsification are overlapping, and it is unclear why we even need both when either would do quite well. If one makes up data (i.e., fabricates data), then by definition those data are false; they were not generated by the experiment as outlined in the paper or presentation. Conversely, if one manipulates data to generate results that are different than would have been generated had the experiment been properly conducted and the data properly recorded, it is difficult to see how that falsification is not also fabrication. A subtle distinction, however, can be made between the two. Fabrication is making up data out of whole cloth. Falsification is changing data that in fact exist or manipulating inappropriately specimens, equipment, or data so that the research results are not accurately represented in the research record. Fabrication, though, does involve falsification because the statements that are being made about the data are in fact not true. The term "research record" is defined as

> the record of data or results that embody the facts resulting from scientific inquiry, including but not limited to, research proposals, laboratory records, both physical and electronic, progress reports, abstracts, theses, oral presentations, internal reports, journal articles, and any documents and materials provided to HHS or an institutional official by a respondent in the course of the research misconduct proceeding.[63]

Thus, it would appear that a postdoc who orally reports data or results during an informal laboratory meeting has affected the "research record" because this informal talk is an "oral presentation" or an "internal report," or both.

Fabrication or falsification can take a host of forms limited only by human inventiveness. Sometimes there is no question but that data have been fabricated or falsified. For example, a clinical coordinator reported having conducted follow-up interviews with two patients on two specific dates. The only problem was that the patients had each died before the interviews took place. The ORI viewed these transgressions as "fabrication." Interview dates for two other patients were incorrect. ORI viewed these two errors as falsifications.[64] In another recent case, a psychology graduate student at UCLA falsified and fabricated "data and statistical results for up to nine pilot studies on the impact of vulnerability on decision making from Fall 2000 to Winter 2002 as a basis for her doctoral thesis research."[65]

In other instances, there may a significant dispute over whether data or results were falsified or fabricated. For example, a University of Illinois at Chicago professor was found by the university and the government to have committed research misconduct by using a figure in an NIH grant application that the government believed had been either falsified or fabricated. The researcher denied the allegations, claiming that his original data were missing as a result of "involuntary relocation of his laboratory." Ultimately, the case settled. The researcher, without admitting guilt, agreed to certain sanctions, namely not submitting grant applications without supervision and not serving on any study sections for a specific period.[66] Finally, in another case a University of North Carolina professor was found to have "engaged in scientific misconduct by falsifying DNA samples and constructing falsified figures for experiments done in his laboratory to support claimed findings of defects in a DNA repair process that involved rapid repair of DNA damage in the transcribed strand of active genes, included in four grant applications and in eight publications and one published manuscript."[67] The professor contested the charges, arguing that "a systematic error was introduced into the experiments in question and he recognizes that it could have influenced or accounted for the results."[68] He further stated that he "entered into a Voluntary Exclusion Agreement [under which he would not undertake PHS-sponsored research for five years] because he cannot sustain the significant financial burden of a legal proceeding to resolve the disagreements between his position and that of HHS."[69] As will be discussed later, the researcher's claim that he could not pursue further legal avenues because of the expense is not an idle excuse. The cost to mount a full challenge to an adverse government finding can easily exceed $500,000, making it beyond the means of many, if not most, researchers.

As noted earlier, most cases (more than 80 percent) of alleged misconduct end with a decision in favor of the accused (see table 1). Some of these cases are reported by the government, but without the investigator's name and without sufficient detail to understand what the real issues were.[70] In many cases, the alleged transgression, even if it occurred, would not constitute research misconduct. For example, disputes over authorship priority are a constant source of misconduct allegations. Whether someone is first or third author on a paper does not fall within the definition of misconduct. In other cases, the alleged fabrication or falsification never occurred; the reported data turned out to be correct. In one case, for example, a postdoc, after unsuccessfully attempting to replicate a series of experiments by a researcher involving sister chromatid exchange, filed misconduct charges against the researcher and the heads of the laboratories involved. The matter ultimately made its way into federal court, and, after years of litigation, it was discovered that the postdoc had been unable to replicate the studies because she had consistently reversed two steps in the protocol.[71] The postdoc gave up science and became a lawyer (not a joke). In other cases, it cannot be determined one way or another whether the results are correct or not. Those tend to be factually troubling cases, but not as difficult to resolve as those cases in which everyone agrees the data or results are erroneous and the disagreement focuses on the underlying intent: Were the errors inadvertent or due to sloppiness or due to some malevolent intent? Inadvertence and sloppiness are not punished, at least by the government (except FDA); evil intent, though, can be.

Intent is not relevant when it comes to research under FDA's jurisdiction (i.e., clinical research aimed at testing a new drug or device). FDA relies on the data from clinical trials to decide whether to approve a new product for a specific use and, if it does approve it, the warnings and contraindications that ought to appear on the labeling. As a result, FDA has adopted a relatively inflexible standard—if a researcher is sloppy or if someone working under the researcher's supervision is sloppy, then the researcher can be punished or sanctioned. FDA's tolerance for error can be relatively low. Specifically, a researcher can be disqualified from receiving test articles (i.e., unapproved drugs, biologics, or devices) if he or she "repeatedly or deliberately failed to comply with the [regulatory] requirements . . . or has submitted to FDA or to the sponsor false information in any required report."[72] In egregious cases, investigators have been criminally prosecuted for failing to keep complete and accurate records.[73] Recently, for instance, a court of appeals reinstated fifteen counts of an indictment against a clinical researcher whom FDA had charged with failing to keep complete and accurate records of her clinical research.[74] The researcher, Maria Carmen Palazzo, a psychiatrist, had been retained by SmithKline Beecham Corporation (SKB) to conduct clinical trials concerning the safety and efficacy of Paxil in children and adolescents. The agreement and FDA regulations required Palazzo to maintain records of the trials and report her results to SKB, which, in turn, was required to report to FDA. The indictment alleged that Palazzo had failed to comply with the study protocol and had failed to maintain appropriate records. Palazzo had already been convicted of Medicare and Medicaid fraud.[75]

INTENT: WHAT STATE OF MIND UNDERLIES FABRICATION OR FALSIFICATION?

In cases in which the data or results are plainly wrong and there is an allegation of misconduct, one has to determine whether the errors amount to fabrication or falsification. Errors, though, occur all the time in science, especially in laboratory work where many things can and usually do go wrong. How does one determine whether a researcher had the requisite intent that transforms an error—or usually a set of errors—into fabrication or falsification? The definition of misconduct actually incorporates two elements of intent. First, the actions have to be sufficiently egregious to support a finding of evil intent, and second, it must be established that the errors were done "intentionally, knowingly, or recklessly."

Significant Departure

To separate run-of-the-mill laboratory errors from conduct that should be sanctioned, the government has injected the requirement that the error or the way in which it occurred has to deviate significantly from the norm. Thus, for example, if one tosses out four outliers and all are 5 standard deviations from the mean, that might well be acceptable, even if the data cleansing made the results publishable when they otherwise might not have been. On the other hand, if one tossed out two outliers that were 1.5 standard deviations from the mean, that might not be acceptable, especially if it transformed a nonpublishable data set into a publishable one. The concept of "significant departure" does not mean that one can intentionally falsify a small amount of data; falsification, by definition, incorporates the requisite *mens rea* to sustain a misconduct finding. The concept of "significant departure" recognizes that science is error-prone and that neither the government nor universities can or even should police scientific competence—that is the purpose of the peer review system. It also recognizes that neither the government nor universities should attempt to punish nonconformity or promote conformity. Science has made some of its greatest strides when individuals have had the courage and intellect to challenge scientific orthodoxy. These types of challenges should be encouraged and not discouraged. The concept of "significant departure" is designed in part as a recognition that the last thing the government wants to do is to stifle creativity or to even chill it.

Intentionally, Knowingly, or Recklessly

The three terms that form the basis of the intent necessary to support a misconduct finding have very different meanings. The term "intentionally" normally means that the actor intended the consequences of his or her act, namely that he or she intended to falsify or fabricate data or results. It is a term usually reserved for a specific intent offense. The term "knowingly" tends to be reserved for a general intent offense. All one has to do is knowingly perform the acts, even if one does not fully appreciate the consequences of those acts. However, it is difficult to see how one can "knowingly" generate erroneous data without having a specific intent to do so. Thus, in the context of scientific misconduct, it would appear that the terms "knowingly" and "intentionally" have similar meanings. By contrast, the term "recklessly" is a term that really denotes gross negligence or willful indifference. Thus, a researcher can be found to have committed misconduct if his or her errors were the result of significant departures from accepted practices of the relevant research community and the researcher was grossly negligent. Suppose a researcher permits an untrained lab assistant to perform a complicated experiment. The lab assistant had no idea of what he was doing, and his results were due to serious laboratory errors that someone with even a modicum of training would not have made. The senior researcher publishes the results without further ado. Has the senior researcher exhibited the requisite intent necessary to establish research misconduct, even though he had no intent to publish incorrect data and did not know that the data were wrong?

Honest Error

The concept of "honest error" is the antithesis of "intentionally, knowingly, or recklessly" generating incorrect data or results. While neither the government nor universities are required to prove a lack of honest error to sustain a finding of misconduct, in fact the only way that they can ever truly show the requisite intent is by disproving that the error was honest. Interestingly, honest error, when used as defense to a charge of misconduct, is treated as what lawyers call an affirmative defense. That means that the respondent scientist has the burden of introducing evidence to show that the admitted error was honest and not result of an evil intent. What then is an honest error? The Sharma case, discussed above, and the first one to go through a full hearing before the DAB, actually involved what the board concluded was nothing but a single typographical error repeated many times. In that case, the respondent was able to show how he mistyped the subscript.[76]

What Is Plagiarism?

WHAT IS THE THEORY BEHIND PUNISHING PLAGIARISM?

Plagiarism is decidedly different from either fabrication or falsification. Whereas fabrication or falsification involves the dissemination of incorrect information, plagiarism usually involves the overdissemination of correct information. Fabrication or falsification undermines the validity of the scientific record leading unsuspecting researchers down the wrong path or even worse, leading policymakers to rely on scientifically infirm information to make decisions affecting significant segments of the population. Plagiarism, in contrast, is an offense affecting attribution or credit. Putting aside the federal definition of misconduct, plagiarism, in and of itself, is not legally wrong—civilly or criminally. How should one define plagiarism? Plagiarism is not the same as copyright infringement, which is discussed in chapter 7. Plagiarism involves the failure to attribute; copyright infringement involves using someone's materials even with attribution, but without their consent.

In the late eighteenth century, three men writing under the pen name Publius took turns writing letters to the editor of the *Independent Journal*, a semiweekly newspaper in New York City. Publius was a plagiarist, of sorts. Shortly after their original publication, the letters were compiled into a book. However, the identity of the authors remained a closely guarded secret. This enabled one of them, in a note discovered after his death, to lay claim to having written sixty-three of the letters, some of which were plainly authored by one of the other two. The actual authorship of the letters would likely have engendered little curiosity were it not for the fact that the three authors were Alexander Hamilton (the deceased claimant), James Madison (later Secretary of State and president), and John Jay (the first Chief Justice of the United States). Their letters (published in 1787–88), collectively known as

The Federalist Papers, were extraordinarily influential discourses extolling the virtues of the United States Constitution, as yet unratified.

Owing to the importance of *The Federalist Papers*, Hamilton's note, according to one historian, "touched off a tortuous dispute that went on for generations between [Hamilton's] political heirs and those of Madison over the authorship of . . . various papers."[77] It was not until the mid-1950s that the late, renowned American historian Douglass Adair (husband of the late American poet Virginia Adair), using a variety of stylistic indicators, unraveled the mystery. He concluded that Madison had written twenty-six, Jay had written five, three were jointly written by Madison and Hamilton, and Hamilton had written fifty-one, or nine fewer than he claimed to have written, counting those that were jointly written. Few, other than historians, actually cared whether Hamilton had written fifty-one, fifty-four, or sixty-three of the letters; he was not seeking either tenure or grant money, but rather the presidency. Was Hamilton a plagiarist?

In his *The Little Book of Plagiarism*, University of Chicago law professor Richard A. Posner, also a judge on the United States Court of Appeals for the Seventh Circuit, notes that many judges do not write the opinions that bear their names.[78] They are written by their young law clerks, and the names of those clerks appear nowhere on those opinions. Does this mean that most of the federal bench is populated by plagiarists? Lawyers, not infrequently, expropriate for their own briefs good chunks of briefs written by other lawyers. While there are many reasons why people think ill of lawyers, should we add plagiarism to this list?

Let's examine each of these three cases in reverse order. Lawyers are paid by the hour. The longer it takes a lawyer to write a brief, the more it costs the client. Using words written by other lawyers saves lawyers' time and their clients' money. The clients are therefore happy—indeed overjoyed if the brief takes less time to write. What about the judges who read the briefs? Why should they care who wrote the text? After all, they are probably text borrowers themselves. In the law, all that really matters, from the judge's perspective, is who signed the brief and is therefore taking responsibility for its accuracy. Also, in large law firms, portions of briefs that were not borrowed from briefs written by other lawyers are likely written by junior associates in the law firm; their names may not appear on the brief. The reader, therefore, has no expectation that the person who signed the brief actually wrote the brief. The same is true for judges. No one who reads a judicial opinion actually believes that it was written in its entirety by the judge who signed it. Again, there is no expectation that the author actually wrote the words. The same is true for Hamilton and *The Federalist Papers*. They were published under a pseudonym, so there was no expectation of authorship from the get-go; no reader could have been deceived.

Many, including Posner and myself, have long argued that plagiarism is a form of fraud or deceit requiring that the reader or listener actually believe that the text was written by the named author or speaker. When there is no such expectation, there is no deceit, and it is difficult to find an injury. That does not mean that plagiarism, when it occurs, injures only the reader. In the classroom environment, when a student plagiarizes, the primary victims are those other students against whom the plagiarist is competing for that "A." When an author of a novel or history plagiarizes, the readers may be one set of victims, but the real victims are those whose works have been plagiarized and who are competing against the plagiarist for book sales. In sum, plagiarism involves more than merely copying another's words or ideas without attribution. It involves deceit—at the very least, the readers have to be deceived. When there is no deceit, as in the cases of Hamilton, judges, or lawyers, there is no plagiarism.

One curious case that does not fit the normal mold involves Vice President Joseph Biden, previously a long-serving Senate Democrat from Delaware who ran unsuccessfully for the presidency in 2008 and, twenty years earlier, in 1988. Biden was forced to drop out of the 1988 presidential race after it was revealed that a significant chunk of one of his speeches had been purloined from a speech by Neil Kinnock, the then-head of the British Labour Party. Ironically, Biden had used the material before and had correctly credited Kinnock on those other occasions, but the one time his speech was recorded (apparently by aides of one of his opponents), he had neglected to do so. Was Biden a plagiarist?

How could Biden be a plagiarist? How many voters believe that politicians write their own speeches? The White House employs a cadre of speech writers, as do senators and even lowly members of the House. Presidential candidates are no different; their speeches are ghost written like any other politician's speech. Biden was nailed not because he plagiarized, but more likely because one of his ghost writers plagiarized and because Biden stated that he just thought of the comments on his way to give the speech. The latter was a clear lie.

I stated earlier that plagiarism, by itself, is not a civil or criminal wrong, but it can give rise to a civil or criminal case. For example, plagiarism can constitute copyright infringement if the work is not in the public domain. Take the case of Doris Kearns Goodwin, the presidential and baseball historian seen most frequently on the Public Broadcasting System (PBS). A number of years ago, the *Weekly Standard*, a political magazine of the right, accused Goodwin of plagiarizing from another historian, Lynne McTaggart, portions of Goodwin's 1987 bestseller, *The Fitzgeralds and the Kennedys*.[79]

According to news accounts and Posner, Goodwin settled with McTaggart by paying her money, ostensibly for copyright infringement. However, when material is in the public domain, either because it was never copyrighted (e.g., Shakespeare's works) or because the copyright has expired, (e.g., Dickens's works), then there can be no copyright infringement.

Plagiarism can also lead to suits for breach of contract and for unfair competition. Kaavya Viswanathan, the Harvard undergraduate who purloined significant chunks of her novel from the work of another published author, faced legal liability not only from the other author for copyright infringement, but also from her own publisher, Little, Brown & Co., for breach of contract. Upon learning of the plagiarism, Little, Brown was forced to recall the novel and therefore suffered significant damages in addition to the $500,000 advance it had paid the young author for a two-book deal.

Interestingly, most suits that are based on plagiarism are actually brought by the alleged plagiarizer, usually challenging a university's decision to expel the student plagiarizer. In one case, a law school graduate was denied admission to the bar because he had plagiarized a paper while in law school. The state supreme court, while recognizing that "plagiarism, the adoption of the work of others as one's own, does involve an element of deceit, which reflects on an individual's honesty,"[80] nonetheless concluded that the applicant's actions in copying nearly verbatim twelve pages of a thirty-page paper did not "demonstrate such lack of character."[81] The court ordered that he be admitted to the practice of law in Minnesota.

WHAT IS THE FEDERAL DEFINITION OF PLAGIARISM?

Up to this point, we have focused more on the rough boundaries of plagiarism. We have not, for example, analyzed the intent necessary to support a claim of plagiarism or the precise type of conduct (e.g., verbatim copying versus paraphrasing versus appropriation of ideas). Nor have we focused on the federal definition, which defines plagiarism as

> the appropriation of another person's ideas, processes, results, or words without giving appropriate credit.[82]

Under the federal definition in the research misconduct rule, the conduct at issue must represent a significant departure from the accepted practices in the relevant research community and must be committed intentionally, knowingly, or recklessly.

The federal definition easily accommodates practicing judges, lawyers, and politicians. In their respective communities, copying the words of others without attribution is no sin and thus would not represent a departure, significant or otherwise, from that which is commonly accepted. Judges and lawyers who are writing for law reviews or other academic journals are likely not to be afforded this same leeway. They would be judged by normal academic standards, which arguably do not tolerate copying someone else's words.

Verbatim Copying versus Paraphrasing

In theory, there should be little difference between how one treats verbatim copying and a near-verbatim paraphrase. It is much more difficult, though, to prove that someone paraphrased someone else (without attribution) than it is to prove verbatim copying. When someone paraphrases someone else's writing, for instance, they could claim that they did so inadvertently or came up with wording independently; verbatim copying is far less susceptible to this defense.

Plagiarism of verbiage is normally established through a subjective probability assessment—it is extremely unlikely that two people, working independently, would generate text that is identical in every way. Also, psycholinguistic research shows that it is extraordinarily difficult for someone to remember text word for word, especially subconsciously, but much easier for someone to remember the "gist" of the text. As one moves away from cases involving verbatim copying to paraphrasing, plagiarism becomes more difficult to establish. However, where paragraph after paragraph in one text matches the gist of paragraphs in another text, the defense of "inadvertence" becomes less plausible, even though the two texts are not identical. This was the case with Doris Kearns Goodwin. Here are some excerpts as they appeared in the *Weekly Standard*:

McTaggart, for example, writes that

> "her [Kathleen's] closest friends assumed that she and Billy were 'semiengaged.' On the day of the party reports of a secret engagement were published in the Boston papers. . . . The truth was that the young couple had reached no such agreement." (p. 65)

The corresponding passage in Goodwin's book differs by just a few words:

> "her [Kathleen's] closest friends assumed she and Billy were semi-engaged. On the day of the party, reports of a secret engagement were published in the Boston papers. . . . The truth was that the young couple had reached no such agreement." (p. 586)

McTaggart:

> "Hardly a day passed without a photograph in the papers of little Teddy, taking a snapshot with his Brownie held upside down, or the five Kennedy children lined up on a train or bus." (p. 25)

Goodwin:

> "Hardly a day passed without a newspaper photograph of little Teddy taking a snapshot with his camera held upside down, or the five Kennedy children lined up on a train or bus." (p. 523)

McTaggart:

> "Mrs. Gibson gave a tea in her honor to introduce her to some of the other girls—hardly a routine practice for new recruits." (p. 130)

Goodwin:

> "Mrs. Harvey Gibson gave a tea in her honor to introduce her to some of the other girls—hardly a routine practice for new recruits." (p. 666)[83]

Even when copying is involved, plagiarism may be difficult to establish. How much does one have to copy to be labeled a "plagiarist"? Do five words constitute plagiarism? What about twenty-five words? What about paraphrasing the words of another in describing a well-known laboratory technique? What about a verbatim copy of another's description of a well-known laboratory technique where credit is given to the person who developed the technique but not to the person whose description was copied? For example, suppose that scientist A develops a given laboratory technique that becomes widely used. Scientist A's description is written in remarkably pedestrian prose: short sentences, all ordered subject, verb, object. Some years later, scientist B comes along and describes scientist A's technique in far more literary fashion, using less wooden language. Scientist C copies verbatim scientist B's description of scientist A's technique, but credits only scientist A. Has scientist C committed plagiarism? If in fact the discipline at issue places some independent value on the literary value of a given write-up, then perhaps scientist C is a plagiarist. If, however, the discipline is one that places little value on a turn of phrase, it may be difficult to argue that scientist C is a plagiarist.

Compare this with two real cases. In the first case, a senior researcher submitted a grant proposal to NSF containing "several paragraphs in its literature review that were identical or substantially similar to material in an article published by two other scientists."[84] The scientist argued in his defense "that when he submitted the proposal, he believed the grammatical changes he made in the original text rendered the use of quotation marks inappropriate, and that he 'may have erred by using parts of sentences verbatim without proper citation.'"[85] The scientist's university found that he had plagiarized, and the NSF inspector general concurred in that finding.

In the second case, "a review panel member alleged that a researcher included in an NSF proposal a paragraph describing a laboratory procedure that was practically identical to a paragraph in a published article written by another scientist. Further inquiry revealed two additional instances in which the subject had incorporated this paragraph into proposals without proper citation."[86] In the end, the researcher's university concluded that the transgression was not sufficiently serious to warrant a finding of misconduct under the NSF definition, and the Office of the Inspector General (OIG) concurred. The OIG reported that it believed that to constitute misconduct under the NSF definition, the conduct had to be "a serious deviation from accepted practices" and that

> several factors, no one of which alone would disqualify an act from being misconduct, mitigated the seriousness of what the subject did. Among these were the following facts:
>
> The copied paragraph occurred in proposals in which the article was frequently cited.
> The subject made clear the source of the ideas. The only originality of the passage that the subject copied lay in its original combination of words.
> The passage itself was only one paragraph long.
> The subject was an inexperienced investigator with a limited command of the English language who had been trained in another country.[87]

There is some conduct that scientists may tentatively label "plagiarism" which by definition cannot be plagiarism. One is permitted to copy verbatim one's own prior work without attribution: One cannot plagiarize from oneself. "Self-plagiarism" is an oxymoron. What happens, though, when a paper is jointly authored by five scientists, and one of those scientists copies a couple of paragraphs from the paper for use in another article, all without attribution. Under normal rules of copyright law, all authors are deemed jointly to own the article, and each can independently use it or license it to another without the consent of the coauthors.[88] Since under copyright law each has the right to full use of the article as if he or she were the sole author (other than an obligation to split any resulting royalties), use of a portion of the article by one of the authors without attribution is really a form of "self-plagiarism" and hence, not plagiarism at all.

In the end, plagiarism, other than its most obvious manifestations, may be difficult to prove. Plagiarism, like falsification and fabrication, must represent a significant departure from what is commonly accepted in the respective scientific community, and it must be accomplished with the requisite intent.

Words versus Ideas

The federal definition diverges from what most would consider plagiarism in one important respect—under the federal definition, an idea or process can be plagiarized. For many reasons, this aspect of the federal definition is counterproductive. Ideas, unlike lengthy passages of text, can be developed independently. In science, it happens all the time. Were Isaac Newton, Charles Darwin, and Howard Temin plagiarists because each developed an idea at about the same time as others did? Newton and Gottfried Leibniz developed calculus at about the same time, but independently. Darwin and Alfred Russel Wallace developed the theory of evolution at about the same time, but independently. And Temin and David Baltimore each discovered reverse transcriptase at about the same time, but independently.

Second, ideas build on one another. Often it is difficult to discern when one idea blends into another or is used as a stepping stone for someone's major discovery. Sometimes, the scientists themselves may not even remember what they relied on in developing their theories. Albert Einstein, for example, gave conflicting accounts when asked whether he had relied on Albert Michelson's and Edward Morley's work in developing his theory of special relativity.

Third, there is a real question as to whether ideas deserve any protection at all. Part of the scientific process is the ability to throw ideas around informally. Does one have to keep copious records of what everyone says at one of these bull sessions to protect oneself from a charge of plagiarism? Because unadorned ideas are virtually impossible to protect, our patent or copyright laws make clear that an idea, by itself, is not subject to protection. There is no reason to believe that those at a university or at ORI would be any better at discerning who developed which idea first than those at the Patent and Trademark Office or the Library of Congress, which has jurisdiction over copyrights. In the course of discussing the near-simultaneous discoveries by Newton, Leibniz, Darwin, and Wallace, Posner states that the "most important distinction between plagiarism of verbal passages . . . on the one hand, and plagiarism of ideas, on the other—a distinction that suggests that much copying of ideas isn't plagiarism at all—is that old ideas are constantly being rediscovered by people unaware that the ideas had been discovered already."[89]

What Is Not Research Misconduct?

While fabrication, falsification, and plagiarism capture most misconduct in science, there are actions that may be equally troubling but that do not fall within the narrow definition and hence are not misconduct; not all bad behavior in a laboratory is research misconduct. We discuss some of the more common types of behavior or misbehavior that do not meet the regulatory definition of misconduct. Some of the conduct is just poor form; some, however, while not research misconduct, can give rise to criminal liability. For example, there have been more than a few cases in which competition in a single laboratory has gotten out of hand, especially among postdocs and junior researchers, to the point where one researcher sabotages another's research. While this childish conduct does not constitute fabrication or falsification, it is likely a crime in most states (e.g., destruction of property) and may well be a federal crime if the experiment is being supported by NIH or NSF. Correspondingly, spending grant money on lavish vacations or purloining scientific equipment, while not research misconduct, can constitute grand theft or other state or federal crimes for which scientists have been prosecuted criminally.

AUTHORSHIP DISPUTES

The laboratory, like any other workplace, is the site of many bizarre disagreements and antisocial behavior, most of which, while improper, is not research misconduct. The most common form of nonsubstantive disagreement among scientists involves issues of authorship—who is or ought to be an author and who should be the first author, second author, and so forth. There are some basic, commonsense rules governing authorship that have been published both by NIH and by the International Committee of Medical Journal Editors (ICMJE). However, authorship policies and customs vary dramatically from field to field, making a single set of policies unworkable; furthermore, custom (e.g., the actual practice) deviates from the policy (e.g., the written guidelines). In some disciplines, where formal collaborations are the norm (e.g., high-energy physics), the rules governing authorship are set out in a collaboration agreement, a form of contract. Where formal collaboration agreements are lacking, as is most often the case, researchers should be guided by university policy and common sense. Because authorship disputes have become endemic, many institutions have formal policies and appeals procedures that they expect their faculty, graduate students, and staff to follow and use. One of the more detailed sets is published by the University of Pennsylvania, which lists authorship policies by discipline.[90] Another, published by the ICMJE, recommends the following:

- Authorship credit should be based on 1) substantial contributions to conception and design, or acquisition of data, or analysis and interpretation of data; 2) drafting the article or revising it critically for important intellectual content; and 3) final approval of the version to be published. Authors should meet conditions 1, 2, and 3.

- When a large, multi-center group has conducted the work, the group should identify the individuals who accept direct responsibility for the manuscript. These individuals should fully meet the criteria for authorship defined above and editors will ask these individuals to complete journal-specific author and conflict of interest disclosure forms. When submitting a group author manuscript, the corresponding author should clearly indicate the preferred citation and should clearly identify all individual authors as well as the group name. Journals will generally list other members of the group in the acknowledgements. The National Library of Medicine indexes the group name and the names of individuals the group has identified as being directly responsible for the manuscript.
- Acquisition of funding, collection of data, or general supervision of the research group, alone, does not justify authorship.
- All persons designated as authors should qualify for authorship, and all those who qualify should be listed.
- Each author should have participated sufficiently in the work to take public responsibility for appropriate portions of the content.[91]

Under the ICMJE guidelines, providing either a reagent or the funding is insufficient, by itself, to warrant authorship. The ICMJE guidelines, as well as the guidelines for most disciplines, disapprove of honorific authorship (e.g., the head of the laboratory is always last author, even if he or she had nothing to do with the research other than provide the funding). The reality (and hence custom), though, tends to be the opposite. In most large laboratories, the head of the laboratory is usually the last author; in some cases it is justified under the ICMJE guidelines, and in other cases it may not be.

Although authorship problems are difficult to resolve, they are not intractable. There is a variation on the theme that is more difficult to resolve—it involves the inverse of a claim of authorship and is known as "authorship extortion." This occurs when one putative author, dissatisfied with his or her authorship position, refuses to sign the journal's copyright license form or form authorizing publication unless his or her authorship position is improved. These problems are difficult to resolve, although in most cases, the recalcitrant author is removed altogether from the authorship list and left with an acknowledgment and a reputation that may make it difficult for him or her to find collaborators in the future.

On rare occasions, authorship disputes end up in court. In one recent case, Roberto Romero, the chief of the Perinatology Research Branch of the National Institute of Child Health and Human Development, an institute within NIH, sued a former colleague, Irina Buhimschi, for breach of contract and defamation arising out of an authorship dispute. The research at issue sought to identify biomarkers present in amniotic fluid that was associated with inflammation or infection. The court found that Romero had maintained a bank of amniotic fluid that was used in the research, that he had developed significant aspects of the methodology for the research, including criteria for classifying whether the fluid came from a woman who had an infection or inflammation, and that he developed the idea for two additional aspects of the study "which substantially increased its scientific value."[92] Finally, "Romero provided significant assistance to Buhimschi in preparing the manuscript" and was involved in editing and revising the manuscript. The original manuscript was submitted to *The Lancet* with Romero as one of the authors. However, following a disagreement between the two, when Buhimschi resubmitted the article to address certain concerns, she removed Romero's name as an author. "When Romero learned of this, he contacted *The Lancet*, which had been under the impression that he had agreed to the removal of his name, and which subsequently informed Buhimschi that it would not publish the work until the authorship dispute was resolved. Buhimschi withdrew the manuscript, and some months later submitted a revised version to the British Journal of Obstetrics and Gynaecology ("the BJOG"), again without Romero's knowledge and without listing him as an author."[93] After the BJOG published the article, Romero complained and asked that his name be added as an author. The BJOG refused, but did indicate that authorship was disputed. Thereafter, Romero instituted suit for breach of contract, arguing that an oral contract existed under which he would assist in the research activities, including preparing the manuscript, and in exchange, he would be one of the authors. The court, in granting summary judgment in favor of Buhimschi, held that in order for a contract to exist there must be "consideration." Usually, the consideration would be Romero's efforts in helping with research and preparing the manuscript. However, here the court reasoned that Romero, as a government employee, was already being paid to do research and therefore, as a matter of law, there could not be an enforceable contract between the two: "All this is to say that a government employee who has discretion in his work is not, as a matter of contract law, free to make side bargains with third parties that will control the exercise of that discretion."[94] The ramifications of the court's reasoning is troubling because it treats government scientists differently than other scientists, when in fact all scientists are paid to do research no matter who their employers may be.

To reiterate, authorship disagreements (unless they really involve plagiarism) do not fall within the federal definition of research misconduct.

They are best handled with common sense and composure; where intractable, the authors should seek out a disinterested mediator to help resolve the conflict.

SIMULTANEOUS PUBLICATION AND THE MINIMAL PUBLISHABLE UNIT

There are a host of publication and authorship practices that many may find disquieting but that do not constitute misconduct. For example, is it appropriate for a researcher simultaneously to publish the same results in two journals without advising either journal of the simultaneous publication? Many argue that publishing the same data in two journals is improper for three reasons. First, it wastes journal space and reviewers' time. Second, it gives the false impression that a researcher is more productive than he or she actually is. And third, many journals require that you affirm that the data in the submitted article have not been and will not be submitted to another journal. By doing just the opposite, the researcher has either lied on the journal form or breached the contract that he or she would not submit the data elsewhere.

Although these are legitimate concerns, there are strong policy reasons why there is nothing wrong with simultaneous publication, especially when the author is attempting to reach two diverse audiences. Indeed, sharing one's results with the largest possible audience is part and parcel of the scientific method. The notion that simultaneous publication somehow wastes journal space and takes up reviewers' time would be a legitimate concern were it not for the fact that there are more than 3,500 peer-reviewed journals in the biomedical sciences alone, many of which are hungry for good papers. The notion that simultaneous publication somehow gives a false impression about one's productivity merely caters to the idea that it is proper to evaluate a person's scientific productivity by counting publications in a curriculum vitae. In fact, if one were to take the trouble to read the publications, there would be no false impression conveyed. Finally, if simultaneous publication leads to a breach of contract, then it is a private dispute between the author and the journal editors. Scientific journals, after all, are businesses, and the articles that they carry are in many instances legally nothing more than advertisements.[95]

Another concern is the tendency among some researchers to bolster their publication lists by seeking the minimal publishable unit (MPU). This is accomplished by breaking a single experiment down into a series of distinct subexperiments and publishing the results of each as separate articles. What might have been a single large article is now five smaller ones.

The practice of publishing the MPU has drawn much attention and, ironically, has been the subject of many journal articles. Should one count as a publication an article decrying the publication practices of others? Despite all the concerns about "salami slicing," another term for MPU, it is not scientific misconduct.

VIOLATING HUMAN SUBJECTS OR ANIMAL WELFARE RULES

As we discuss later, there are a host of rules governing the treatment of human subjects, their data, and specimens, as well as the welfare of laboratory animals. However, violation of these regulations and statutes does not amount to research misconduct. The rules governing human subjects are discussed in chapter 4; those governing animal welfare appear in chapter 8.

SEXUAL HARASSMENT AND LABORATORY TRYSTS

Laboratory romances are more common than many are willing to admit. When the relationship involves researchers of unequal stature (e.g., married laboratory head and single postdoc), the opportunity for disappointment is ever present. Many allegations of scientific misconduct have their origins in these types of relationships, once they have soured. However, aside from one unusual case, sexual harassment and laboratory romances gone bad do not constitute scientific misconduct. The one exception occurred under the old definition in which misconduct was defined as "fabrication, falsification, plagiarism, or other practices that seriously deviate from those that are commonly accepted within the scientific community for proposing, conducting, or reporting research." This definition was adopted by both NIH and NSF.[96] In 1993, the NSF inspector general concluded that sexual harassment of a junior researcher by a more senior one constituted "other practices that seriously deviate." In the NSF inspector general's view, sexual harassment in the laboratory or in an archaeological dig, as was the actual case, constituted misconduct in science. It was a silly decision and is precluded by the current definition. The senior researcher's conduct, though, was equally silly.[97]

RESOLVING ALLEGATIONS OF RESEARCH MISCONDUCT

An allegation that someone engaged in research misconduct can be a nightmare for all parties: for the accused scientist, whose reputation and career are threatened; for the accuser, who may be viewed as an untrustworthy troublemaker (perhaps making him or her unemployable in the future); for the vice president for research, dean, or department chair, who must balance the interests of the institution and those of the other faculty while simultaneously overseeing a process that must be fair to all concerned; and for the faculty members, who will be called upon to sit in judgment of one of their peers.

The process for resolving allegations of misconduct actually involves five discrete steps: (1) complaint, (2) inquiry, (3) investigation, (4) government determination (in the case of NIH or NSF), and (5) DAB hearing (in the case of NIH). We examine the first three aspects of the process from the vantage point of the relevant actors. The process that unfolds below is one in which everything is done properly, everyone is competent, and everyone's motivations are relatively pure.

The Complaint

THE COMPLAINANT

A "misconduct play" has up to five possible acts and a large, but constant, cast. The central character in the first act is the complainant, the person who is alleging that someone engaged in research misconduct. Complainants come in all sizes and shapes; some may be motivated by the belief that "it's the right thing to do," while others may have less noble aims (e.g., revenge). Normally, a complainant's motivation is not relevant unless that person's credibility becomes important at some point during the proceeding. From a purely practical perspective, the most important questions that a complainant should ask are rarely verbalized: Will there be retaliation, and will my "dropping the dime" adversely affect my career? The importance of these questions is a function of the status of the complainant relative to the accused. For example, a postdoctoral fellow is likely to feel considerably more vulnerable should he accuse the head of his laboratory of misconduct than if it were the other way around. In fact, studies confirm these commonsense expectations. Swazey, Anderson, and Louis mailed questionnaires to 2,000 doctoral students and 2,000 faculty in chemistry, civil engineering, microbiology, and sociology, and found that more than half in each group believed that they would suffer retaliation if they were to file a misconduct complaint. The authors also found, not surprisingly, that 60 percent of faculty indicated that they would feel safe in filing charges against a graduate student, but only 35 percent would feel safe in filing charges against a colleague.[98] A much more ambitious study by Research Triangle Institute (RTI) under contract with ORI focused on actual whistleblowers and the perceived consequences that they experienced. Some of the findings were surprising. For example, the RTI team found that in the cohort of whistleblowers examined (n=68, 34 with tenure), about 75 percent had negative experiences, and it did not matter much whether the allegations were substantiated or not.[99] However, only 7 out the 68 interviewed indicated that they would not blow the whistle again.[100]

Let's assume that the whistleblower in our little drama is John Whitaker Chambers, a twenty-seven-year-old postdoctoral fellow in his second year at a major research university working for an established researcher whom he both likes and respects. Chambers received his PhD in microbiology from one of the preeminent institutions in the country, has an excellent publication record, both in number and significance, and has every expectation of landing a tenure-track position at another major research university. Things are going along remarkably well until, one day, he goes to the lab building late at night and is surprised to find his mentor, Arthur Hiss, busily at work. He is about to announce his presence when he observes Hiss generating data without performing the experiment. Our postdoc quietly leaves the building without saying a thing, his presence apparently undetected.

The following week at a lab meeting, Hiss discusses the results of a new experiment that he has just completed and that he is quite excited about. Chambers is almost certain that the results under discussion were the ones that he had seen created out of whole cloth, but he can't be certain. Over the next few weeks, he secretly tries to replicate his mentor's results—the ones discussed during the lab meeting. He cannot. He doesn't know what to do, so he calls up his dissertation advisor and lays out the problem. His former advisor recommends that he not act precipitously, and that he try again to replicate the experiment; if he can't, he really has a moral obligation to report the matter to the university's vice president for research. Over the following two weeks, working secretly in the middle of night and using his mentor's lab notebook, he carefully tries again to replicate the experiment. His results are marginal, not quite what his mentor got, but significantly closer than the last time he ran the experiment. He is now in an even greater quandary; he is less certain than before, but is positive that he saw his mentor generating data without doing the experiment. The only thing he is not certain about is whether those make-believe results are the ones that were reported during the lab meeting.

He is about ready to call his advisor again, when all of sudden he is asked by his mentor to review a grant application to NIH. The application sets out, as preliminary findings, the results reported during the lab meeting. To further complicate things, Chambers sees that a significant percentage of his time is going to be covered by this grant, if it is awarded. He meets with his mentor and asks if he has replicated the work. The mentor states that that is the purpose of the grant; he does not have time and, as it is, "I had to do the work late at night in the lab all by myself." That statement clinches it for John. He meets with the VP, lays out what he knows and what he does not know, and hands over the results of his two attempts to replicate. Chambers is scared; he is shaking. The VP is not "Mr. Warm and Fuzzy." He admonishes Chambers not to discuss the matter with anyone, especially his mentor. But, before ending the meeting, tells John not to worry because the institution will act appropriately and there are laws that protect whistleblowers under precisely these conditions.

THE INSTITUTION: VP FOR RESEARCH ET AL.

The VP for research, Dr. Charles McCarthy, has a number of decisions to make and little time to make them. The first thing he does is precisely what he ought to do. He calls the university's general counsel, Harry Palsgraf. Palsgraf, like McCarthy, is relatively new but has the advantage of having handled a misconduct case in the past. He recognizes that any misconduct proceeding can be easily compromised either by overreacting and treating the accused as a pariah or by underreacting and sweeping the allegations under the rug. He counsels McCarthy that a rational middle course is best; he also knows that in these types of matters time is your enemy. First, Palsgraf wants to know whether the preliminary research that formed the basis of the grant application was federally funded and, if so, by what agency. If it were funded by NIH, then the NIH process would govern, and the university might have some reporting obligations. If it is not federally funded and the grant application is not forwarded to NIH, then there is no federal jurisdiction, and the matter does not have to be disclosed to anyone outside the university. McCarthy believes that Hiss's entire operation is NIH funded, and therefore the experiment would necessarily have been done using NIH money. He tells Palsgraf, however, that he will double check. However, he can't do that without talking to Hiss and looking over records. McCarthy asks whether the mere preparation of an NIH application, even if it is not submitted, is enough to trigger federal jurisdiction. Palsgraf indicates that unless the application is submitted or unless the research reported in the application and discussed during the lab meeting is federally funded, there is no federal jurisdiction.

Second, Palsgraf wants to understand better whether, from a scientific perspective, the allegations make sense; he explains that under the rules governing NIH-supported research, an institution is obligated to conduct an "inquiry" if the allegation, if true, would amount to "research misconduct" and "is sufficiently credible and specific so that potential evidence of research misconduct may be identified."[101] McCarthy ponders the matter for a few moments and then indicates that he believes that Chambers's allegations are sufficiently credible and specific to warrant an inquiry.

As a result, Palsgraf recommends that McCarthy ask the medical school dean and Hiss's department chair to join them for a meeting immediately. At the meeting, McCarthy lays out what occurred and indicates that the institution is obligated under federal law to conduct an inquiry. Palsgraf adds that an inquiry is a preliminary fact-finding procedure designed to determine whether there is a reasonable basis to believe that the allegations (1) fall within the HHS definition of misconduct, (2) involve PHS funding, and (3) may have substance. If all three criteria are satisfied, Palsgraf explains, we go to the next stage, which is to conduct a formal investigation.

Palsgraf indicates that the university is obligated to conduct an inquiry, but first, "we have to inform Hiss and make every effort to secure all of the relevant evidence without destroying his laboratory. We need to leave him with copies of all of the notebooks that we take and an inventory listing everything that we take. I want this done correctly the first time. The worst thing is to keep going back to gather more evidence. That makes us look incompetent and also makes us look like we are predisposed to finding Hiss guilty." Palsgraf goes on to tell the three that he "will draft up a letter to Hiss from the department chair outlining the allegations, indicating that the university has to take possession of all relevant laboratory notebooks and physical evidence, and that a three-person inquiry committee will be appointed, as required by the university's internal rules."

Palsgraf's overview of what must be done by the university mirrors the federal requirements. Those requirements, though, provide the university with significant latitude in deciding how to conduct an inquiry. It can be conducted by a committee or an individual. In fact, according to an ORI-contracted survey, about 58 percent of the universities reported using a three-person inquiry committee, whereas the rest either use a single person or determine the number necessary based on the circumstances.[102] The form of the inquiry is largely up to the university: It can be conducted through a series of interviews at which the accused is not permitted to be present, or the interviews can be open to the accused, or transcripts of those interviews can be made available to the accused. The central requirement is that every recipient of PHS or NSF funding must have a written policy setting out its procedures for resolving allegations of misconduct, including conducting inquiries and investigations.

THE ACCUSED

Hiss is having a good week; his postdocs are productive, and his research is coming along better than he expected. To boot, his wife just learned that she is pregnant. He is a happy fellow. He is somewhat surprised, though, when his department chair tells him that they have to meet this afternoon to discuss something important. Hiss has no idea what's in the offing, but being a good faculty member he promptly shows up at 4:00 p.m. for his meeting with his department chair. Hiss is surprised to see Palsgraf also in the office; he has met him on few occasions, but really doesn't know him. The department chair looks like he is going to be sick. He fumbles with papers and then haltingly tells Hiss that someone has alleged that he has engaged in scientific misconduct, and, in particular, the preliminary data in his as-yet-to-be submitted NIH grant application are alleged to have been fabricated. Hiss is dumbfounded. He can't believe what he is hearing and indeed, after about thirty seconds, he sees his department chair's lips moving but is unable to process anything that is coming out of the man's mouth. The chair hands Hiss a letter that spells everything out in detail, including the facts that a three-person inquiry committee will be assembled, Hiss will be given an opportunity to be interviewed by the committee, and, in the interim, the chair will need to gather up all of Hiss's relevant records, including lab notebooks. Hiss asks if they will take the originals and is told that they will, but will make complete copies of everything so that Hiss can both defend himself and continue with his work.

Palsgraf then suggests that Hiss may wish to retain a lawyer, although the university cannot reimburse him for those expenses. They walk back to Hiss's laboratory. The department chair asks Hiss to identify all his records and lab notebooks relevant to the experiment in question. Hiss points to them; the department chair catalogs them, writes out a receipt for Hiss, and leaves with an armful of records. Before the chair and Palsgraf leave, Hiss is beginning to regain his composure and asks who could have made such an allegation. Palsgraf responds that at this point the identity of the complainant is not relevant. Hiss decides that he needs the comfort of his wife and home, along with a belt of good single-malt scotch.

The Inquiry

THE INSTITUTION

Palsgraf, working with McCarthy and Hiss's department chair, quickly identifies three candidates for the three-member committee. All are tenured science faculty, none from Hiss's department, but each with enough knowledge in fields closely related to Hiss's research to enable him or her to understand the science and the allegations. Also, none of the three has ever worked or socialized with Hiss, and none knows the complainant.

Palsgraf then calls each committee candidate and asks to meet with each in McCarthy's office. After the three members arrive, McCarthy and Palsgraf provide them with a quick summary of the allegations, a copy of the university's procedures, the relevant laboratory notebooks, and the letter that was handed to Hiss. Palsgraf explains that while under federal law they have up to sixty days to complete the inquiry, he would like it done within a few weeks.[103] There is some complaining by two of the three about other obligations, but McCarthy puts up his hands and says, "This is one of the more important roles you will ever play as a faculty member, and it is one that no one will be permitted to shirk." Palsgraf also explains that he will be sitting in on all of the interviews and will be able to answer any questions that they have and may also be able to suggest questions that they might ask.

Palsgraf then drafts a letter informing both the complainant and the accused that an inquiry committee has been assembled and providing the identities of the members. In keeping with the university's written procedures, Hiss is given an opportunity to object to any member on the grounds that that person is prejudiced against him or has some other conflict that might make it difficult for that person to render an impartial decision. The letter ends by informing the complainant and the accused that the committee may be contacting each party shortly to arrange for an interview.

THE COMMITTEE

The three committee members, Joan Townsend, Marc Goldberg, and Larry Usher, decide to meet on their own without university administrators present. They want to go over the scientific record and try to figure out what to ask each witness. According to Palsgraf's notes, Chambers stated that on February 25, at about 11:30 p.m., he went to the laboratory to do some work. As he was walking down the hall, he noticed that the door to Hiss's lab was open, the lights were on, and Hiss was writing in a laboratory notebook what appeared to be data. None of the equipment was on, and no experiments were actively being conducted. According to Palsgraf's notes, he asked Chambers how far from Hiss he was at this time, and Chambers responded, "about 35–40 feet." "How was it possible for you to have seen what Hiss was writing in his notebook," Palsgraf asks, "if you were that far away?" Chambers responds that Hiss's practice, ever since Chambers had been in the lab, was that "Hiss only wrote data in notebooks with short cross-references to the actual procedure or protocol. And since he was writing for a few minutes, it had to be data." After reading Chambers's account, Townsend notes that while she wants to keep an open mind, she feels that the allegation is flimsy at best; Hiss could have been writing anything in his notebook. "Well, let's take a look at Hiss's notebook and see what's there for February 25," Usher suggests. They flip through one of notebooks, see that on February 25 the only entries are data; there are a few cryptic cross-references to procedures. However, Goldberg notices that a time entry at the end of the data trail reads: 2/25/08—22:00. Goldberg ponders a bit, and says, "I'm not too good at military time, but I believe that 22:00 hours is 10:00 p.m. Something odd is going on here, and I don't know what it is."

They then review printouts and Hiss's write-up of his preliminary results that were later inserted into the grant application. After about four hours, they are no closer to reaching any conclusions than when they had started. Before packing it in for the night, the committee agrees on two things: that there isn't much more information that Chambers can provide other than his two attempts at replicating Hiss's work, and that they want to hear Hiss's side of the story. Goldberg e-mails Palsgraf and asks him to set up a meeting with Hiss so that the committee can interview him and find out what is going on.

THE ACCUSED

Hiss is lucky in one respect. His wife Helen is a lawyer who works for a relatively large law firm. However, no one in the firm has ever handled a case like this. One of Helen's senior partners does some calling around and finds that no one has much experience handling these cases; the attorneys who did a lot in the 1990s did them primarily pro bono and aren't doing them anymore. He does speak with one of those attorneys, who advises that ideally the attorney should know how a university works and how science is practiced. Finally, Helen locates a former university general counsel, Chester Vanderbilt Hall, now retired, who agrees to represent Hiss. Hall reviews Hiss's notebooks, discusses them with him, and reviews the charge letter, which just states that someone saw Hiss writing out data at 11:30 p.m. on February 25. Hiss acknowledges being in lab at that time, but maintains that he was not writing out data; he finished his last experiment at about 10 p.m. give or take, as indicated in lab notebook, and he can't remember what he was doing at around 11:30.

In the meantime, Hall has called Palsgraf to let him know that Hiss is represented by counsel, that Hall intends to be present with Hiss if and when he is called to testify, that all communications with Hiss should be directed to Hall, and that no one should attempt to speak with his client about the charges. Hall also confirms that this matter must remain confidential; Hall drafts up a letter to Palsgraf confirming their conversation.

The following day, Hiss receives an e-mail from the committee asking him to present himself in two days at 2:00 p.m. Hiss forwards the e-mail to Hall, who is furious. Hall immediately drafts a short note reminding Palsgraf that all communications to Hiss must be made through him.

THE INQUIRY HEARING

The inquiry committee, at Palsgraf's suggestion, decides to tape-record the proceeding and also to allow Hall to appear with Hiss. Between their last meeting and the first interview with Hiss, the committee has had a better opportunity to understand the experiments and what the real issues are.

Hiss and Hall enter the room and shake hands with Townsend, Goldberg, Usher, and Palsgraf. Townsend, acting as chair, goes over the allegations and asks Hiss to explain his side of the story. Hiss recounts that he had spent about one month trying to get an experiment to run properly, and finally, on the 25th, it worked. He discusses the purpose of the experiment and indicates that it took him all day to set it up and that he began running and collecting data in the early evening and finished at about 10 p.m., give or take a few minutes. He can't remember for the life of him what he was doing thereafter in the lab. Goldberg points to some data points that Hiss had reported in the putative grant application and asks where in the notebooks one can find those data. Hiss picks up the notebooks and begins turning pages and pages; finally he looks up and says, "My God, they're not here; I may have written them on a piece of paper." Goldberg continues to ask questions and Hiss answers them, but the missing pages or paper spread a pall over the entire process.

After Hiss leaves, Usher remarks that "this case makes absolutely no sense. The fact that Chambers had difficulty replicating Hiss's work doesn't trouble me because Hiss had lots of difficulty getting the experiment to run; his own records indicate that it took him a month." "What's equally mystifying," Townsend remarks, "is that Hiss has some of the data written in his notebook, but about a third is missing, and uncharacteristically written on some loose pieces of paper, according to Hiss." Goldberg nods in agreement and continues that the "timing thing is also screwy; the notebook indicates that he was done writing data at 10:00 p.m., yet Chambers sees him ninety minutes later. The only inference we can draw is that Chambers saw him writing the mysterious missing data, but that inference would be clearly incorrect because Chambers saw him writing in a notebook, not on a piece of paper. It was the notebook writing that piqued Chambers's curiosity to begin with. Nothing about this case makes any sense."

The three of them turn toward Palsgraf and, almost in unison, say, "Okay, what are we supposed to do now?" Palsgraf reaches toward his mouth for his nonexistent pipe—having given up smoking about a decade earlier—and shakes his head. "I think that there are just too many unanswered questions, and the real problem is that there are missing data. I recommend that you reluctantly find that there is some evidence pointing toward misconduct and that an investigation ought to be convened."

Palsgraf goes on to tell the committee that under federal law, they have to write up their findings in a report that must be shared with the accused. They may also share relevant portions of the report with the complainant, if they wish.[104] Over the next week the committee prepares its report going over the reasons why they believe that an investigation is warranted.

The Investigation

THE COMPLAINANT

Chambers plucks the envelope out of his university mailbox and immediately notices that it is from McCarthy's office. He waits until there is no one around him and quickly tears it open, pulling out the single sheet of paper that reads: "This is to inform you that the misconduct complaint that you filed has proceeded to the investigation stage and that you will be called upon to give evidence. We have enclosed a copy of the inquiry committee report, which you may comment upon if you wish. In the interim, you are not to discuss this matter with anyone. The matter remains confidential, and we anticipate that you will treat it accordingly."

Having not heard from anyone about anything since he filed his complaint, he was worried that no one was taking him seriously. He's gratified to see that that isn't the case, but he is also concerned about Hiss. He really likes the guy and feels some empathy for him. But Hiss should have thought of that before he started to play fast and loose with data.

THE ACCUSED

Hiss learns from Hall that the powers that be have decided to take the matter to an investigation. Hall explains that he has a copy of the inquiry committee's report, which he will scan and e-mail to Hiss. Under the federal rules, Hall says, the committee was obligated to complete a report and the university was obligated to share that report with Hiss and may, but is not required to, share that report or portions of it with the complainant.[105]

Hall goes on to explain that the investigation is a far more formal process than the inquiry. Under the university's rules, it is conducted by a three-member committee that may consist of the same faculty who conducted the inquiry; that is up to McCarthy. Palsgraf explains that "there are a number of differences between an inquiry and an investigation. First, before the investigation can even start, the university has to inform ORI that it will be conducting an investigation into an allegation of misconduct against you and will have to share with ORI a copy of the inquiry report. All and all, it is not a good thing."

"Second, the investigation, which under federal rules has to start within 30 days of the completion of the inquiry and should take no more than 120 days, is a more thorough and formal process than the inquiry. The committee is obligated to interview the complainant, the accused, and anyone who may have any information relevant to the case; all the interviews have to be either recorded or transcribed; each recording or transcription has to be provided to the interviewee for correction.[106] The good part, though, is that unlike the inquiry, where there really isn't any burden of proof, the investigation committee can find you guilty of misconduct only if the finding is supported by a preponderance of the evidence."[107]

"Is preponderance of the evidence like proof beyond a reasonable doubt?" asks Hiss. "No, it is not. In criminal cases, the prosecutor has to prove the government's case beyond a reasonable doubt, meaning that the evidence favoring guilt has to be substantially greater than the evidence pointing toward innocence. The 'preponderance of the evidence' standard, which is used in most civil cases and administrative hearings, is a much lower standard. The evidence of guilt only has to be slightly greater than the evidence of innocence. If the evidence pointing to guilt is equal to the evidence pointing toward innocence, you must be acquitted. In other words, in civil cases, including yours, our society is willing to permit an incorrect outcome nearly 50 percent of the time.[108] In other words, our legal system, at least in the civil area, operates at a level of validity only slightly better than a coin toss."

"However, what makes your case somewhat more troublesome is the missing data. The university is actually required by the federal rules to draw a negative inference from the missing data." "What does that mean?" "Well, it's not good; what it means is that the unexplained absence of critical research data can be viewed as evidence of misconduct."[109]

After meeting with Hall, Hiss feels as if his entire career is about to implode. He can't believe that this is happening to him. During the past couple of weeks, he has been extremely depressed and has hardly gone into the lab. His wife is really worried about his mental state and tries to cheer him up, but whenever she does, he goes from being depressed to being on the verge of rage. He wants to know who would do this to him and why. And "why couldn't the idiots on the committee have seen that everything that he's done is honest?" he rants.

THE INSTITUTION

Once the committee has recommended that the matter should proceed to an investigation, the institution has a variety of obligations. First, it has to provide ORI with a copy of the inquiry report, it has to identify the PHS funding involved, and it must identify the complainant and the respondent (i.e., the accused). Once this is done, the institution has to decide whether it will empanel a new committee or permit the inquiry committee to handle the investigation as well. Federal rules do not require the institution to do one or the other, and the university's rules are equally flexible. Ultimately, McCarthy and Palsgraf decide to proceed with the same committee. They advise the committee, the complainant, and the respondent of this decision.

THE INVESTIGATION HEARING

The committee has decided to call not only Chambers and Hiss, but also others in the laboratory. Rumors have been spreading throughout the department, and the committee feels that it has to move quickly to quell them. The first witness whom they hear from is Chambers. He retells what he saw and when he saw it. He admits that he couldn't tell what Hiss was writing. All he could say for sure was that Hiss was writing in a standard lab notebook. He also admits that Hiss's experiment, at least the one Hiss reported about, was hard to get to run. He thinks that he blew it the first time, but is fairly sure that his second effort was pretty faithful to the protocol.

He reiterates for the committee his belief that his inability to replicate fully is evidence of misconduct. Goldberg asks Chambers whether he has ever failed to replicate one of his own experiments. Chambers states that that has never occurred. Goldberg pulls out a lab notebook and asks Chambers if he can identify it. Chambers states that "yes, it is one of my notebooks from last year." Goldberg then turns to a page and asks Chambers, "is this experiment one that you published just recently in a fairly good journal?" Chambers proudly replies "yes." Goldberg then turns a few pages in the notebook and asks Chambers if this is the same experiment, only run a few days later. Chambers looks at the pages and begins to look uneasy, as he answers "yes." Goldberg then asks if in fact this later experiment was his first attempt to replicate the previous experiment. Chambers replies that it was. Goldberg then asks whether he succeeded in replicating the experiment, to which Chambers responds, "no, but I managed to replicate it a couple of weeks later." "That is not the point," states Goldberg. "Should we infer from your inability to replicate that you in fact committed misconduct?" "No, of course not. I just mean that here things are different."

The committee then hears from others in the laboratory, all of whom testify to the fact that Hiss had been secretive about one set of experiments he had been trying to run. They all got the impression that Hiss had been unable to get it to work and that it took a long time before he perfected the technique. They all testified about Hiss's practice of writing data in lab notebooks, but were able to testify about little else.

Finally, the committee hears from Hiss. Hiss tries to appear confident, but it is clear to the committee that he is anything but. Hiss still can't remember what he was doing between 10 p.m. and 11:30 p.m., and he still maintains that he had absentmindedly written part of the data on a piece of loose paper. He states that he has turned his lab, his clothes, and his house inside out, but can't find the missing data. He maintains that the data that are recorded in the lab notebook are strongly supportive of his conclusions.

THE COMMITTEE AND ITS DELIBERATIONS

The entire hearing lasts two days. The members of the committee weigh and reweigh the evidence. There is certainly evidence of misconduct: the testimony of Chambers that he saw Hiss writing in a lab notebook while not performing an experiment and that he couldn't replicate Hiss's results, at least not the first time around; with the second attempt, his results were equivocal. There were also the lab notebooks themselves, which contained only a fraction of the data that comprised Hiss's so-called preliminary results. And then there was the elephant in the room—namely the missing piece of paper that supposedly contained the missing data. There was also evidence of no misconduct. Although Chambers eye-witnessed an event—Hiss writing in a lab notebook—there is only marginal evidence that that event is relevant. If Hiss were committing misconduct, he certainly would have recorded the entire experiment and not put himself in the position of having to argue that some data were on a missing piece of paper. However, this all assumes that the "10 p.m." notation in the lab notebook is false. But why would Hiss have falsified the time? He had no apparent motive to do so.

The committee members are going around in circles, looking at the same evidence over and over. Townsend and Usher believe that scales have tilted in favor of a misconduct finding; for them the missing piece of paper is the critical factor. Goldberg, on the other hand, argues that he is unwilling to torpedo someone's career because of a missing piece of paper. Everyone agrees that without that "missing piece of paper," there would literally be no evidence of misconduct. Townsend suggests that they leaf through the notebooks one more time and then call it quits for the day. Townsend picks up the three relevant notebooks and hands one to Goldberg and the other to Usher; she keeps the third. Goldberg looks over and asks her, "What about that fourth notebook, the one that everyone says is really not relevant? Shouldn't we look at it?" Townsend rolls her eyes, picks the notebook up by one of its covers, and, as she is handing it to Goldberg, a piece of paper falls out of the notebook. Townsend picks it up, and there on the sheet of paper, dated 2/25/08 10:15 p.m., are data that match perfectly the missing data in Hiss's preliminary report.

"Boy, did we dodge a bullet," suggests Usher. The group takes a quick vote and then calls Palsgraf and McCarthy with their findings. They are told to write it up just as it happened so that the university has a defensible position when ORI reviews their work. McCarthy immediately picks up the phone and calls Hiss with the good news.

While our make-believe case ended happily for all concerned, many cases do not. When a university determines that an accused has committed misconduct, the university is free to impose whatever sanctions are appropriate under the university's policies, including termination.

THE GOVERNMENT'S ROLE IN MISCONDUCT CASES HANDLED BY UNIVERSITIES AND OTHER GRANTEES

Further Government Review

At the completion of any investigation involving research funded by PHS or NSF, the institution is required to forward its report and findings to ORI, in case of PHS funding, or to the NSF inspector general, in the case of NSF funding. Some universities do not end the process with the investigation. These institutions permit an accused who has been found to have committed misconduct to have his or her case reviewed by the university president or similar official. The university can delay submitting its report until that final review has been completed.

ORI or NSF, depending on the funding source, is supposed to review the university's findings to ensure that those findings are consistent with the evidence and that the institution followed the relevant federal rules. In recent years, both agencies have been remarkably reluctant to overturn a university finding either way.[110] However, when the university finds misconduct, the government has an added obligation—it not only reviews and confirms the findings, but it also may impose sanctions of its own, including debarment. The fact that a person may have already been punished by his or her university does not preclude the government from imposing its own sanctions.

ORI sanctions tend to involve at least two restrictions. First, the culpable researcher is usually debarred for a period of time (see chapter 2). That means the researcher may not work on any federally funded grant, contract, or cooperative agreement. The researcher's name is placed on a government-wide debarment list. Thus, if a researcher is debarred by HHS for three years, that means that he or she cannot work on a Department of Defense research grant or contract. Second, ORI also bans those found culpable of misconduct from serving on a study section or other HHS advisory committee and attempts to get the researcher to retract those publications that contained tainted data. ORI can also seek repayment of grant funds, but that route is considerably more problematic, and, for constitutional reasons, it is doubtful that repayment can be sought simply through the administrative process. In rare instances, HHS, usually operating through its inspector general, can refer a case for criminal prosecution. This occurred most recently in the case of Eric Poehlman, a former tenured professor of medicine at the University of Vermont School of Medicine, who pled guilty to falsifying research data in seventeen grant applications involving $2.9 million in funding from NIH and the U.S. Department of Agriculture between 1992 and 2000. On June 28, 2006, Poehlman was sentenced to one year plus one day of jail time for research fraud.[111]

While the government requires that institutions act with dispatch in resolving misconduct cases (60 days to conduct an inquiry and 120 days to conduct an investigation), ORI imposes no time constraints on its review process. ORI reports that it handles cases more quickly than the institutions. In its 2005 annual report, for instance, it notes that while it takes institutions on average 8.4 months to take cases through investigation ($n=22$), it takes ORI only 5.8 months to review cases. What is troubling is not the mean, but rather the large standard deviation: ORI processing time ranges from 1 month to 24 months, with a mode of 2. One would think that a federal agency ought to be able to review a case in at least the same time that it took the university to conduct its investigation. The fact that ORI has been unwilling to impose time constraints on itself can only further erode its stature in academic circles.

Final Review: Hearings and the Like

As noted earlier, once ORI seeks to impose sanctions, the researcher has the right, under limited circumstances, to have the matter essentially retried by an administrative law judge (ALJ) within the DAB. An ALJ is a career employee of the Department of Health and Human Services. The ALJ is permitted to retain a scientific expert to assist in the trial. DAB review, while available, is almost never used, for a number of reasons. First, as a practical matter, few if any scientists will have the resources to seek full review by the DAB. Since 1992, when the DAB was first given appellate jurisdiction over ORI findings, only nine cases have been appealed to it, and no case has been taken through hearing since 1999. Second, recent changes in the regulations have made an appeal to the DAB less attractive. The original three-judge panel was replaced by a single ALJ, possibly assisted by a scientist, and access to an appeal is no longer automatic. To qualify you must now specify those aspects of the ORI findings that are factually incorrect and why they are incorrect. Moreover, even if you were to prevail at the DAB, the ALJ decision is no longer a true ruling as in the past, but now "constitutes a recommended decision to the Assistant Secretary for Health."[112]

The NSF system does not provide either an independent or a trial-like review. Rather, the recommendations of the OIG are forwarded to the deputy director, who reviews the recommendations and may confirm the recommendations or call for additional evidence. The regulations also permit an appeal from the deputy director's decision to the director, but there is no indication what criteria are to be used by the director.[113]

The decision of the Assistant Secretary for Health, in the case of HHS, or the director, in the case of NSF, is "final agency action."[114] This means that a dissatisfied researcher can challenge the decision of either in a U.S. district court under the Administrative Procedure Act.[115] However, courts are required to defer to the agency's findings and may overturn those findings only if they are arbitrary or capricious, which is difficult to establish. Thus, while judicial review may theoretically be available, as a practical matter it is not. To my knowledge, no scientist has ever successfully challenged a misconduct finding in federal court.

In the United States, allegations of misconduct are relatively rare, given the universe of those who are active scientists. When they arise, though, they can throw lives and institutions into turmoil. Therefore, it is important for institutions to act rationally and fairly with all concerned and to move the process along. It is also important, especially when federal funds are involved, for institutions to scrupulously apply the relevant federal definitions and procedures. Those federal definitions and procedures trump institutional policies that may differ. When the research at issue is not federally funded, institutional definitions and policies apply. Above all, misconduct is a legal and not a scientific concept. Attempting to resolve a case through science, in my experience, is a recipe for disaster.

CASE STUDIES AND PROBLEMS

Case 1: The Mystery of the Missing Data

(Use applicable PHS rules.)

Professor Claudia Powers ran a large laboratory at the University of East-West. At any time, Powers had at least ten postdoctoral students. One of those students, Harrison Green, was an intense young man bent on getting as many publications under his belt as possible during his two-year stay at the laboratory. Green was on an NIH training grant. During his first year, he had published a number of papers, all in mediocre journals. Indeed, during that year he had submitted six articles for publication, and none was earth shattering. Quantity appeared to be Green's primary objective.

Recently, Green had been attempting to develop a new antisense technique that would more easily enable one to shut down the expression of a particular gene. At first, Green was having difficulty getting his technique to work. As a result, he had gone five months without generating publishable data. Suddenly, during a laboratory meeting, Green announced that he had gotten his technique to work and produced data showing a statistically significant reduction in the amount of the target protein expressed by cells that had been subjected to his new technique. Specifically, during his presentation, he set out data from a series of twenty-nine experiments using independent controls (fifteen experiments) and treatments (fourteen experiments). His data, as presented during the lab meeting, are set out in Attachment 1. Although the results were not spectacular, they were, in Powers's opinion, worthy of publication. Green indicated that he would set about writing up his results immediately.

ATTACHMENT 1

	CRUMPLED PAPER FOUND BY DR. POWERS		PAGE PRESENTED AT LABORATORY MEETING	
	Compilation of Data		*Compilation of Data*	
	Treatment	Control	Treatment	Control
	48	79	48	79
	54	85	54	85
	59	80	59	80
	62	105	62	105
	65	92	65	92
	59	76	59	76
	109	100	66	100
	66	100	55	100
	55	81	65	81
	65	89	49	89
	49	90	98	90
	98	72	65	72
	65	80	78	80
	78	32	12	32
	12	36		36
Mean	62.93	79.80	59.64	79.80
t-test (two-tailed)	$p > 0.05$ "No good"		$p = 0.00561$	

That evening, while in the laboratory near where Green worked, Powers received a telephone call giving her directions on how to get to a social event planned for the next evening. She needed some paper to take down the directions and, without thinking, reached into a trash can and pulled out some crumpled paper; she took down the directions on the reverse side of the paper.

About a week later, Powers was going through her pockets and found the crumpled-up paper with the directions. She was about to toss it out when she noticed that there were data on the other side. The data appeared to be almost identical to those presented by Green, and indeed, she recognized the writing as belonging to Green. She studied the data and remembered that the compilation that Green had presented during the lab meeting had one more control than treatment, whereas these data had an equal number of controls and treatments. She pulled out Green's loose-leaf notebook and compared the two. She immediately noticed that the scratch paper had an extra data point that Green had apparently deleted from the treatment group. With that extra data point included, Green's results appeared no longer to be statistically significant. Indeed, this very fact was noted on the scratch paper: "$p>0.05$. No Good."

The next day, Powers asked Green why he had deleted the data point. Green responded that it was an outlier and had been deleted because he had botched the experiment that day and felt justified in tossing the point out. Green said that the statement "no good" indicated that he had botched an experiment. Powers asked to see the raw data for that experiment. Green got out his notebook and leafed through it. The raw data for the other twenty-nine experiments (fifteen control and fourteen treatment) were present. The raw data page for the discarded experiment was missing. Green had no explanation.

That evening, while pondering what to do, Powers reanalyzed Green's data and found, much to her surprise, that Green had miscalculated the t-value the first time around and that even including the discarded treatment value of 109, Green's original results were actually statistically significant at the 0.05 level.

What should Powers do, if anything? If you were a university administrator judging the case, how would you resolve it?

Case 2: Revenge of the Disgruntled Postdoc

(Use applicable PHS rules.)

Felicity Frank had been a professional postdoc, having spent five years in John Mulligan's laboratory. During her first few years at the lab, Frank had helped Mulligan write a number of successful NIH grant applications. Moreover, Mulligan and Frank had, at one point, been lovers. However, their personal relationship soured, and pretty soon that began to affect their professional relationship. One day, Mulligan announced that he thought it would be a good idea if Frank looked for a tenure track position. Frank, who wasn't particularly interested in leaving the lab, paid no attention to Mulligan's suggestion. Frank had noticed, though, that Mulligan had been submitting grant applications without Frank's assistance and further, that Frank was not included in the budgets for these applications.

One day, Mulligan handed Frank a letter, which read as follows:

> As we have discussed, it is critical for your career development that you obtain a tenure track position at a major research university. Your ability to do so is likely to be adversely affected if you remain in this lab as a postdoc for much longer. Accordingly, I have decided, after consulting with the dean, not to renew your contract. As you know, your contract is due to expire at the end of next month.
>
> *Sincerely yours,*
> *John Mulligan, Ph.D.*

Frank was furious. She tried to discuss the matter with Mulligan, but he refused to talk to her. That evening, she returned to the laboratory and photocopied all of the grant applications that Mulligan had submitted after their relationship had soured. None of those applications included Frank. During the next few weeks, she examined each application in detail. Finally, in one application she noticed that the 300-word method section for one of the proposed experiments was a verbatim copy of the method section that she had written for one of her articles. The method section in the application did not reference the article. In that article, Frank was the first author; Martin and Goldstein, other researchers in the lab, were the second and third authors, respectively; and Mulligan was the last author. Frank had developed the method and had written up the entire paper. Although Mulligan did no work on the paper, he was included as the senior author because it was custom to include the head of the lab on all papers generated by researchers in that lab.

Frank also noticed that in discussing another proposed experiment, the application contained the following statement: "We have already run some of these experiments on rabbits and they demonstrate that our method is viable." Frank remembered that at the time the application was submitted, Mulligan had been unable to get the rabbit experiment to work. However, two weeks before the study section met, he had succeeded in getting the experiment to work.

The next day, Frank files with the dean charges of scientific misconduct against Mulligan. The dean convenes an inquiry panel. You are a member of that panel. How would you resolve this case?

Case 3: The Problem of the Prolific Professors

(Use applicable PHS rules.)

Dr. William Peters is a well-respected researcher in biophysics. Originally trained as a theoretical physicist, Peters learned biology on his own, more than thirty years ago, while a graduate student. Since then he has pioneered many innovations in the field. Peters has always had difficulty aiming his publications at the "right" audience. Much of his work, whether theoretical or empirical, can be written for at least two audiences. As a general rule, biologists will not understand the physics-laden component of his research, while physicists have difficulty grasping the more technical biological component of his work. Frequently, Peters finds that the best way to solve this problem is to write for a general science audience, publishing in journals like *Nature* or *Science*. Occasionally, however, he attempts to address multiple audiences.

Recently, he and his colleague, Robert Roberts, completed an experiment that yielded surprising results. Given the likely impact of the experiment, Peters submitted the article reporting on the experiment to a general science journal similar to *Nature* or *Science*. Unbeknownst to Peters, his coauthor, Roberts, had submitted a similar article to a physics journal. Part of the confusion stemmed from the fact that both were in a rush to go on their vacations, and neither consulted the other prior to submission. Consistent with past practices, both of their names were included as authors on each of the two submissions. However, neither was aware of the other's submission until both articles had been accepted.

At that point, the two decided to modify the physics article to make it more physics oriented. Neither needed additional publications; they both had tenure. They were both concerned, however, that the general science article had lost some of its impact when the physics had been edited out so as to appeal to the more general audience. They therefore decided to permit both articles to be published, even though the underlying data were identical and would appear simultaneously in both journals.

Soon after publication, the editors of both journals realized the similarity. Acting together, the editors wrote both Peters and Roberts scathing letters criticizing them for violating the journals' policies against duplicative publication and demanding an apology from both. Peters and Roberts were both taken aback by the letters, but wrote the apology, thinking that that would end the matter.

Unfortunately, the editor of the general science journal was still miffed. He called the federal agency that had funded the research and filed a formal misconduct charge against both Peters and Roberts. The agency has referred the matter to the university for appropriate action. You are the dean of the school of arts and sciences. What should you do?

Case 4: The Case of the Understanding Urologist

Dr. Hans Cutter, a renowned surgeon and researcher, is the principal investigator of a large NIH-funded study comparing various treatments for prostate cancer: surgery, radiation, and a new chemotherapy, ProSTAT, just approved by FDA, which shuts down tumor cell angiogenesis. About one hundred centers nationwide are participating in Cutter's protocol, including the Brigham Institute of General and Focused Experimental Exercises (BIG-FEE), part of Haryvale University. Dr. Sebastian Archer is the chief of oncological urology at BIG-FEE and is the principal investigator at that site.

There is a detailed protocol with rigid entrance criteria (e.g., between ages of 45 and 50, primary tumor no larger than 8 mm in diameter). Patients who meet the criteria and who agree to participate are randomized into one of three arms—(1) surgery plus radiation, (2) radiation plus drug, or (3) drug only. Because surgeons, as opposed to researchers, are the primary entrance points, the protocol is set up to accommodate an error rate of 10 percent.

The new drug is extraordinarily expensive ($50,000 per year), and many insurance companies have declined to place it on their formularies, i.e., they won't pay for the drug. Many of Archer's patients cannot afford the drug and do not meet the entrance criteria. To resolve this problem, Archer falsifies patients' birthdays so that they qualify—they then have a two-thirds chance of getting the drug for free. Archer enters about eighty patients, and he jiggers the entrance criteria in seven or eight cases; all of his other cases are perfect.

After Cutter accumulates the data from all the centers, he writes up his results and submits them for publication. Three weeks later, he learns about Archer's conduct, but decides to go forward with the publication anyway.

Has Cutter committed misconduct? What if he notes the data problems at BIG-FEE, but retains Archer's data? Analyze under both the PHS and FDA definitions.

NOTES

1. The statistics are difficult to compile only because ORI's Annual Reports are less than models of consistency or clarity. The number of "new cases" reported in table 1 is based on table 9 in ORI Annual Report 2007. The number of those cases ending with findings of misconduct has been culled from each ORI Annual Report. Because of the lag between the time an allegation is first reported and a case ends, yearly data are difficult to interpret. The averages over the thirteen-year horizon noted in table 1 are more pertinent than the yearly numbers.

2. The author or his friends and former colleagues at a large Washington, D.C., law firm defended many of the scientists or universities involved in the major misconduct cases during the 1990s, including Rameshwar K. Sharma (Charrow and Tom Watson), Robert Gallo (Joe Onek and Watson), Thereza Imanishi-Kari (Onek and Watson), Margrit Hamosh (Charrow), Bernard Fisher (Onek and Charrow), and the University of California (Charrow and Watson). While this may give the author a firsthand view of these cases, it also may bias that view. The names in the parentheses indicate the lawyers involved in each case. Mikulas Popovic was represented by Barbara Mishkin and her team at another large Washington, D.C., law firm.

3. The charges against Summerlin had been revealed six weeks earlier in a *Times* story that noted that some in the scientific community had likened the incident to a "medical Watergate." *See* Jane E. Brody, *Charge of False Research Data Stirs Cancer Center Scientists at Sloan-Kettering*, N.Y. TIMES, Apr. 18, 1974; Jane E. Brody, *Inquiry at Cancer Center Finds Fraud in Research*, N.Y. TIMES, May 25, 1974.

4. *Id.*

5. *Id.*

6. Instances of data falsification were known to have occurred prior to 1974. Perhaps the most notorious example is the embarrassing case of the "Piltdown Man," a collection of head bones unearthed near Piltdown, a village in East Sussex, England, which were thought to be an unknown form of early man—part ape and part hominid. The "missing link" was named *Eoanthropus dawsoni* after the amateur paleontologist, Charles Dawson, who in 1912 claimed to have discovered the bones. The bones were displayed at the British Museum. Finally, in 1953, the bones were proven to have been assembled from a human skull, the lower jaw of an orangutan, and chimpanzee teeth. It is unclear who was responsible for the hoax or whether it was intended as an actual fraud or merely as a prank to embarrass the scientific establishment.

During the First World War, James Shearer, an American physician serving in the British Army, published an article in the *British Medical Journal* claiming to have developed a method for tracing the path of a bullet in the body. The work proved to be a fabrication; the physician was tried and convicted before a military court and sentenced to death. His sentence was commuted, and he died in prison a few years later. *See* S. Lock, *Misconduct in Medical Research: Does It Exist in Britain*? 297 BRIT. MED. J. 1531 (1988); R. Charrow, *Scientific Misconduct: Sanctions in Search of Procedures*, 2 J. NIH RES. 91 (1990).

In 1967, the Food and Drug Administration established a unit (later known as the Office of Scientific Investigations) in the then-Bureau of Drugs (now the Center for Drug Evaluation and Research) to investigate clinical investigators "believed to have submitted false data . . . or to have conducted research in a manner that did not offer the maximum protection to the subject." Frances O. Kelsey, *Biomedical Monitoring*, 18 J. CLIN. PHARM. 3 (1978). The following year, FDA adopted its first rule providing procedures for dealing with researchers "charged with submitting false data. . . ." 33 Fed. Reg. 8338 (June 5, 1968) (codified at 21 C.F.R. § 130.3).

7. In 1979, according to the *Monitor*, a publication of the American Psychological Association, the Senate Committee on Labor and Human Resources had compiled "a list of 31 cases of research violations first made public in 1979, which included 13 instances of altered or false lab work and eight false patient records." Kathleen Fisher, *The Spreading Stain of Fraud*, 13 MONITOR 1, 7 (1982). It would appear that the list referenced in the article was FDA's listing of researchers who had been disqualified from receiving investigational new drugs.

8. For a more in-depth discussion of these four cases, see Barry D. Gold, *Congressional Activities Regarding Misconduct and Integrity in Science*, *in* 2 RESPONSIBLE SCIENCE ch. 6

(1993). For an excellent discussion of the Soman and Long cases and how they fit into the evolution of federal laws regulating scientific "integrity," see DANIEL J. KEVLES, THE BALTIMORE CASE 104, 106 (1998).

9. AAAS/ABA, REPORT ON WORKSHOP II—PROJECT ON SCIENTIFIC FRAUD AND MISCONDUCT 77 (1989). Felig retracted many articles that he had jointly authored with Soman. *Id.*

10. Philip Hilts, *Cancer Researcher Punished in Fraud Case*, WASH. POST, May 20, 1982, at A10.

11. *See* FDA Records F83–2746, provided in response to Freedom of Information Act request.

12. *See* Nils J. Bruzelius, *Cancer Research Data Falsified; Boston Project Collapses*, BOSTON GLOBE, June 29, 1980; Consent Agreement in the Matter of Marc J. Straus, M.D., FDA Records F83–2746 (May 17–18, 1982) (recounting that FDA formerly conveyed the allegations to Straus on August 27, 1980).

13. HHS News Release (n.d.); Hilts, *supra* note 10.

14. Seven months after being debarred, Straus was again in the news. The October 1982 issue of *Cancer Treatment Reports*, a National Cancer Institute (NCI) journal, published an article by Straus. The journal, aware of Straus's debarment, agreed to publish the article if and only if the data reported in the article were independently verified. NCI, however, learned from Congressman L. H. Fountain (D-NC) that the "independent review" might not have been sufficiently independent or rigorous. *See* Marjorie Sun, *Barred Researcher Publishes a Paper*, 219 SCIENCE 270 (1993). Although there is no evidence that the article contained falsified data, the fact that it came to the attention of a member of Congress may indicate that some in Congress were concerned about the integrity of the scientific enterprise.

15. *Confronted with Evidence of Faked Data, MGH Doctor Resigned in January*, BOSTON GLOBE, June 29, 1980.

16. Robert Charrow & Michael Saks, *Legal Responses to Allegations of Scientific Misconduct, in* RESEARCH FRAUD IN THE BEHAVIORAL AND BIOMEDICAL SCIENCES 35–36 (David J. Miller & Michel Hersen eds., 1992) (referencing Patricia Woolf, *Deception in Scientific Research*, 29 JURIMETRICS J. 67 [1988]).

17. *Fraud in Biomedical Research: Hearings before the Subcomm. on Investigations and Oversight of the H. Comm. on Science and Technology*, 97th Cong. 58 (1981).

18. *See id.* at 10 (testimony of Philip Handler, President, National Academy of Sciences); *id.* at 26 (testimony of Donald S. Frederickson, Director, NIH); *id.* at 65 (testimony of Ronald Lamont-Havers, Chief of Research, Massachusetts General Hospital); *id.* at 165 (testimony of William Raub, Associate Director, NIH); *cf. id.* at 343 (statement of Patricia Woolf, Princeton University).

19. Eugene Braunwald, *Cardiology: The John Darsee Experience, in* RESEARCH FRAUD, *supra* note 16, at 61.

20. *See* Robert Leavey, *Harvard Panel Finds 3 Cases of Falsified Research Data*, BOSTON GLOBE, Jan. 26, 1982.

21. Arnold Relman, *Letters from the Darsee Affair*, 308 N. ENG. J. MED. 1415, 1415–16 (1993).

22. *Id.*

23. *See* H.R. 5919, 97 Cong. (1982) (proposing to add a new section 402(f) to the Public Health Service Act).

24. *See* Marcia Angell, *Betrayers of the Truth*, 219 SCIENCE 1417 (1983) (book review).

25. WILLIAM BROAD & NICHOLAS WADE, BETRAYERS OF THE TRUTH: FRAUD AND DECEIT IN THE HALLS OF SCIENCE 87 (1982).

26. Public Health Service Act § 493, 42 U.S.C. § 289b.

27. H. Rep. No. 158, at 38–39 (1985); *see* H. Rep. No. 99-309, at 83–84 (1985) (Conf. Rep.).

28. *Id.*

29. 15 GUIDE FOR GRANTS AND CONTRACTS No. 11 (July 18, 1986).

30. *Id.* at 5. *See also* PHS Grants Policy Memorandum No. 49 (Sept. 1, 1988); PHS Grant Application Form PHS 398, at 7 (Oct. 1988).

31. Daniel Koshland, Editorial, *Fraud in Science*, 235 SCIENCE 141 (1987).

32. For one of the most thorough and even-handed accounts of the events of this era, see KEVLES, *supra* note 8.

33. *See* April 12, 1988 hearing (Baltimore Hearing), May 4 and May 9, 1989 hearings, May 14, 1990 hearing, and August 1, 1991 hearing.

34. *See Fraud in NIH Grant Programs: Hearings before the Subcomm. on Oversight and Investigations of the Comm. on Energy and Commerce*, 100th Cong., Serial No. 100-189 (1988).

35. The late Ted Weiss (D-NY) also held a scientific fraud hearing on April 11, 1988. *See Scientific Fraud and Misconduct and the Federal Response: Hearing before the Subcomm. on Human Resources and Intergovernmental Relations of the H. Comm. on Government Operations*, 100th Cong. (1988).

36. At that time, all Public Health Service agencies (e.g., NIH; Centers for Disease Control and Prevention; Food and Drug Administration; Alcohol, Drug Abuse, and Mental Health Administration (now Substance Abuse and Mental Health Services Administration); Indian Health Service; Health Resources and Services Administration; and a few others) reported to the HHS Assistant Secretary for Health.

37. *See* 53 Fed. Reg. 36,344 & 36,347 (Sept. 19, 1988).

38. *See* 54 Fed. Reg. 11,080 (Mar. 16, 1989).

39. *See* 54 Fed. Reg. 32,446 (Aug. 8, 1989).

40. Breuning pled guilty to one count of violating the criminal False Claims Act (18 U.S.C. § 287) and one count of violating the False Statements Act (18 U.S.C. § 1001). *See* United States v. Breuning, No. K88-0135 (D. Md. Nov. 10, 1988); *Scientist Given a 60-Day Term for False Data*, N.Y. TIMES, Nov. 12, 1988.

41. Hallum had served on an Institute of Medicine committee that was studying scientific misconduct, and he was an early advocate of a nonadversarial, collegial approach to resolving misconduct cases.

42. Abbs v. Sullivan, 756 F. Supp. 1172, 1177 (W.D. Wis. 1990).

43. Robert Charrow, *Scientific Misconduct Revisited: OSI on Trial*, 2 J. NIH RES. 83 (1990).

44. *Id.*

45. *Id.* Crabb found in favor of the government on Abbs's other claims, and the following year, the court of appeals concluded that Crabb lacked jurisdiction to decide anything because Abbs had tried to short-circuit OSI's decision making. Under well-established principles of administrative law, one cannot sue the government until one is aggrieved; Abbs, according to the court, could not be aggrieved until the OSI investigation had ended with a finding that he had committed misconduct. *See* Abbs v. Sullivan, 963 F.2d 918 (7th Cir. 1992).

46. *See* Bernadine Healy, *The Dangers of Trial by Dingell*, N.Y. TIMES ON THE WEB, July 3, 1996, http://www.nytimes.com/books/98/09/20/specials/baltimore-trial.html.

47. Bernadine Healy was one of the very few government officials with the courage to stand up to Dingell and his staff and to challenge their motives, competence, and fairness.

48. *See* 57 Fed. Reg. 24,262 (June 8, 1992).

49. *See* 57 Fed. Reg. 53,125 (Nov. 6, 1992). Prior to being given responsibility over misconduct cases, the DAB adjudicated mind-numbing cases about the cost principles governing federal grants and the amount due states under the Medicaid program. *See also* 45 C.F.R. pt. 16, app. A, § B(a)(1).

50. *See In re* Sharma, No. A-93-50, Dec. No. 1431 (DAB Aug. 6, 1993), *available at* http://www.hhs.gov/dab/decisions/dab1431.html.

51. *In re* Popovic, No. A-93-100, Dec. No. 1446 (Nov. 3, 1993), *available at* http://www.hhs.gov/dab/decisions/dab1446.html. Shortly after the Popovic decision, ORI dismissed its cases against Robert C. Gallo (the lead author on Popovic's *Science* paper) and Margrit Hamosh, a Georgetown University medical researcher. The charges against Gallo were similar to those against Popovic. In contrast, Hamosh's case was more akin to Sharma—it turned on the meaning of a single word, "presently." In English, as used in England where Hamosh was trained, the word "presently" means "in the future," the opposite of its meaning in American English.

52. *See* Thereza Imanishi-Kari & David Baltimore, *Altered Repertoire of Endogenous Immunoglobulin Gene Expression in Transgenic Mice Containing a Rearranged Mu Heavy Chain Gene*, 45 CELL 247 (1986).

53. *In re* Thereza Imanishi-Kari, No. A-95-33, Dec. No. 1582 (June 21, 1996), *available at* http://www.hhs.gov/dab/decisions/dab1582.html. Page references are not available because the Web version is not paginated.

54. *Id.*

55. *See* 65 Fed. Reg. 30,600, 30,601 (May 12, 2000).

56. *Id.* at 30,601.

57. Jacobellis v. Ohio, 378 U.S. 184, 197 (1964) (Stewart, J., concurring).

58. *See* United States v. Apfelbaum, 445 U.S. 115, 131 (1980) ("In the criminal law, both a mens rea and a criminal actus reus are generally required for an offense to occur.").

59. BLACK'S LAW DICTIONARY 179 (5th ed. 1979).

60. *See* RESTATEMENT (SECOND) OF TORTS § 13 (1965).

61. Scholars and courts disagree on the meaning of both specific intent and general intent. I use the terms here to illustrate two significant points on the spectrum of intent. *See* 1 WAYNE R. LAFAVE, SUBSTANTIVE CRIMINAL LAW § 5.2(e), at 355 & n.79 (2d ed. 2003) (noting the term "specific intent" is ambiguous and that the courts do not use it consistently); United States v. Hernandez-Hernandez, 519 F.3d 1236, 1239 (10th Cir. 2008) ("[W]e have sought to follow the thrust of modern American jurisprudence and clarify the required mens rea, often by reference to the Model Penal Code's helpfully defined terms, rather than persist in employing opaque common law labels that sometimes blur the line between distinct mental elements."). *Cf.* Carter v. United States, 530 U.S. 255, 268–69 (2000) (discussing distinction between specific and general intent crimes). The Model Penal Code § 2.02 uses a four-tiered approach in describing *mens rea*—(i) intent or purpose, (ii) knowledge, (iii) recklessness, and (iv) negligence.

62. *See* 45 C.F.R. pt. 689 (NSF rules on research misconduct).

63. 42 C.F.R. § 93.224.

64. *See* 71 Fed. Reg. 70,966 (Dec. 7, 2006).

65. 71 Fed. Reg. 8304 (Feb. 16, 2006).

66. *See* 71 Fed. Reg. 67,870 (Nov. 24, 2006).

67. 71 Fed. Reg. 33,308–9 (June 8, 2006).

68. *Id.*

69. *Id.*

70. These cases are available at http://ori.dhhs.gov/misconduct/cases/ under the heading "Summaries of Closed Inquiries and Investigations Not Resulting in Findings of Research Misconduct."

71. *See* United States *ex rel.* Milam v. The Regents of the Univ. of Cal., 912 F. Supp. 868 (D. Md. 1995).

72. 21 C.F.R. § 312.70(a).

73. FDA conducts data audits to ensure compliance with its regulations. *See* Martin Shapiro & Robert Charrow, *The Role of Data Audits in Detecting Scientific Misconduct: Results of the FDA Program*, 261 J. Am. Med. Ass'n 2505 (1989); Martin Shapiro & Robert Charrow, *Scientific Misconduct in Investigational Drug Trials*, 312 New Eng. J. Med. 731 (1985). For a critical evaluation of FDA procedures, see Robert Charrow & Edward Goldman, *Regulation of Clinical Trials by the FDA: A Process in Search of Procedures*, 3 Med. Res. L. & Pol'y Rep. 107 (2004).

74. *See* United States v. Palazzo, No. 07-31119 (5th Cir. Feb. 6, 2009).

75. *See Failure to Meet FDA Reporting Requirements Can Lead to Criminal Liability for Investigator*, 8 Med. Res. L. & Pol'y Rep. 128 (2009).

76. *See In re* Sharma, No. A-93-50, Dec. No. 1431 (DAB Aug. 6, 1993).

77. The Federalist Papers xi (Clinton Rossiter ed., 1961).

78. *See* Richard A. Posner, The Little Book of Plagiarism (2007).

79. *See* Bo Crader, *Doris Kearns Goodwin's Borrowed Material*, Wkly. Standard, Jan. 28, 2002.

80. *In re* Zbiegien, 433 N.W.2d 871, 875 (Minn. 1988).

81. *Id.* at 877.

82. 42 C.F.R. § 93.103(c).

83. *See* Posner, *supra* note 78, at 83–88 (discussing the Goodwin case and noting at 88 that "Goodwin did footnote her references to McTaggart's work, but footnotes acknowledge the sources of ideas; they do not acknowledge copying the source's words.").

84. National Science Foundation, Office of the Inspector General, Semiannual Report to Congress October 1, 1994–March 31, 1995 (Aug. 7, 1995), *available at* http://www.nsf.gov/pubs/stis1995/oig12/oig12.txt.

85. *Id.*

86. *Id.*

87. *Id.*

88. In *Thomson v. Larson*, 147 F.3d 195, 199 (2d Cir. 1998), the court stated as follows:

> Joint authorship entitles the co-authors to equal undivided interests in the whole work—in other words, each joint author has the right to use or to license the work as he or she wishes, subject only to the obligation to account to the other joint owner for any profits that are made. *See* 17 U.S.C. § 201(a); Childress v. Taylor, 945 F.2d 500, 508 (2d Cir. 1991); Community for Creative Non-Violence v. Reid, 846 F.2d 1485, 1498 (D.C. Cir. 1988) ("Joint authors co-owning copyright in a work are deemed to be tenants in common, with each having an independent right to use or license the copyright, subject only to a duty to account to the other co-owner for any profits earned thereby."), *aff'd without consideration on this point*, 490 U.S. 730 (1989).

89. POSNER, *supra* note 78, at 100.

90. *See* UNIVERSITY OF PENNSYLVANIA, POLICY ON FAIRNESS OF AUTHORSHIP CREDIT IN COLLABORATIVE FACULTY–STUDENT PUBLICATIONS (Oct. 8, 1998), *available at* http://www.upenn.edu/grad/authorpolicy_alpha.html.

91. INTERNATIONAL COMMITTEE OF MEDICAL JOURNAL EDITORS, UNIFORM REQUIREMENTS FOR MANUSCRIPTS SUBMITTED TO BIOMEDICAL JOURNALS: WRITING AND EDITING FOR BIOMEDICAL PUBLICATION (Oct. 2008), *available at* http://www.icmje.org/#author. Some journals now also request that one or more authors, referred to as "guarantors," be identified as the persons who take responsibility for the integrity of the work as a whole, from inception to published article.

92. *See* Romero v. Buhimschi, 2009 U.S. Dist. LEXIS 2943, at *5–9 (E.D. Mich. Jan. 14, 2009). The court's findings were made for the purpose of evaluating the legal propriety of Romero's claims where all doubts were resolved in favor of Romero.

93. *Id.* at *10–11.

94. *Id.* at *18.

95. Under United States Postal Service regulations, if one pays for the privilege of publishing an article, then that article is viewed as an "advertisement." Many journals require authors (or their institutions) to pay a publication or page fee, usually referred to as a "page charge," thus transforming the article into an advertisement. *See* Tom Scheiding, *Paying for Knowledge One Page at a Time: The Author Fee in Physics in Twentieth-Century America*, 39 HIST. STUD. NAT. SCI. 219 (2009) (discussing page charges by physics journals); United States Postal Service Customer Support Ruling PS-033 (Oct. 1996), *available at* http://pe.usps.gov/cpim/ftp/manuals/csr/ps-033.pdf. ("Postal regulations in Domestic Mail Manual (DMM) C200.4.4, require that editorial or other reading matter contained in publications entered as Periodicals matter must be plainly marked 'advertisement' by the publisher if a valuable consideration has been paid, accepted or promised for its publication." If the page charge, though, is voluntary and not imposed as a condition of publication, then the article is not considered an advertisement by the USPS). *See also* http://pe.usps.gov/cpim/ftp/manuals/csr/CSR%20TOC.pdf (listing Customer Support Rulings as of May 2005).

96. 42 C.F.R. § 50.102 (1990) (PHS); *see* 42 C.F.R. § 689.1(a)(1) (1992) (NSF) (catchall is "other serious deviation from accepted practices").

97. *See* Deborah Parrish, *The Scientific Misconduct Definition and Falsification of Credentials*, PROF. ETHICS REP., Fall 1996, *available at* http://www.aaas.org/spp/sfrl/per/per7.htm; David Goodstein, Conduct and Misconduct in Science (2002), http://www.its.caltech.edu/~dg/conduct_art.html.

98. *See* J. M. Swazey, M. S. Anderson & K. S. Louis, *Ethical Problems in Academic Research*, 81 ACAD. SCI. 542 (1993).

99. *See* J. S. LUBALIN, M. E. ARDINI & J. L. MATHESON, CONSEQUENCES OF WHISTLEBLOWING FOR THE WHISTLEBLOWER IN MISCONDUCT IN SCIENCE CASES 30 (table 10) (1995), *available at* http://www.ori.dhhs.gov/documents/consequences.pdf.

100. *Id.* at 50.

101. 42 C.F.R. § 93.307(a).

102. *See* OFFICE OF RESEARCH INTEGRITY, ANALYSIS OF INSTITUTIONAL POLICIES FOR RESPONDING TO ALLEGATIONS OF SCIENTIFIC MISCONDUCT at 6-4 (table 6-1) (2000).

103. *See* 42 C.F.R. § 93.307(g).

104. *See* id. § 93.308.

105. *See id.* § 93.308(a) ("The institution must notify the respondent [i.e., the accused] whether the inquiry found that an investigation is warranted. The notice must include a copy of the inquiry report and include a copy of or refer to this part and the institution's policies and procedures adopted under its assurance.").

106. *See id.* § 93.310(g).

107. *See id.* §§ 93.103, 93.106.

108. *See* Richard O. Lempert, *Modeling Relevance*, 75 MICH. L. REV. 1021 (1977).

109. *See* 42 C.F.R. § 93.106(b).

110. I am familiar with only one instance where an institutional finding of misconduct was reversed by ORI. That case involved a relatively small medical school and a Veterans Administration hospital.

111. *See* United States v. Poehlman, No. 2:05-CR-38-1 (D. Vt. June 28, 2006).

112. 42 C.F.R. § 93.523(a).

113. *See* 45 C.F.R. §§ 689.9–689.10.

114. Other agencies have procedures that are similar to those used by NSF or ORI. *See, e.g.*, 10 C.F.R. pt. 733 and § 600.31 (Dept. of Energy Misconduct Rules); 14 C.F.R. pt. 1275 (NASA Misconduct Rules); 48 C.F.R. § 952.235 to .271 (Dept. of Energy Misconduct Rules for Procurement Contracts); 48 C.F.R. § 1252.235 to .270 (Dept. of Transportation Misconduct Rules for Procurement Contracts).

115. *See* 5 U.S.C. §§ 702–706. In reviewing the outcome of an administrative hearing, a variety of standards come into play depending on the issue and the nature of the hearing. For example, if a plaintiff claims that the outcome is just wrong, courts ask whether the agency's decision is supported by "substantial evidence." 5 U.S.C. § 706(2)(E). The term "substantial evidence" merely means that there is some evidence—it does not have to be very much—to support the ALJ's decision. In short, overturning the results of a hearing is difficult unless the ALJ made an error of law. Technically, the substantial evidence standard applies to hearings that are required by statute. A scientific misconduct hearing is not required by statute, but there is no reason to believe that a federal court would use a different standard to gauge the evidentiary propriety of the outcome.

CHAPTER 4

Protecting Human Subjects: Bits, Bytes, and Biopsies

Research involving[1] human subjects is tightly regulated by the federal government, as well as by some states. Clinical researchers tend to be familiar with those rules; bench researchers usually are not. After all, what relevance do clinical research rules have for those who never see patients and rarely even set foot in a hospital? In fact, a decade ago, clinical research rules had little bearing on nonclinical scientists, but today, that is not the case. Now, the bench researcher who is using tissue specimens to link a trait with a gene and the psycholinguist who is using a paper-and-pencil test to study how the mind processes certain grammatical constructions may be conducting "regulated" research to the same extent as the clinician who is studying competing treatments for a given disease.

This chapter discusses how and why the government and universities got involved in regulating clinical research. It also explains the five types of rules that govern research involving human subjects:

1. The common law, which pertains to virtually anything anyone does to another anywhere at any time
2. The Common Rule (which has little to do with the common law), which applies to all federally funded or federally regulated research and university and state law counterparts
3. The special Food and Drug Administration (FDA) rules that apply when studying an unapproved new drug, biologic, or device in a human
4. The federal privacy laws, including the Health Insurance Portability and Accountability Act of 1996 (HIPAA), which regulate the use, transmission, and storage of patient-specific information collected by hospitals and physicians
5. The special federal guidelines affecting the manipulation of genes[2]

The common law differs from the other sources of regulation in two important respects. First, the common law is usually not enacted as a statute or issued as a regulation, but rather is made by the courts in the course of deciding cases. Second, it provides a remedy only to an injured party, and therefore directly applies only once something has gone awry. Its primary focus is retrospective (i.e., correcting a past wrong), as opposed to prospective (i.e., preventing a wrong). However, this is not to say that the common law does not have a profound prospective influence. The threat of having to pay damages influences people's conduct and leads them to take precautions that they might not otherwise have taken. Take the case of the Ford Pinto, a 1970s automobile with a propensity to explode following rear-end collisions. In one case, a new 1972 Pinto stalled on a southern California freeway; apparently the car had had numerous mechanical problems. The Pinto was immediately rear-ended and burst into flames because the car's gasoline tank was located too near to the rear bumper and the rear bumper provided inadequate protection. The driver was killed and the thirteen-year-old passenger badly burned over most of his body. The passenger and the driver's estate instituted suit against Ford: The jury determined that the Pinto was defectively designed and that Ford knew about these defects but chose to go into production without addressing them. The trial court awarded the passenger $2.5 million in compensatory damages (e.g., medical expenses, pain, and suffering) and $3.5 million in punitive damages. The driver's estate was awarded $559,000. On appeal, the California Court of Appeal affirmed the judgment.[3] The court's award, though, directly affected only the injured passenger and the driver's estate; it had no direct effect on the way in which Ford designed its cars. However, the specter of large punitive damage awards and significant adverse publicity should lead a rational manufacturer to modify its design, even though it is not required to do so by existing law. This happened in the case of the exploding Pinto, where public pressure and large punitive damage awards forced Ford to redesign the car's fuel system.

As we will see shortly, the common law was in the process of developing rules of informed consent when events overtook it and led to the Common Rule, state rules, and university rules, i.e., rules two through four. The FDA rules, while similar in many respects to the Common Rule, have significantly different reporting and paperwork requirements, and substantially more serious ramifications for violations. The HIPAA rules are decidedly different in many significant respects from the Common Rule and apply irrespective of whether the research is federally funded. The rules affecting gene research apply to any institution that receives National Institutes of Health (NIH) funding for recombinant DNA research.

THE COMMON LAW—THE BEGINNING OF INFORMED CONSENT

We've all seen movies and television programs in which the macho cop character tells the crusty desk sergeant to book the miscreant for "assault and battery." Have you ever wondered precisely what "assault and battery" are? They are distinct but related offenses. Battery is the intentional and nonconsensual offensive touching of another. In contrast, assault is the placing of another person in imminent apprehension of a battery. Thus, if I punch you in the nose without your consent, I have committed a battery. I may have also committed an assault, but not necessarily. For example, if I punched you in the nose while you were asleep, I have committed a battery, but not an assault, because you were unaware that you were about to be slugged. The law of battery developed during the Middle Ages in England and was largely intended to provide a remedy for unprovoked physical attacks by one person on another. In the United States, battery took on another role—it became the vehicle of choice for early medical malpractice cases, especially those in which the physician deviated from the treatment that he had originally discussed with the patient. This was so because battery necessarily presupposes the absence of consent.

Let's venture back to Minnesota at the beginning of the twentieth century where a patient agreed to undergo surgery to her right ear to correct a hearing problem (although one has to wonder whether it was possible to surgically correct a hearing problem at the start of the twentieth century). While the patient, Mrs. Mohr, was under a general anesthetic, the surgeon realized that the left ear was more in need of treatment than the right one. Without reviving Mohr, which would have been risky especially in 1900, the surgeon proceeded to operate on her left ear rather than the right. The surgery was an apparent success, but the patient nonetheless instituted a battery suit against the surgeon arguing that he did not have her consent to operate on her left ear; the court agreed that the surgeon had committed a technical battery.[4] Cases like *Mohr*[5] emphasize the importance of consent; without a patient's consent, a physician could easily commit a battery, even when things work out well.[6]

Battery deals with the simple concept of "consent," not informed consent. The notion of "informed consent" did not enter the legal equation until relatively recently, starting in the 1960s, and then only tentatively.[7] Things changed quickly, though, soon after the Court of Appeals for the District of Columbia Circuit decided *Canterbury v. Spence*.[8] *Canterbury* was one of those nightmarish cases in which everything that could go wrong in a hospital did. The court's summary of the case tells it all:

> The record we review tells a depressing tale. A youth troubled only by back pain submitted to an operation without being informed of a risk of paralysis incidental thereto. A day after the operation he fell from his hospital bed after having been left without assistance while voiding. A few hours after the fall, the lower half of his body was paralyzed, and he had to be oper-

ated on again. Despite extensive medical care, he has never been what he was before. Instead of the back pain, even years later, he hobbled about on crutches, a victim of paralysis of the bowels and urinary incontinence. In a very real sense this lawsuit is an understandable search for reasons.[9]

The court went on to hold that a patient's consent is meaningless unless it is an informed consent, one based on the potential risks and benefits of a proposed course of treatment as well as an understanding of the treatment options. The court held that it is the physician's responsibility to provide the patient sufficient information so that he or she can make an informed decision. The physician who treats a patient without having the patient's informed consent can be held liable for common law battery.

It is likely that, over time, the common law notion of "informed consent" articulated in *Canterbury* would have led universities, research institutions, and hospitals to adopt internal rules for informed consent both for treatment and research purposes and to put in place formal processes for approving clinical studies. But in politics, like the stock market, timing is everything. And the ethics involved in giving (or denying) patients informed consent were about to move to center stage. About three weeks after the *Canterbury* decision became final, the *New York Times* reported on the Tuskegee syphilis study, which, according to the paper, was "the longest non-therapeutic experiment on human beings in medical history." From 1932 to 1972, as part of a Public Health Service cooperative study, treatment was denied to 399 poor African American sharecroppers in Macon County, Alabama, who had been diagnosed with syphilis. This was done to permit government scientists to study the natural course of the disease. The revelations in the press shocked the conscience of the nation.

THE COMMON RULE AND ITS FDA VARIATIONS

Responding to the Tuskegee study, Congress, in 1974, enacted the National Research Service Award Act, which required any entity[10] applying for funding under the Public Health Service Act to submit with its application "assurances satisfactory to the Secretary [of Health and Human Services] that it has established . . . a board (to be known as an 'Institutional Review Board') to review biomedical and behavioral research involving human subjects. . . ."[11]

The new law, though, proved too narrow because it failed to cover most agencies that funded basic research. In response, Congress created the President's Commission for the Study of Ethical Problems in Medicine and Biomedical and Behavioral Research to review the adequacy and uniformity of government-wide policies for protecting human subjects.[12] In its First Biennial Report, the Commission recommended that uniform federal regulations be adopted.[13] On June 18, 1991, the White House Office of Science and Technology Policy published its Model Federal Policy for Protection of Human Subjects (Model Federal Policy), which has since been adopted by the relevant agencies with appropriate changes reflecting the individual needs of each agency involved.[14] The Model Federal Policy is usually referred to as the Common Rule. (A link for the Department of Health and Human Services [HHS] version of the Common Rule, as amended, is set out in appendix C, along with links for the versions adopted by other agencies.)

An Overview of the Common Rule

The Common Rule revolves around two central themes—informed consent and independent review. First, no one should be asked to participate in a clinical study unless he or she is fully aware of the risks and potential benefits, if any. And second, a group of individuals not directly associated with the research should review the proposed research and decide whether its benefits outweigh its risks. If they do, the group would approve the protocol; if not, they would disapprove it.

INFORMED CONSENT

Research involving human subjects usually can be conducted only with a duly approved informed consent form that meets the regulatory requirements and includes, among other things, the risks and benefits of the proposed study, appropriate alternative procedures, an acknowledgment that the subject can cease participating whenever he or she wishes, and a discussion of the possible significance of the research.[15]

INSTITUTIONAL REVIEW BOARD

The Model Federal Policy, which is mirrored in the regulations adopted by most agencies,[16] requires that an awardee (e.g., university, research institute) assure the agency (e.g., NIH) that it will comply with the agency's policies on human subjects.[17] In general, this means that the awardee will create an Institutional Review Board (IRB) to review the risks and benefits of any proposed research involving human subjects. Each IRB must have at least five members, with varying backgrounds; at least one member must be a nonscientist and at least one member must not be affiliated with the institution.[18] The IRB must review, discuss, and approve any research involving human subjects before that research can be funded by the government. As part of its evaluation, the IRB also reviews and approves the informed consent form that usually must be signed by each human subject. Research involving "minimal risk" can be approved, on an expedited basis, by a single designated member of the IRB (usually the chairperson); the research need not be reviewed or approved by the full board.[19] In the course of considering a protocol, the IRB must identify and weigh the risks of the proposed research against its potential benefit. At least annually, the IRB must review and approve any application for continuing the research.

RECORD RETENTION

Each IRB is required to maintain its records, including the grant application, informed consent forms, minutes of meetings, and disposition, for whatever periods the agency specifies in its regulations. For example, the HHS rules require that IRB records be maintained for at least three years after completion of the research at issue.[20]

SANCTIONS

Funding agencies reserve the right to sanction grantees found to have violated an agency's human subjects regulation. The method and severity of the sanction varies across agencies. At the National Science Foundation (NSF), for example, violations of that agency's human subjects regulation are investigated by the NSF inspector general. In contrast, at HHS such investigations are conducted by the Office for Human Research Protections (OHRP).[21] At HHS, a grant may be terminated or suspended if there is a finding that there has been a "material failure to comply with the terms of [the HHS human subjects] policy."[22] If the problems are deemed systemic, OHRP may suspend all federally funded human subjects research at the institution except for patient follow-up. For example, on Thursday, July 19, 2001, following the death in June of a healthy volunteer, OHRP suspended all HHS-funded human trials at Johns Hopkins University because of what it perceived to be systemic violations of the Common Rule.[23] The following Monday (July 23, 2001), OHRP lifted its ban. Many believed that the quick turnaround was due to two factors. First, OHRP's action was out of proportion to the problems at Hopkins, and those problems were unrelated to the death. And second, OHRP folded under the political pressure brought to bear by both Maryland senators, Paul Sarbanes (D-MD) and Barbara Mikulski (D-MD); Mikulski was and is a member of the appropriations committee and could have held up funding for OHRP.

In extraordinary cases, a federal agency such HHS can seek to debar the institution, the investigator, or both.[24] In addition to administrative sanctions, failure to adhere to human subjects regulations can subject researchers and their institutions to actions for assault, battery, and negligence, and can jeopardize intellectual property interests that are based on tissue removed from a patient as part of the study.[25]

The Who, What, and When of the Common Rule

In deciding the extent to which a proposed project is regulated under the Common Rule, you have to ask and answer a number of questions. First, does the project involve "human subjects research"? Second, if it does involve human subject research, does it qualify for one of the regulatory exemptions? Third, if it does not qualify for one of the exemptions, is it of a type of research that poses so little risk that it qualifies for expedited IRB review? Fourth, if it does not qualify for expedited review, are the subjects nonimprisoned, nonpregnant adults? If so, then the research qualifies for normal IRB review. Otherwise, when the subjects are children, pregnant women, or those who are incarcerated, a more focused, specialized review is required.

THE COMMON RULE APPLIES TO MOST TYPES OF "HUMAN SUBJECTS RESEARCH"

As with most regulatory schemes, the rules that govern research that involves humans hinge on the definitions of the key terms "research" and "human subjects." If the activity does not involve both "research" and "human subjects," then the panoply of regulations, including IRB review and approval, does not apply. Not infrequently, the IRB, aided by counsel, must make this determination.[26] Deciding whether something is research or whether an individual is a subject can be far from simple. Note that most of these issues arise when no federal funding is involved; universities often voluntarily apply the Common Rule to all research activities irrespective of their funding sources. When an individual seeks federal research funding, this very act should logically preclude the individual from later claiming that his or her now-funded activities are not really research.

What Is Research?

Francis Bacon, Immanuel Kant, Karl Popper, and other great philosophers have devoted tomes to defining science and research.[27] The Common Rule is more terse and significantly less philosophical; it defines research as

> a systematic investigation, including research development, testing and evaluation, designed to develop or contribute to generalizable knowledge. Activities which meet this definition constitute research for purposes of

this policy, whether or not they are conducted or supported under a program which is considered research for other purposes. For example, some demonstration and service programs may include research activities.[28]

It is unlikely that any philosopher of science would embrace the Common Rule's definition of "research." That may be beside the point. Regulators tend to be more concerned with the practical and less with the theoretical. They seek to balance two competing interests: a desire, on the one hand, to define terms broadly so that "bad behavior" does not escape scrutiny and, on the other hand, to define terms in such a way that the line separating regulated conduct from nonregulated conduct is relatively crisp. Regulators usually dislike uncertainty. As illustrated below, the strain between "breadth" and "certainty" is evident in the Common Rule, especially when attempting to decide whether a proposed endeavor is "research" or "nonresearch."

Is a case study research? In some instances, deciding whether certain activities qualify as research can be murky. This is especially so for case studies, which populate the medical literature. For example, Dr. Bell, a professor of internal medicine at a highly respected medical school, after examining a patient with an unusual syndrome, decides to write it up as a case study and submit it to a major medical journal. Assuming that Dr. Bell's institution applies the Common Rule to all research irrespective of the funding source, do any of Dr. Bell's activities constitute research, thereby requiring IRB approval? Most physicians would properly argue that Dr. Bell's physical examination of the patient was a "systematic investigation." Furthermore, by publishing the results, Dr. Bell apparently hoped to "contribute to generalizable knowledge." Under the definition of "research," though, the intent to contribute to generalizable knowledge must coexist temporally with the "systematic investigation." Here, the examination was apparently undertaken solely as a prelude to treating a patient; there is no evidence to suggest that Dr. Bell intended to craft a case study before or even while conducting the examination. Accordingly, because the "systematic investigation" was not "designed to develop or contribute to generalizable knowledge," the physical examination does not constitute research, even though the physician decides post-examination to write it up and submit it for publication.

Change the timing and the results will change. Suppose, for instance, that Dr. Bell let it be known that he was interested in examining any patient who presented with a given set of symptoms and that one of the purposes of examining the patient was to provide additional information for an article that he was writing. In such a setting, an IRB should conclude that Dr. Bell's activities constitute research requiring IRB review and approval.

Is a new surgical procedure research? Dr. Cutter, a noted thoracic surgeon at a famous academic medical center, has been performing a particular procedure for years. He has always entered the patient's chest cavity by making an incision in the patient's back. One day, while performing the procedure, he decides to try entering from the front and finds that it is much easier and quicker to perform the surgery in that way. He writes up his findings about his experimental procedure and continues to perform the procedure entering from the front. Has Dr. Cutter engaged in research? Merely because a surgeon might label a procedure as "experimental" does not mean that he is engaging in research when he performs that procedure. It is important not to blur the line that separates treatment from research. After all, many would argue that neither IRBs nor OHRP should be in the business of deciding "best medical practices."[29]

Did Dr. Cutter engage in a "systematic investigation"? It would be difficult to label his innovative procedure as "systematic." It has none of the indicia that one normally associates with research: There is no experimental design; there is no control group; there is no hypothesis capable of being disproved. The fact that he writes up his experience, even labeling it as experimental, does not alter this conclusion. The act of putting pen to paper cannot transform a set of surgical procedures into a "systematic investigation."

Naturally, if Dr. Cutter actually decided to measure differences between patients' outcomes depending on which procedure he used, then that would transform his purely surgical work into research, and IRB review and approval would be necessary.

Science versus Journalism. One of the most fascinating cases highlighting the limitations of the Common Rule's definition of "research" involved two psychologists, Elizabeth Loftus, then at the University of Washington, and Melvin Guyer, at the University of Michigan School of Medicine. Both were interested in legal cases that involved so-called repressed and recalled memory, when a witness (usually a young adult) is prodded by a psychologist or psychiatrist into recalling an assault (usually sexual abuse by a parent) many years after the fact, a traumatic event that the witness had totally forgotten (i.e., repressed). Both Loftus and Guyer doubted the reliability of repressed memory testimony, and both had agreed to jointly author an article in a magazine called the *Skeptical Inquirer*, which is devoted to debunking "claims of paranormal, fringe-science, and pseudoscientific phenomena."[30] No federal funding was involved.

They focused on a single case that had been written about by two psychologists, David Corwin and Ema Olafson, in an article that appeared in the journal *Child Molestation*. The gist of the article was that one of psychologists, Corwin, had tape-recorded a 1984 interview that he had conducted with a six-year-old girl whose parents were in a custody battle. The court had asked Corwin to conduct a psychological evaluation of the young girl.

In response to questions during that 1984 interview, the girl recalled that her mother had sexually abused her. Fast forward to 1995. The psychologist who conducted the 1984 interview contacts the now seventeen-year-old girl and her stepmother, seeking permission to reinterview the girl. During this later interview, the girl at first does not remember her earlier interview, but then, the article reported, she has a "spontaneous return of that reportedly unrecallable memory." The article contained a transcript of the interviews. Any inconsistencies between the 1984 interview and the one eleven years later Corwin believed were inconsequential.

Loftus and Guyer, however, believed that the inconsistencies were telling and wanted to learn more about the case. This could have proven to be impossible because Corwin and his coauthor were careful in their writings to hide any information that could be used to identify the girl, whom they consistently referred to as "Jane Doe." However, in speeches, Corwin was not as careful, and the girl's first name and town of residence slipped out on at least one occasion. Armed with this limited information, Loftus and Guyer "searched legal databases and found a published appellate court case relating to allegations that Jane's father had failed to comply with visitation orders. . . . That case provided additional factual details about Jane Doe's family. Furthermore, the disclosure of the father's first name and last initial led to a successful search for the father's identity, and, according to the authors, 'from there we uncovered the full history of the custody dispute and the abuse allegations.'"[31] Armed with the identities of the parties, Loftus and Guyer interviewed Jane Doe's natural mother, her foster mother, and her stepmother. During that interview, the stepmother was reported to have "volunteered that the way [Jane's father and stepmother] got Jane away from Mom was 'the sexual angle.'"[32]

In their article, Loftus and Guyer questioned many of the legal findings, including whether the original allegations against the natural mother were true. They also questioned "whether 17-year-old Jane's memory of an alleged prior event was, in fact, a recovered memory."[33] The Loftus and Guyer article (which appeared in two parts) did not identify Jane Doe or the other parties.

Loftus and Guyer, though, had initial difficulties publishing their article. Apparently, one or more of the parties involved in the case complained at first to a University of Michigan IRB. The IRB initially concluded that Guyer's work was not research, reversed itself one month later, and later reversed itself again, finally concluding that the work was not research. Loftus, armed with the decision of the University of Michigan, went forward with publication. However, the University of Washington's IRB then undertook an investigation of its own; that investigation lasted twenty-one months before exonerating Loftus. In the end, Loftus, who had been the star of the University of

Washington's psychology department, was offered a chaired professorship at the University of California, which she promptly accepted, in part because of the way in which the University of Washington had handled her case.

Both universities had to decide whether Loftus and Guyer's foray into the almost-lay press constituted scientific research on human subjects. In resolving this question, they probably examined two factors. First, how different was the nature of the work from that of an investigative journalist? Loftus and Guyer, like journalists, unearthed information about people who would have preferred to remain anonymous. Loftus and Guyer, like journalists, sought to discover whether a court had done justice more than one decade earlier in awarding custody to the father and stepmother. And Loftus and Guyer, like journalists, published in a magazine rather than in a science journal. Had the article been written by a member of either university's journalism department, no one would have raised an eyebrow. Second, both universities had to consider that, as state institutions, they were severely constrained under the First Amendment in how they regulate their faculty. That is especially the case when the faculty are speaking out on controversial issues that affect the body politic. Indeed, many respected legal scholars believe that state universities that require IRB review of social science research, especially interviews, where no federal funding is involved are likely violating the First Amendment.[34] State universities, unlike private institutions, are bound by the First Amendment, which prohibits state, local, or federal governments from conditioning or interfering with speech, especially academic speech.[35]

You may have noticed that much of the history underlying the Loftus–Guyer case was gleaned from a court opinion. In fact, Jane Doe ultimately sued just about everyone involved for defamation and various other causes of action, some stemming, in part, from statements that were allegedly made at a professional meeting; none of those statements ever identified her. The case went all the way up to the California Supreme Court, which held that all but one of her claims had to be dismissed on First Amendment grounds.[36] The case illustrates the risks associated with certain types of research; on occasion, the researcher may be in greater danger than the subject.

Calibration versus Experimentation. On occasion, laboratory equipment needs to be calibrated, and human volunteers are sought either to provide bodily fluids or sometimes aspects of their humanity—their voices, their hearing. Does this type of activity qualify as human subjects research? Certainly, subjects may be placed at risk or made to feel uncomfortable. However, the elimination of risk is neither possible nor the goal of the Common Rule. It is aimed only at reducing those risks incurred in the course of research. Is calibrating an instrument research? Probably not, because there is no effort to obtain generalizable information. To the contrary, calibration runs in the opposite direction, namely individualizing a piece of machinery. There is another way of looking at it. Suppose that a piece of laboratory equipment could be calibrated only with great difficulty requiring two sets of hands, one to hold the equipment in place and the other to get under the equipment and tighten a bolt. Would the young research assistant who aids the professor in adjusting the equipment by shimmying under the equipment with wrench in hand be considered a human subject? Clearly not. The fact that a person is being asked to give blood rather than sweat should not affect whether the activity is research.

Providing Specimens versus Conducting Research. Not infrequently, one group of researchers (group A) may provide another group (group B) with specimens that have been derived from humans. Are those in group A who collected, maintained, and transmitted the specimens to the other researchers engaged in research involving human subjects? What about those in group B? If the specimens were "not collected specifically for the currently proposed research" by group B and are coded so that those in group B cannot link the name of patient with a given specimen, then those in group A are not conducting research on humans by transferring the specimens; correspondingly, those in group B are not conducting research involving humans.[37] If, however, the two groups collaborate on the research (e.g., group B collects the data and group A analyzes those data), then both may be viewed as conducting human subjects research. This is so because the research teams (group A and group B) when taken together have the ability to decode the specimens and link a patient with a specimen. It is not legally feasible to separate the two groups of researchers since they will be jointly responsible for any joint publication.

When Does Research Begin?

Usually, it is relatively easy to decide *whether* a project involves research, especially when the project is funded by NIH or falls under the aegis of FDA. However, it is not always clear *when* that research begins. This can lead to "gotcha" moments—instances when a researcher fails to get the necessary informed consent because he or she (mistakenly) believes that the clinical research has not yet begun. Figuring out when research begins is critical because research involving human subjects may legally proceed only after informed consent is obtained. Suppose, for example, that a researcher is testing a new drug aimed at treating certain types of angina. Only patients who have been admitted to the hospital suffering from severe chest pains are eligible to participate. The entrance criteria for the study also require a specific white blood cell count and a negative human immunodeficiency virus (HIV) test. Blood work is performed routinely by the hospital on all new admissions, but an HIV test is not routinely performed. Does the study begin at the time of hospitalization, at the time the blood is drawn, at the time the HIV test is ordered, or at some other time?

As a general rule, screening or eligibility tests are considered part of the study and may not be performed unless informed consent is first obtained. However, tests that are part of a patient's treatment or are routinely performed on all patients similarly situated, whether they participate in a study or not, should not trigger the start of a study, even if the results from those tests are used to determine eligibility. Conversely, laboratory tests that would not have been performed but for the study are sufficient to trigger the start of a study.[38]

A second aspect of screening potential subjects involves recruiting them. Researchers who use mass media (e.g., radio, television, newspapers, phone banks) to recruit potential study subjects are usually required to have the IRB review and approve the advertisement or script. If potential subjects are going to be screened over the telephone, the script should contain an oral informed consent that tells subjects that they are being screened to participate in research, that the questions might call for embarrassing information, that the participants do not have to answer any questions they do not want to answer, and the like.

The privacy provisions of HIPAA also restrict a researcher's ability to review medical records at his or her own institution to identify potential subjects. It is possible that there may be two temporal dividing lines between research and nonresearch, one for the Common Rule and FDA, and another for HIPAA. HIPAA is discussed later in this chapter.

Who Is a Human Subject?

It would seem that anyone, other than a researcher, who participates in "research" is a "human subject." Before jumping to conclusions, let's examine precisely how the term is defined in the Common Rule:

> *Human subject* means a living individual about whom an investigator (whether professional or student) conducting research *obtains*
> (1) data through *intervention* or *interaction* with *the individual*, or
> (2) identifiable private information.
> *Intervention* includes both physical procedures by which data are gathered (for example, venipuncture) and manipulations of the subject or the subject's environment that are performed for research purposes.

> *Interaction* includes communication or interpersonal contact between investigator and subject.
> *Private information* includes information about behavior that occurs in a context in which an individual can reasonably expect that no observation or recording is taking place, and information which has been provided for specific purposes by an individual and which the individual can reasonably expect will not be made public (for example, a medical record). Private information must be individually identifiable (i.e., the identity of the subject is or may readily be ascertained by the investigator or associated with the information) in order for obtaining the information to constitute research involving human subjects.[39]

The definition is so broad that it is relatively easy to transform the Common Rule into a constitutionally infirm regulation.[40] For example, the definition of "human subject" would seem to encompass just about anyone, including those interviewed by political pollsters. Pollsters, after all, are seeking generalizable knowledge—e.g., how will the body politic vote on election day based on the opinions of 1,500 registered voters recorded well before the election? Indeed, under the Common Rule definition, CEOs of publicly traded companies could be construed as "human subjects" and Wall Street analysts as "investigators." Wall Street analysts are paid to predict how a company's stock will do based on bits and pieces of information; some of that information is gained by interviewing a company's officers. The analyst is collecting data through systematic investigation of a company, including "interaction" with its CEO, which is designed to contribute to generalizable knowledge. As we can see, the regulatory definition is overly broad because included within its ambit are subjects who are not real subjects and researchers who are really not researchers, in the scientific sense.

Would the target of an investigative journalist be a "human subject" if the journalist were deemed to be conducting research? Suppose, for instance, that the journalist went through the subject's trash and gathered private information. It would seem that that would transform the target of an investigative reporter into a human subject if, in fact, the journalist were deemed to be conducting research.

Can Third Parties Be Subjects? Surveys are usually not controversial, and the actors involved are easy to categorize. The person asking the questions is the researcher, and the person giving the answers is the subject. Suppose, however, that the subject is being asked questions about his or her father, mother, and siblings, and suppose those questions relate to drug use or sexual relations or other inherently private matters. Do the father, mother, and siblings suddenly become "human subjects," and if so, are their consents required before the survey can take place? This is not an idle question, nor is it one with a simple answer.

In 1999, a father opened an envelope addressed to one of his adult daughters; the daughters were participating in a study about identical twins. The envelope contained a survey instrument, which among things sought information about the twins' parents, including whether the father had suffered from depression or had abnormal genitalia. The father complained to OHRP that the survey effectively transformed him into a subject, yet he had not consented to permit his daughters to provide the information. OHRP took the complaint seriously and conducted an investigation. One has to question the agency's real-world experience. Anyone with an adult child knows that what he or she tells others about you is not subject to moderation or parental controls; the father wanted the government to step into this breach, and OHRP obliged. Also, one has to wonder why OHRP ignored the father's possible legal transgression—it is a federal crime to open mail addressed to another without his or her permission and to abstract the contents.[41]

OHRP, following its investigation, apparently determined that the father was a third-party subject from whom informed consent was needed. The publicly available documents do not reveal how OHRP reached this conclusion: The documents are remarkably devoid of any analysis.[42] This is not surprising. In my dealings over the years with OHRP, legal analysis (whether sound or otherwise) has normally been lacking. Cogent legal analyses are not easy to prepare; they require thought, time, and skill. However, they should be an essential aspect of any decision making that depends on federal regulations.

In response to OHRP's findings, its advisory committee studied the issue and effectively concluded, albeit without the benefit of any reasoning and using remarkably waffley language, that there really is no such thing as a third-party subject:

> Neither reference to a third party in a research design, nor the recording of information about a third party in research records suggests that a third party must be regarded as a research subject. Nevertheless, investigators, in designing and proposing research projects and IRBs . . . should consider how the research design might focus not only on the identified human subjects, but on other persons.[43]

Some suggest that the Common Rule does not address third-party subjects.[44] I believe that there is no such thing as a third-party subject. Specifically, an individual becomes a subject either because the researcher has interacted with the individual or has obtained identifiable private information about that person. The Rule defines private information to include

> information about behavior that occurs in a context in which an individual can reasonably expect that no observation or recording is taking place, and information which has been provided for specific purposes by an individual and which the individual can reasonably expect will not be made public (for example, a medical record).[45]

Under the Common Rule, information is private only if the third party has a reasonable expectation of privacy. Thus, if someone were observed bagging cocaine by the police through a window with the shade up, it would be difficult for him to argue that he had a reasonable expectation of privacy. If he wanted privacy, he should have dropped his window shades.[46] Correspondingly, if a person tells another, albeit a family member, about certain conduct or if the family witnesses that conduct, then one could argue that any expectation of privacy necessarily vanishes. How can you have a reasonable expectation of privacy when you reveal your secrets to a third party, especially to a son or daughter? How can you have a reasonable expectation of privacy when you undertake conduct in such a way that it can be easily observed by a third party? In contrast, a third party would have a reasonable expectation of privacy if he or she confided in a healthcare provider or attorney. Those individuals are precluded by law from revealing confidences.

Interestingly, many large studies of the ways in which physicians treat various ailments involve third-party information. In the typical study, physicians are asked certain questions about the first patient with a specific ailment they saw on the third Tuesday in the month. The physician provides the surveyors with patient-specific information, but does not reveal to the surveyors any identifying information about the patient. In these studies of clinical practices, each physician is the subject; the patients are not subjects because the physician provides no identifying information about them to the surveyor. As such, the surveyors have no way of linking medical information with any patient.

Ultimately, it will be up to the IRB to determine who are subjects and who are not. Some IRBs may be so risk averse that they opt to treat everyone as subjects. While this is ill advised and, in my view, not required by the regulations, no bureaucrat ever got into trouble by saying "no" or by requiring a researcher to jump through additional hoops. Other IRBs may adopt a more reasoned approach and recognize that if third parties are transformed into subjects, then so might fourth or fifth parties be.

Can a Researcher Be His or Her Own Subject? This brings us to an interesting question: Is self-experimentation subject to the Common Rule? This too is not an idle question. A few years ago, a senior researcher at a major research institution proposed conducting some genetic studies on himself. The research was not federally funded, but the institution voluntarily adhered to the Common Rule and was concerned that the IRB would need to approve the research. The senior researcher argued that even though he was conducting research, he was not a "human subject." Both the language and the history of the Common Rule supported his view. I contacted OHRP, and they told me that the researcher was a human subject. When I asked for their analysis, none was forthcoming. In the absence of any government analysis, let's try the following abbreviated analysis. As noted earlier, a "human subject" is "a living individual about whom an investigator . . . obtains," through interaction or intervention, data or identifiable private information. "Obtain" means "[t]o succeed in gaining possession of [something] as the result of planning or endeavor; [to] acquire."[47] Thus, one can no more "obtain" something from oneself than one can rob oneself. A similar result is reached when the words "interaction," "intervention," and "private information" are analyzed; none of these words carry a reflexive meaning.

The notion that self-experimentation does not come within the Common Rule is consistent with the purpose and genesis of the Rule. The Common Rule, as noted above, owes its genesis, in part, to a report issued by the National Commission for the Protection of Human Subjects of Biomedical and Behavioral Research, an advisory committee created by the National Research Act of 1974. This report, commonly referred to as the Belmont Report (because it grew out of a conference at the Smithsonian Institution's Belmont Conference Center), sets out three basic ethical principles to guide researchers—respect for persons, beneficence, and justice.[48] According to the report, respect *for persons* involves a recognition of the personal dignity and autonomy of individuals, and special protection of those persons with diminished autonomy.[49] This ethical principle forms the basis of informed consent, namely that individuals are in no position to exercise independent judgment unless the researcher provides them with all necessary information to make an informed decision.

However, this ethical principle also effectively precludes regulation of self-experimentation. Individual autonomy necessarily implies the ability to do something to oneself that others may not approve. While the IRB and informed consent processes are designed to ensure individual autonomy when applied to human subjects, they have precisely the opposite effect if applied to researchers who wish to engage in self-experimentation. In short, self-experimentation, which is not regulated by the plain text of the Common Rule, ought not to be regulated if one wishes to remain true to the ethical principles underlying the Common Rule.

This is especially true for genetic research where the primary risk involves dissemination of private information.[50] Privacy remains the hallmark of genetic research. In that regard, privacy has been variously defined as the "concept of separating self from society" or the "ability to control access to information about oneself."[51] Regulating, even benevolently, the extent to which an individual may freely publish his or her genetic information blurs the separation between self and society and interferes with one's control over one's personal information; any such regulation necessarily undermines rather than promotes privacy. Thus, in the area of genetic information, interference with self-experimentation represents an arguably inappropriate erosion of one's privacy.

RESEARCH THAT IS EXEMPT FROM THE COMMON RULE

The Common Rule does not regulate all "human subjects research." Those who wrote it recognized that the Rule included activities that either posed no risk or were not real science, or that raised potential constitutional issues. Therefore, the Rule's drafters included six exemptions:

1. Research involving educational practices (e.g., comparing instructional techniques, curricula)
2. Research involving use of educational tests (e.g., SAT, MCAT, LSAT, GREs), surveys, interviews, or observations of public behavior, unless the subjects are identifiable and disclosure of their responses could subject them to criminal or civil liability or affect their reputations
3. Research involving educational tests, surveys, interviews, and the like that does not qualify for exemption 2, is nonetheless exempt if the subjects are public officials or, if not, the researchers are prevented by federal law from revealing the identities of the subjects
4. Research involving existing data or specimens either if gathered from a publicly available source or if the investigator reports the information in a way that the subjects cannot be identified
5. Research conducted by federal agencies to assess the effectiveness of federal benefit programs
6. Consumer acceptance studies of foods

A few of these exemptions warrant special attention. Exemptions 2 and 3 relate to surveys, interviews, and other relatively benign forms of research. This type of research is exempted from the Common Rule, even if subjects are identified but nothing is disclosed about their responses that "could reasonably place the subjects at risk of criminal or civil liability or be damaging to the subjects' financial standing, employability, or reputation." The exemption was not artfully drafted, and sometimes it is difficult to figure out who is permitted to know the identity of the subject and who is permitted to know embarrassing details about the subject's life. Let's take a look at the precise wording of exemption 2, which exempts

> (2) [r]esearch involving the use of educational tests (cognitive, diagnostic, aptitude, achievement), survey procedures, interview procedures or observation of public behavior, unless:
> (i) Information obtained is recorded in such a manner that human subjects can be identified, directly or through identifiers linked to the subjects; and
> (ii) any disclosure of the human subjects' responses outside the research could reasonably place the subjects at risk of criminal or civil liability or be damaging to the subjects' financial standing, employability, or reputation.[52]

To better understand exemption 2, let's look at an example. Suppose a survey asks subjects for their names and other pertinent demographic information, and then asks whether they have ever shoplifted. The information is dutifully recorded, but the researcher does not reveal in any publication which subjects shoplifted and which did not. The fact that the researcher has chosen not to reveal the embarrassing information is not relevant, however, and the exemption does not apply. What governs is not what *will* occur but what *could* occur, namely that sufficient information has been brought together (e.g., name plus history of sticky fingers) to permit someone to reveal who has shoplifted.

Suppose that the survey is conducted by a physician using his patients as subjects. Does this change our analysis? It may, but not necessarily. Exemption 3 applies to research that fails to satisfy the conditions of exemption 2, but only if "federal statute(s) require(s) without exception that the confidentiality of the personally identifiable information will be maintained throughout the research and thereafter."[53] I have no idea what this exemption means. There is no federal statute that protects "personally identifiable information" "without exception." Every federal privacy statute, including the privacy provisions of HIPAA, contains exceptions allowing, for instance, law enforcement personnel to execute a search warrant or to obtain the information using a simple subpoena.[54]

Exemption 4, too, is not very clear. It exempts research

> involving the collection or study of existing data, documents, records, pathological specimens, or diagnostic specimens, if these sources are publicly available or if the information is recorded by the investigator in such a manner that subjects cannot be identified, directly or through identifiers linked to the subjects.[55]

The "publicly available" proviso is relatively easy to understand. The problem occurs with information that is not publicly available. An example will highlight the problem. Suppose that a hospital provides a researcher with tissue specimens along with the names, addresses, ages, and other information about each patient who provided each specimen. All of the patients suffer from a specific form of cancer, and the researcher is interested in seeing if she can find a marker or set of markers associated with that form of cancer. To qualify for the exemption, does the researcher have to remove the patient's name from each specimen container, replace it with a number, and then destroy the list of names so that she cannot at a later time link a name with a specimen? Or does it mean that the researcher can qualify for the exemption if she removes the names, replaces them with numbers, but keeps the document linking names with the numbers a secret? The problem lies with the phrase "cannot be identified." We are not told who precisely is precluded from identifying the subjects. What if the hospital keeps a split sample of each specimen along with the names? Should that change anything?

The most reasonable way of interpreting exemption 4 is that it applies only when the researcher destroys the document linking the name and number before the research begins or never creates a code sheet in the first place. If the hospital provides the specimens without any names or other identifying information, the researcher would not even be conducting human subjects research because the specimens would have been collected by a third party for treatment and not research purposes (i.e., no intervention by a researcher) and the researcher would not have acquired any personally identifiable information. A good analysis of this issue is provided by OHRP.[56]

The problem with existing specimens has been exacerbated by HIPAA and is discussed later in this chapter. However, one point is worth noting: The Common Rule and its FDA variations apply only to the living. A corpse cannot be a human subject. Furthermore, once someone dies, the privacy protections of the Common Rule die with him or her. That is not the case with HIPAA; its implementing regulations apply to anyone whether in heaven, hell, or in between.

An Overview of FDA's Variant on the Common Rule

As emphasized above, the Common Rule applies only when the research is federally funded, when the grantee voluntarily agrees to abide by the Common Rule, or when the research is under the jurisdiction of FDA.[57] Under the Food, Drug, and Cosmetic Act (FDCA), FDA regulates the approval, manufacture, distribution, and sale of drugs, medical devices,[58] and biologics (e.g., blood, vaccines, and genetic therapies).[59] Broadly, a drug is anything that either is listed as a drug in the *United States Pharmacopeia* or is intended to diagnose, cure, mitigate, treat, or prevent disease in man or beast.[60] A device is defined similarly, except that a device is not metabolized and does not primarily involve chemical reactions within the body.[61]

Whether something is a food, drug, device, or something else frequently turns on the intent of the seller and how the product is marketed. Take an ordinary lemon. It is a food when it sits idly in the supermarket's fruit and vegetable section. It can also be a drug, if, for example, a lemon seller were to tout it as a "great source of vitamin C which will cure colds and other maladies." It can also be a medical device if one sells it as a source of citric acid, "a natural sterilant for medical devices."

In the United States, it is illegal to distribute drugs, devices, or biological products that have not been approved or cleared by FDA. But in most cases, FDA will not approve or clear these articles for marketing without some clinical evidence that the articles are safe and effective for their intended use.[62] How, then, can one conduct clinical trials of new drugs, devices, or biologics if distributing the unapproved articles is illegal? It is possible because the FDCA contains exceptions that allow researchers to test new drugs, devices, and biological products on humans. For drugs and biologics, an article that is being clinically tested is called an IND, or Investigational New Drug; devices are clinically tested under an Investigational Device Exemption, or IDE. The IND and IDE processes differ in some respects.

First, a researcher has to decide whether the IND or IDE requirements even pertain. Some clinical trials can be performed without either an IND or IDE. (This does not mean that the Common Rule does not apply. The opposite is the case, as we shall see.) A researcher can conduct a clinical trial of a drug or biologic without FDA clearance: (1) if the article is already approved by FDA, (2) if the data from the trial will not be used to support an application to FDA to expand the article's intended uses, (3) if any change in the way the article will be used will not significantly increase its risk (e.g., route of administration, dosage, patient population), and (4) if the clinical trial complies with the Common Rule and FDA's special requirements. Thus, for example, a researcher can conduct a clinical trial of an approved drug using a higher-than-approved dose or a different route of administration, provided the higher dose or different route does not significantly increase the risk to patients; IRB approval and informed consent would still be necessary, though. Indeed, one of the functions of the IRB is to confirm that the clinical trial is exempt from IND requirements.

Second, if a researcher decides that the research can be conducted only under an IND, she then has to decide what role she, as a researcher, wishes to play in that research. Most clinical trials have four sets of actors: (1) the sponsor, (2) the investigator (along with his or her employing institution or site), (3) the contract research organization (CRO), and (4) the data safety monitoring board.

The sponsor is the person or entity that provides the experimental drug, device, or biologic being tested, and in theory, pays for all aspects of the clinical trial. The sponsor is really the "producer," in the cinematic sense, of the clinical trial; aside from financing the trial, the sponsor is responsible for getting the IND approved, recruiting clinical sites and investigators, developing the case report forms, preparing a brochure or "how-to book" for the investigators, training the investigators, monitoring all sites to ensure compliance with FDA requirements, and filing reports with FDA, including adverse event reports and annual reports. Normally, the sponsor is a drug, device, or biologic manufacturer.

The investigator differs from the other roles in that only an individual may serve. (While individuals are not foreclosed from acting as either a sponsor or CRO, both require significant resources and expertise, which most individual faculty members lack.) The investigator, in FDA parlance, is called the principal investigator of a clinical trial at a given site and is responsible for actually giving or prescribing the test articles to subjects. For this reason, the investigator must be licensed in that state to prescribe the drugs, devices, or biologics. This usually means that the investigator is a physician, podiatrist, or dentist.

The investigator is like a lead actor, and like an actor, he or she is required to sign an agreement with the company overseeing the production, in this case FDA. FDA Form 1572 (for trials of drugs and biologics) requires the investigator to do the following (among other things):

- Personally conduct or supervise the investigation
- Ensure that all associates, colleagues, and employees assisting in conducting the study are informed about their obligations
- Conduct the study in accordance with the protocol
- Comply with all requirements regarding obligations of clinical investigators
- Inform subjects that the drugs are being used for investigational purposes and ensure that informed consent and IRB requirements are met
- Report adverse events to the sponsor so that the sponsor can fulfill its responsibilities
- Read and understand the investigator's brochure

FDA clinical trials are not designed to be performed by scientists who wish to improvise or to test out their own theories. Clinical trial work is repetitive science that requires meticulous obedience to the protocol; departures, when required because of patient safety, must be carefully documented. In addition, it is the investigator's responsibility to obtain IRB approval at his or her institution and to continually update the IRB and the sponsor about adverse events. If data are recorded incorrectly, informed consents not appropriately obtained and documented, test articles not appropriately accounted for, or protocol not followed, the investigator will be held responsible. FDA disciplines investigators, even though the errors were committed by others on their team. To help ensure that this does not occur to you or to faculty at your institution, make sure that data are reviewed meticulously against the patient's medical records and that the sponsor (usually through its CRO) regularly monitors the data and informed consent forms to ensure that things are being done correctly.

Because sponsors frequently do not want get involved in the day-to-day management of a clinical trial, they usually look to unload some of their responsibilities onto a CRO, which in exchange for a nice payment agrees to perform certain aspects of a clinical trial. The FDA regulations permit a sponsor to cede by contract its responsibilities to a CRO. The CRO is like a film director. In most multisite clinical trials, the CRO recruits the sites and investigators, sets up a meeting where investigators and their staffs are taught (1) the protocol and the informed consent form and process, (2) how to collect and record the data, and (3) the known adverse effects associated with the test article.

The final actor is the data safety monitoring board (DMSB), a group of independent individuals (usually experts in the field) who are hired by the sponsor or CRO. The DMSB regularly reviews all the data unblinded to see if trends are developing that would warrant the protocol being terminated, suspended, or modified. Usually, it is the DSMB that first spots the trend that the test article is less effective than an existing therapy or is causing an inordinate number of serious side effects. If the test article is less effective than an existing alternative treatment, for instance, it would unethical to continue to deny subjects access to the existing treatments. In such a case, the DSMB would recommend that the trial be stopped. Then there is the rare event—the lottery of clinical trials—dreamt of by every sponsor, namely that the DSMB recommends that the trial be stopped because the test article is so effective that denying patients access to it would be unethical. This is occurred with the first AIDS drug, azidothymidine (AZT), in 1986, and now, most multicenter trials have "stop rules" built into the protocol so that if certain good or bad things occur, the study is stopped.

Clinical trials are complicated, and even small ones require extensive monitoring and scads of paperwork. Most research universities lack the infrastructure and resources necessary to comply with all the duties of a sponsor. Most even lack the resources necessary to file a clean IND application. Drug companies sometimes want to conduct small trials of their approved products at different dosages, but do not want to act as the sponsor. They may offer a researcher at a university a sizable chunk of change to act as both the sponsor and the investigator, and even promise to help the researcher get an IND approved. (The IND is necessary because the drug company intends to submit the data to FDA to support a labeling change.) Unless you have significant experience acting as a sponsor and have the resources and knowledge to fulfill all of the FDA requirements, this is an offer that you should run from as quickly as possible.

Suppose that you have satisfied yourself and university officials that you are up to the task of acting as a sponsor. How do you get FDA approval for your clinical research? To obtain approval of an IND, you normally have to satisfy FDA that the drug is safe in animal studies, that it has some proven efficacy in bench studies, that there is some underlying biological plausibility for its theoretical action, and that the protocol for the clinical trials is well designed and uses appropriate end points.

Clinical research involving new devices can be a little easier only because FDA does not require a sponsor to file for an IDE if the device and the way it will be used pose a nonsignificant risk. However, if you guess wrong and a patient dies during the trial, and the device contributed to the death, FDA will invariably conclude that the new device posed a significant risk and you should have sought express FDA approval for the IDE. To provide a researcher with some cover, FDA requires that an IRB independently determine that the clinical trial of the device poses a nonsignificant risk.

The most complicated part of acting either as a sponsor or investigator is reporting adverse events. I discuss this in detail below, both with respect to studies conducted under INDs and IDEs and to studies that are subject only to the Common Rule.

The Structure of Informed Consent

The Common Rule and the FDA rules set out in great detail what must be in an informed consent form and what cannot be.[63] Instead of laboriously going over that list, I highlight some of the more important issues. First, an informed consent form should be comprehensible to the average patient or subject and at the same time provide an accurate discussion of the research, the risks, benefits, and alternatives. Balancing detail and accuracy with comprehensibility can be challenging. Some survey research suggests that informed consent forms are written on average at the tenth-grade reading level, while the average subject reads at the sixth- to eighth-grade level.[64] But if you simplify the consent form too much you may lose some scientific accuracy; also, if there is too much detail you may lose your subject—he or she may not understand, may be too embarrassed to ask, and may "tune out." Unfortunately, there is no simple and reliable way of measuring the comprehensibility of an informed consent form.[65] Second, the consent form should avoid exculpatory language (e.g., the subject agrees not to sue the investigator). Third, owing to the complexity of many protocols, it is important that the consent form is not the sole means of communicating the details of the proposed study to the potential subject. Finally, an informed consent form should be approved by the IRB, signed by the subject or his or her guardian, and witnessed.

AN INFORMED CONSENT FORM MUST CLEARLY ARTICULATE RISKS, BENEFITS, AND ALTERNATIVES

An informed consent form ought to highlight all "reasonably foreseeable risks or discomforts," should describe alternative treatments, and should note that there are unforeseeable risks.[66] Most litigation arising out of clinical trials focuses on these three items.

Lenahan v. University of Chicago,[67] for example, concerned foreseeable risks. It started life as a basic medical malpractice case in which the plaintiff sued after his non-Hodgkin's lymphoma was misdiagnosed and, as a result, he received improper chemotherapy that allowed his disease to progress unchecked. Eventually he was seen at the University of Chicago, whose physicians recommended a "high-dose chemotherapy/stem-cell transplant regimen." In addition, one of the physicians recommended that Lenahan participate in a phase 1 clinical trial at the university. Clinical trials normally come in one of three phases. A phase 1 trial is designed primarily to test the safety of the drug or biologic and learn about dosing. It is not designed to measure effectiveness. For most drugs, other than those to treat cancer and certain other diseases, a phase 1 trial involves a small cohort (usually fewer than twenty) of healthy volunteers. The phase 1 trial at the University of Chicago involved a cancer vaccine made by expanding the patient's own T cells and reinjecting them to fight the cancer. Lenahan signed the informed consent form but died during the course of the experimental treatments. His estate sued the university, the biotech company sponsor, the investigator, and others alleging, among other things, that the informed consent form failed to reveal that there was a 95 percent chance of death during the course of the University of Chicago protocol. It is unclear whether this statement was in fact accurate. The court essentially held that if the expected mortality was 95 percent and that was not mentioned in the form, then a jury could find the university negligent. The appeals court refused to dismiss that aspect of the case, implying that it would be up to the jury to decide whether the consent form accurately portrayed the risks associated with the trial.

Sometimes there is a fight over precisely what alternative treatments ought to have been spelled out in the consent form. For example, in *Stewart v. Cleveland Clinic Foundation*,[68] the plaintiff was diagnosed at the Cleveland Clinic with advanced neck cancer implicating his tongue (i.e., stage IV squamous cell carcinoma originating at the base of the tongue, an oropharyngeal tumor). His oncologist advised him that they were conducting a phase 3 trial of a chemotherapy cocktail (5-fluorouracil and cisplatin) that involved the random assignment to one of two arms. Subjects in the first arm would receive the standard treatment for his type of advanced cancer, which is surgery and radiation. Subjects in the second arm would receive "experimental preoperative chemotherapy," followed by the standard treatment (surgery and radiation). He was advised that he had two options: He could decline to participate, in which case he would receive the standard treatment, or he could participate, in which case he had a 50 percent chance of receiving the experimental treatment plus the standard treatment. Stewart opted to enroll in the study; he was assigned to the first arm—standard treatment only.

Nevertheless, Stewart remained disease-free for more than five years. However, in the sixth year, it appeared that his original cancer had metastasized to his lungs; he died shortly thereafter. Before dying, however, he instituted suit against the Cleveland Clinic and virtually every physician there who had treated him. He alleged, among other things, that the informed consent form was defective for two reasons. First, the phase 2 data strongly supported the added effectiveness of the chemotherapy, and second, he was never advised that he could have received the chemotherapy off-label even if he had decided not to participate in the study. The appeals court refused to dismiss the claim because there was competing evidence concerning the efficacy of the alternative treatment and the adequacy of the informed consent form.

The court never addressed what I perceive to be the real and tough legal issues in the case, namely, does a patient have a right to be informed about off-label use of drugs and the results of an earlier-phase trial. Nothing in the FDCA prohibits a physician from using a drug off-label. Various studies have shown that off-label use of approved drugs is the norm in oncology. However, it is illegal for a manufacturer to promote the off-label use of one of its drugs, and it necessarily follows that a drug manufacturer sponsor cannot lawfully include in its standard informed consent form any reference to the off-label use of one of its products, even though it is aware that it is used off-label. Merely because the sponsor is precluded from inserting a reference to the off-label use in its standard informed consent form does not preclude the site or the investigator from doing so. Therefore, does a clinical trial site and its IRB have an independent duty to discuss various alternative treatments involving the off-label use of an approved drug? There are strong policy arguments on both sides. For example, if the drug company itself were running the trials, it could not lawfully insert language about off-label use. Why, then, should a university have a duty to do so? Alternatively, when the off-label use has become the standard of care, it would arguably be unethical to omit this information from the informed consent form. The greatest difficulty occurs in cases in which many physicians use a drug off-label and the off-label use is being tested, but it is not the standard of care.[69]

The second issue is somewhat easier; results from a phase 2 trial are preliminary and potentially suggestive, but they do not represent the type of proof that would permit, let alone mandate, a researcher to discuss the results in the informed consent for a phase 3 trial. Indeed, it could be argued that referencing positive results from the phase 2 trial in the consent form may be unrealistically optimistic and could form the basis of a lawsuit against the site by a disappointed phase 3 participant should the trial not go well, as is often the case.

AN INFORMED CONSENT FORM SHOULD AVOID EXCULPATORY LANGUAGE

One of the more controversial aspects of the Common Rule and FDA rules, at least in application, is their prohibition on including so-called exculpatory language in an informed consent form. Usually this means that subjects cannot sign away their right to sue. However, in *Moore v. The Regents of the University of California*,[70] a case that is discussed in much greater detail in the next chapter, the California Supreme Court ruled that although subjects did not have a property interest in their discarded tissue samples, they had to consent to allowing their discarded tissue to be used for experimental or commercial purposes. In reaction to *Moore*, research institutions began inserting language into their informed consent forms seeking a subject's consent to use his or her ordinary tissue in experimental or commercial activities. Many of these institutions also inserted into the consent form language along the following lines: "I [the subject] will, unless otherwise agreed by the principal investigator, have no rights to share any profit." OHRP considered this type of language to be exculpatory.[71] In my view, it is not exculpatory, but merely an accurate statement of property law.

OHRP believes that certain phrases are exculpatory and others are not. For example, it believes that this statement is exculpatory: "By agreeing to this use, you should understand that you will give up all claim to personal benefit from commercial or other use of these substances." It believes that this statement is not: "Tissue obtained from you in this research may be used to establish a cell line that could be patented and licensed. There are no plans to provide financial compensation to you should this occur." The difference between the two is without legal meaning because a subject does not have any ownership interest in the intellectual property developed with his or her tissue, and therefore, as a matter of law, there is nothing to waived. That is not always the case. Compare "I waive any possibility of compensation for injuries that I may receive as a result of participation in this research" and "This hospital is not able to offer financial compensation nor to absorb the costs of medical treatment should you be injured as a result of participating in this research." The first purports to be a waiver of one's legal right to collect a certain type of damages, and the second merely reflects the hospital's distaste for providing free medical care.

THE INFORMED CONSENT PROCESS SHOULD EDUCATE, NOT INTIMIDATE, THE SUBJECT

Studies of medical malpractice litigation have consistently shown that patients are far less likely to sue over a bad outcome if they believe that the physician truly cared about them, spent time with them, and answered their questions openly and simply, yet not demeaningly. This is particularly important when the underlying science is complicated and the patient's prognosis is bleak.

One study, for example, compared the ways in which primary care physicians with no malpractice claims and those with two or more claims communicated with their patients. It found that those who had no malpractice claims educated their

> patients about what to expect . . . , laughed and used humor more, and tended to use more facilitation (soliciting patients' opinions, checking understanding, and encouraging patients to talk). No-claims primary care physicians spent longer in routine visits than claims primary care physicians (mean, 18.3 vs 15.0 minutes), and the length of the visit had an independent effect in predicting claims status.[72]

The same is likely to be the case with clinical research. The informed consent form should not be the sole means of explaining the protocol to the subject. Prospective subjects are more likely to feel that they are being properly treated when the physician in charge of the study spends time with them discussing their illness, going over the protocol, raising questions that the average patient usually does not know enough to ask, and answering questions. Principal investigators (PIs) should encourage prospective subjects to take the informed consent form home, discuss it with others including their primary care physician, and do research on the Internet. Prospective subjects usually do not get this warm and fuzzy anti-lawsuit feeling when the "consenting process," as it is called, is relegated to a nurse or even a fellow, or when they feel pressured to sign, or when they are handed the form, asked to read and sign it with little or no explanation. A case in point is *Lett v. Sahenk*,[73] in which a patient instituted suit against an Ohio State University research neurologist. The plaintiff had sought treatment from the neurologist for Charcot-Marie Tooth Syndrome. During the initial visit, however, the physician asked the patient if she would be willing to have a nerve biopsy for research, not for diagnostic purposes. According to her complaint, five months later, while she was sitting on a surgical table awaiting the biopsy, she was presented with a consent form for the first time. At no prior time, according to the complaint, had the researcher discussed the consent form with Lett or shown it to her. The physician ultimately prevailed, but it took nearly four years of litigation. A more complete and timely explanation of the entire process by the physician along with a full assessment of the possible risks might have reduced the risk of suit.

AN INFORMED CONSENT SHOULD BE DOCUMENTED

We have assumed that the informed consent can only be obtained in writing. Generally, "informed consent shall be documented by the use of a written consent form approved by the IRB and signed and dated by the subject or the subject's legally authorized representative at the time of consent. A copy shall be given to the person signing the form."[74] Most standard informed consent forms have signature lines for a witness and for the PI, and dates for each. This is technically not required by either the Common Rule or FDA, but has become accepted practice. Remember, however, the more information that must be filled in, the greater the opportunity that someone will forget to fill in a date or sign the form. And normally, when you agree to adhere to a higher standard than required by the law, the law holds you to that higher standard. Omissions become common in studies that require six pieces of filled-in information (e.g., signature and dates for the subject, the PI, and a witness).

The FDA rules permit the information in the informed consent form to be presented orally when authorized by the IRB. Ironically, the oral presentation method actually requires more paperwork than the ordinary informed consent form. The researcher needs two documents, one a statement to the effect that all of the elements required to be in an informed consent form have been presented orally to the subject. This is supposed to be signed by both the subject and by a third-party witness. In addition, there must a document that actually summarizes the oral presentation; this summary document must be expressly approved by the IRB and must be signed by the witness and the oral presenter.[75]

Establishing and Operating an IRB

An IRB is creature of the Common Rule and the FDA's rules.[76] Books have been written about how an IRB ought to function, how it should keep its records, and the like. By contrast, this section provides an overview of how an IRB should act, what it should require each researcher to provide, and how a researcher ought to interact with the IRB. Two quick points. First, each IRB must develop detailed written procedures for all aspects of its business. The topics that must be addressed are set out in 45 C.F.R. § 46.107, and include, by way of example only, how the IRB will conduct an initial review, how it will conduct continuing reviews, how it will conduct expedited reviews, and the like. Second, the IRB must also register with OHRP either by paper or electronically.

IRB MEMBERSHIP

Central to setting up and running an IRB is ensuring that its membership is sufficiently diverse in terms of substantive expertise so that there will be at least one person on the IRB who will fully understand the science in any given proposal. At research institutions that focus on one type of ailment or one aspect of science, this may be relatively easy to do. At major research universities where the range of research is broad, this may be more difficult to achieve unless the IRB is large. On the one hand, you do not want an IRB that is so large that it is unwieldy, making open yet focused discussion difficult. This is especially so at universities, where it is not unusual to find that everyone has something to add. Nor do you want an IRB that is too small to accommodate the subject-matter diversity necessary for the appropriate review of a proposal. Accordingly, most large research universities have many IRBs, usually at least one for each school that conducts research involving human subjects. Some large medical schools have multiple IRBs.

Each IRB must have at least five members with at least one member who is not employed by the university and one member whose primary concern is nonscientific.[77] IRBs cannot act on a proposal unless there is a quorum present (i.e., a majority of the voting membership); the quorum must contain the nonscientist. Thus, if there is an IRB with nine members, eight of whom are scientists, a quorum consists of five members, one of whom must be the nonscientist member. The quorum must be maintained throughout the entire meeting. Suppose, for example, you have an eight-member IRB and five members show up for a meeting (including the nonscientist), but one of those members is the PI on proposed research scheduled to be considered by the IRB during that meeting. In such a case, that researcher's proposal may not considered by the IRB during that meeting. Because the

PI member must recuse herself from participating as an IRB member when her proposal is being considered, no quorum would be present.

THE MECHANICS OF REVIEWING PROPOSALS

Ideally, each IRB should be assigned an executive secretary (usually a member of the research staff) whose job it is to set the agenda, to keep the minutes, to develop the process by which proposals are to be submitted, to tie up loose ends, and to alert researchers when it is nearly time to renew their IRB approvals. Many university IRBs also have assigned either as a voting or nonvoting member an attorney from the general counsel's office.

IRBs should meet often enough so that proposals can be submitted and reviewed in an orderly fashion. Each IRB should develop a simple form to be completed by the PI asking basic questions. At the very minimum, each IRB member should be provided with (1) the full protocol; (2) a proposed informed consent document; (3) any relevant grant application(s) or, if the research is funded by a drug or device company, a copy of the site agreement; (4) the investigator's brochure, if one exists; (5) the case report form, if there is one; (6) any recruitment materials, including advertisements intended to be seen or heard by potential subjects; and (7) relevant research concerning the risks and benefits.

Most, but not all, IRBs ask the PI to hold him- or herself available to answer questions; other IRBs ask the PI to present the protocol, to answer questions, and then to leave before the discussion and the vote. IRBs are required by the Common Rule to discuss the benefits and risks of the proposed research and to highlight ways of reducing risks without appreciably affecting the scientific integrity of the proposal. The focus of the IRB should be on weighing the risks versus the benefits of the proposed research, and the tenor of that discussion ought to be duly reflected in the minutes of the meeting. The IRB should not attempt to "improve" the protocol's design or change the focus of the research. All too often, PIs come out of an IRB meeting shaking their heads because the questions focus more on the minutiae of the protocol, the end points, or even the proposed statistics than on the protocol's safety. The IRB meeting is not intended to be a "study section" meeting, a faculty meeting, or a symposium, and members ought to remember that their role is limited. This is not to say that the scientific merit of the methodology is never on the table. Clearly, when the risk posed by a study is significant and there are ways of reducing that risk without affecting the study's integrity, then it may be appropriate for IRB members to speak up. In my experience, those occasions are few and far between.

The IRB secretary should take care to make sure that he or she dutifully summarizes the gist of the debate with special emphasis on discussions of the risks, the benefits, and the adequacy of the informed consent document. The IRB must vote on each proposed research project and that vote must be recorded in the minutes, preferably with the names of each member and how that member voted. A simple statement that "the protocol was approved by the IRB" in my view is not sufficient.

For record-keeping purposes, each informed consent form should carry the date that it was approved by the IRB and the latest date on which it will expire (i.e., one year after the IRB's initial approval). It is not unusual for an IRB to approve a protocol conditionally pending certain changes. If the changes, for example to the informed consent, are relatively minor (e.g., not substantive), conditional approval is appropriate provided the modified document is forwarded to the IRB. For tracking purposes, the informed consent should carry a footer or header such as the following: "Modified on 08-20-09 in accordance with IRB Conditional Approval of 08-16-09. *This Form Expires on 08-15-10*." If the conditions for approval require substantive changes or changes that are not minor, then approval must be deferred; the IRB must review the modifications and vote affirmatively to approve the modified protocol.[78] IRB business, especially review of a protocol that is be resubmitted to accommodate IRB concerns, can be conducted via a telephone conference call.

As noted above, an IRB approval lasts no more than one year and must be renewed.[79] This means that a researcher must file a request with the IRB well enough in advance of the one-year deadline so that the IRB can act before the year expires. When does the one year begin to run? Suppose that an IRB met and approved a research project on March 5, 2009, subject to minor changes in the informed consent form. The researcher then transmits the revised informed consent form to the IRB chair on March 15, 2009. Does the one year run from March 5 or March 15? When contingent approval requires only minor changes, changes that do not require the IRB to meet and vote a second time, the one year begins to run from the original conditional approval on March 5. However, if the changes sought by the IRB were substantial, requiring the IRB to meet again on March 15, 2009, and vote on whether the changes satisfied the IRB's original concerns, then the one-year period runs from March 15.

There are differences between the initial review and any continuing review (e.g., review to determine whether to approve the research for an additional year). With the initial review, one can only speculate about the risks posed by the research. With a continuing review, however, there should be some data to better shape the IRB's initial risk assessment. The continuing review is really a Bayesian process, with each year's risk assessment being altered by new data. In conducting a continuing review, OHRP recommends that each IRB member examine the following:

(i) the number of subjects accrued;
(ii) a summary of any unanticipated problems and available information regarding adverse events (in many cases, such a summary could be a simple brief statement that there have been no unanticipated problems and that adverse events have occurred at the expected frequency and level of severity as documented in the research protocol, the informed consent document, and any investigator brochure);
(iii) a summary of any withdrawal of subjects from the research since the last IRB review;
(iv) a summary of any complaints about the research since the last IRB review;
(v) a summary of any recent literature that may be relevant to the research and any amendments or modifications to the research since the last IRB review;
(vi) any relevant multi-center trial reports;
(vii) any other relevant information, especially information about risks associated with the research; and
(viii) a copy of the current informed consent document and any newly proposed consent document.[80]

REGISTERING WITH THE GOVERNMENT

Institutions that receive federal funding for research involving human subjects are required to execute a Federalwide Assurance (FWA), promising to abide by the Common Rule for all of their federally funded research.[81] The assurance program is administrated by OHRP and covers all federal agencies that have adopted the Common Rule.[82] Thus, for example, if an institution has an FWA on file with OHRP, that institution would be eligible to receive funding from the Department of Defense for human subjects research.

To obtain an FWA, an institution must register its IRB with OHRP, provide certain other information, and then indicate whether it will voluntarily agree to extend the protections of the Common Rule to non-federally funded research. It is doubtful that the federal government could enforce a university's promise that it will apply the Common Rule to all research conducted by its employees irrespective of the funding source. However, from a liability perspective, it would be difficult for a university to maintain two sets of rules: one for subjects in federally funded or FDA-regulated research and another less stringent set for all other research. Therefore, it is not surprising that most universities have voluntarily agreed to abide by the Common Rule irrespective of the funding source.

Recently, there has been a move to accredit IRBs. Private accrediting agencies visit an IRB and determine whether its operations meet the particular accrediting body's criteria, whatever those might be. The accrediting bodies are private entities that do nothing other than accredit IRBs for a fee. The Association for the Accreditation of Human Research Protection Programs, Inc., a not-for-profit company, appears to be the largest accrediting body. It charges an application fee based on the number of protocols that the IRB reviews annually and that ranges in 2009 from about $10,200 (1–100 protocols) to $79,500 (more than 7,001 protocols); its annual maintenance fee ranges from $4,700 to $26,000.[83] It is unclear, however, whether accreditation translates into improved subject safety.[84]

Reporting Adverse Events

An investigator's relationship with his or her IRB does not end with the initial approval. In fact, the approval actually marks the beginning of the relationship. First, the IRB must review and reapprove the protocol at least annually. Second, the IRB is responsible for continuously monitoring the results of the protocol and deciding whether it should be modified or suspended.

Reporting adverse events is critical. It enables an IRB and the FDA[85] to monitor continuously whether a protocol is posing greater or different risks than may have originally been thought. It is not uncommon for an IRB to modify the informed consent form to reflect new risks or for an IRB or the FDA to suspend a study after receiving a bevy of adverse event reports. Figuring out when to report adverse events can be tricky, if for no other reason than the definitions are frequently vague and the policy considerations may not be intuitively obvious. For example, one would think that erring on the side of reporting an event would be playing it safe. The opposite may be the case. Overreporting can be just as pernicious as underreporting. Because overreporting can actually mask real problems, FDA and smart IRBs try to discourage it. Striking the perfect balance—the "yin and yang" of adverse event reporting—is the goal. However, the goal can be elusive, indeed impossible to achieve when a researcher is faced with five adverse event reporting systems, no two of which are the same. For example, not infrequently I come across investigators who have to figure how to accommodate the adverse event reporting requirements of the Common Rule, FDA, the university, the protocol itself, and sometimes even the funding agency. Sometimes, everyone's reporting requirements are comparable; other times they are not. An investigator who agrees to operate under five disparate reporting systems is setting him- or herself up to fail.

The FDA reporting requirements alone can be complicated. First, FDA actually has two distinct reporting systems—one for drugs and biologics, and the other for devices. Second, the FDA reporting requirements are more complicated than the generic rules set out in the Common Rule and differentiate between the investigator and the sponsor. However, since the investigator is required to provide the sponsor with all of the information that the sponsor needs to provide to FDA, the investigator is indirectly saddled with the sponsor's reporting requirements as well, but with tighter time constraints.[86] In turn, the sponsor must report to FDA certain types of adverse events and other occurrences, and is required to notify all IRBs in a multicenter trial about increased risks and the like. The investigator has no independent reporting relationship with FDA, unless the investigator is serving as both an investigator and sponsor.

ADVERSE EVENT REPORTING FOR DRUGS AND BIOLOGICS

On the drug and biologic side, an investigator is required to promptly report to the sponsor

(i) all unanticipated problems involving risk to human subjects, or
(ii) any adverse effect that may reasonably be regarded as caused by, or probably caused by, the drug. If the adverse effect is alarming, the investigator shall report the adverse effect immediately.[87]

One of the triggering events—"unanticipated problem"—does not require a link between the drug or biologic and the injury. The other triggering event—adverse effect reasonably regarded as caused by the drug—clearly requires a causal link. Therefore, an anticipated injury only has to be reported when it is causally linked to the test article. All unanticipated injuries, whether caused by the drug or not, have to be reported to the sponsor. This is similar to the reporting threshold under the Common Rule (researcher must report to his or her IRB "any unanticipated problems involving risks to subjects or others"[88]). What precisely is an "unanticipated problem"? How does one gauge whether something is a problem and whether it is unanticipated? Let's look a number of examples. Suppose you have indicated in the informed consent that there is a risk of developing the "dry scritos" (DS), a serious skin ailment, and suppose that a subject develops DS. The event is certainly not unanticipated, and therefore is not a reportable event. However, suppose that fifteen subjects out of the first one hundred accrued develop DS; but the anticipated rate when the study began was about 5 percent. DS is a problem and the rate at which it is developing was "unanticipated" (i.e., greater than originally expected). Accordingly, the fifteen DS events are all reportable and should be carefully analyzed for the IRB by the investigator. Dumping information on an IRB without any explanation is not beneficial. The investigator would also want to report this to the sponsor (assuming it is an FDA trial), and the sponsor, in turn, would report it to FDA and to the IRBs at other sites (if it was a multicenter trial).

Let us assume that rather than seeing fifteen subjects developing DS, one subject's white blood cell count drops significantly. Usually, an event involving only a single subject is not reportable. A data point provides little information by itself. In contrast, suppose that a single subject developed an extremely rare ailment that is normally drug induced, such as Stevens-Johnson syndrome?[89] The single occurrence of an event that is normally rare in the absence of drug exposure is considered an unanticipated problem that ought to be reported to the IRB and the sponsor immediately.[90] It may lead the IRB, the sponsor, or FDA to modify the informed consent forms to highlight Stevens-Johnson syndrome as a risk.

Once the sponsor of an IND receives the information from the investigator, the sponsor must report to FDA any "*serious and unexpected adverse experience that is associated with the use of the drug or biologic*" within fifteen days of receipt of the information or within seven days if the adverse experience is death or life threatening.[91] However, it is assumed that the term "unanticipated problems" is far broader than any of the terms that trigger a sponsor's reporting obligations. The fifteen-day and seven-day windows, though, are important to the investigator because he or she must timely report information to the sponsor so that the sponsor can discharge its reporting obligations to FDA. As a practical matter, any "unanticipated problem" that is neither life threatening nor results in death can be reported by the investigator in less than fifteen days (usually much sooner than that), and an "unanticipated problem" that results in death or is life threatening should be reported immediately. It should also be noted that while the investigator's reporting obligation to the sponsor does not require a cause and effect relationship, the sponsor's reporting obligation does.

Sponsors are also required to advise FDA whenever other research or other findings suggest that the risks associated with the experimental drug are more significant than originally thought.[92] Finally, sponsors are required to summarize annually various events for FDA, including all deaths irrespective of cause.[93] Thus, for example, under the FDA rules, a sponsor (and by implication an investigator) must report all subjects who died as a result of an automobile accident or other causes arguably unrelated to the experimental treatment. A similar regime operates on the device side.

ADVERSE EVENT REPORTING FOR DEVICES

On the device side, investigators are required to submit to their IRB and to the sponsor a report of any "unanticipated adverse device effect (UADE) occurring during an investigation as soon as possible, but in no event later than 10 working days after the investigator first learns of the effect."[94] A UADE is

> any serious adverse effect on health or safety or any life-threatening problem or death caused by, or associated with, a device, if the effect . . . was not previously identified in nature, severity, or degree of incidence in the investigation plan or application . . . or any other unanticipated serious problem associated with a device that relates to the rights, safety, or welfare of subjects.[95]

The device rule differs somewhat from the drug rule in that the investigator must notify both the IRB and the sponsor, and, further, must do so within ten days. If the sponsor determines that an unanticipated adverse device effect presents an unreasonable risk to subjects, it is required to terminate the study within five days of making that determination and not later than fifteen days after the sponsor first becomes aware of the effect.[96]

COMPLYING WITH MULTIPLE SETS OF RULES

What happens when the protocol and the FDA rules or the Common Rule impose different reporting requirements? As a general rule, because the investigator is obligated to follow the protocol, he or she must report adverse events as defined in the protocol in addition to adverse events under the FDA or Common Rule definitions.[97]

Special Circumstances that Justify Departure from the Normal Rules

EXPEDITED REVIEW

The Common Rule permits IRBs to review and approve certain types of low-risk research on an expedited basis. If the research qualifies for expedited review, the IRB is not required to convene a quorum, but instead may delegate the entire review process to a single member of the panel, usually the chair of the IRB. The chair may either conduct the expedited review or designate a member of the IRB with the requisite experience to review the proposal.[98] In theory, expedited review ought to be no different from a full IRB review. But in reality it is, as its name implies. This makes sense as it would be a misallocation of resources to convene an IRB to review a proposal that poses only minimal risks. After the expedited review has been completed, the fact that it occurred and its conclusions must be shared with the other members of the IRB.

Interestingly, an expedited review cannot disapprove the research. Rather, if the single reviewer believes that the research ought not to be approved, it must be referred to the full IRB for complete review. This makes sense because usually when a single IRB member disapproves of an expedited application it is because, in the member's view, the research poses more than a minimum risk, in which case it is not eligible for expedited review.

Expedited review of an initial proposal may be used only when the proposal involves research that fits into one of seven categories specified by the Secretary of HHS, which are as follows:

(1) Clinical studies of drugs and medical devices only if an IND or IDE is not necessary or the device is being used in accordance with its FDA cleared or approved labeling.
(2) Collection of blood samples by finger stick, heel stick, ear stick, or venipuncture under specified conditions.
(3) Prospective collection of biological specimens for research purposes by noninvasive means such as excreta and external secretions, uncannulated saliva collection, placenta removed after delivery, amniotic fluid obtained at time of rupture, but before labor, mucosal and skin cells collected by buccal scraping or swab, skin swab, or mouth washings, and other examples.
(4) Collection of data through noninvasive procedures (not involving general anesthesia or sedation) routinely employed in clinical practice, excluding procedures involving x-rays or microwaves. Where medical devices are employed, they must be cleared/approved for marketing. (Studies intended to evaluate the safety and effectiveness of the medical device are not generally eligible for expedited review, including studies of cleared medical devices for new indications.)
(5) Research involving materials (data, documents, records, or specimens) that have been collected, or will be collected, solely for nonresearch purposes (such as medical treatment or diagnosis).
(6) Collection of data from voice, video, digital, or image recordings made for research purposes.
(7) Research on individual or group characteristics or behavior (including, but not limited to, research on perception, cognition, motivation, identity, language, communication, cultural beliefs or practices, and social behavior) or research employing survey, interview, oral history, focus group, program evaluation, human factors evaluation, or quality assurance methodologies.[99]

Even when a proposal meets the criteria set forth above, this does not mean that it automatically qualifies for expedited treatment. The IRB chair must also find that it involves minimal risk and does not involve prisoners or classified research or clinical trials under an IND or IDE. Therefore, the first thing that the IRB chair ought to do when someone requests an expedited review is to make sure that it is a kind that fits the regulatory criteria. Even if a proposal meets all of these regulatory criteria, the IRB is not obligated to conduct an expedited review. Some IRBs, as a practice, do not permit any proposal to be reviewed on an expedited basis; other IRBs permit an expedited review only after legal counsel has concluded that the research properly qualifies for expedited review.

An IRB may conduct a continuing review on an expedited basis if the original review was conducted on an expedited basis and there is nothing to indicate that the IRB's assessment that the research is minimal risk was in error. Research that was originally approved following a full IRB review can qualify for a continuing expedited review if all that remains is data analysis and no new subjects will be accrued and no additional procedures will performed on existing subjects. Also, research that was originally approved following a full IRB review not involving an IND or IDE may qualify for continuing expedited review if the "IRB has determined and documented at a convened meeting that the research involves no greater than minimal risk and no additional risks have been identified."[100]

COMPASSIONATE USE OF UNAPPROVED DRUGS OR DEVICES OUTSIDE A PROTOCOL

Compassionate Use with the Sponsor's Concurrence

The expedited process noted above is reserved for research that poses only minimal risk to subjects and does not involve INDs or IDEs. At the opposite extreme is research that involves investigational new drugs or devices or preinvestigational articles, all of which may pose significant or unknown risks. Not infrequently, individuals who may benefit from these experimental products, but who do not qualify to participate in a clinical trial, seek special leave to receive the experimental drug or device. For drugs, biologics, and devices, there are two compassionate use mechanisms: (1) an "emergency use" and (2) a "treatment use."[101]

On the drug and biologics side, an emergency use is one in which the drug or biologic has yet to be clinically tested and, of course, has not been approved by FDA. In such a setting, FDA is authorized to permit a sponsor to ship the drug or biologic so that it can be used for a patient in advance of an IND.[102] Usually this is reserved for patients whose conditions are bleak and the drug or biologic is the only available potential course of treatment. Emergency uses are relatively rare and can proceed only with IRB approval and an executed informed consent form. While there is no express "emergency use" mechanism available on the device side, FDA has created one outside of the regulatory process and permits sponsors to make these devices available to a patient (1) who has a life-threatening condition that needs immediate treatment, and (2) for whom no generally acceptable alternative treatment is available. In this setting, FDA permits the chair of the IRB to approve the emergency use of the device, provided that the patient executes an informed consent and the physician provides the IRB with a follow-up report. Furthermore, there are circumstances in which the emergency is so pressing that the physician is permitted to administer or use the test article without prior IRB approval or even without informed consent. These are relatively rare occurrences, and researchers are cautioned to seek to counsel before making this decision on their own.

In contrast to emergency use, a compassionate treatment use of a drug, biologic, or device occurs when the drug or device is being clinically tested (or clinical testing has ended and a marketing application is being pursued) and the subject does not meet the inclusion criteria or triggers one or more of the exclusion criteria. In such a setting, the patient's physician normally contacts a site of the clinical testing, which in turn contacts the sponsor, seeking to use the drug or device for treatment. If the sponsor agrees, it petitions FDA to permit one of the sites to administer the drug outside the protocol. Here, too, FDA approval is conditioned on IRB approval and an executed informed consent form. Treatment use of a nonapproved drug, biologic, or device is reserved for those situations in which, among other things, (1) the "[test article] is intended to treat a serious or immediately life-threatening disease," and "(2) [t]here is no comparable or satisfactory alternative drug or other therapy available to treat that stage of the disease in the intended patient population."[103]

Compassionate Use without the Sponsor's Concurrence

Usually it is the pharmaceutical company that seeks to move its products into clinical testing as quickly as possible and is usually willing to permit its experimental product to be clinically tested outside the protocol, but with FDA authorization. In recent litigation, however, the opposite was true. Patients have sued because they either were denied access to new drugs that had not entered clinical testing or were denied continued access to a drug after a clinical trial in which they had been participating ended. In *Abigail Alliance for Better Access to Developmental Drugs v. von Eschenbach*,[104] the United States Court of Appeals for the District of Columbia Circuit concluded that terminally ill patients do not have a constitutional right to unapproved drugs, even though they may have successfully gone through phase 1 clinical testing. The plaintiffs argued that the normal calculus of risk does not apply to the terminally ill, who have no alternatives and nothing to lose. The court disagreed. FDA is not the only target of these so-called access suits. Sponsors and sites have also been sued. Thus, in *Abney v. Amgen Inc.*,[105] eight patients sued Amgen, seeking to force the company to resupply them with a Parkinson's medication that the patients had received while participating in a clinical trial that had been prematurely terminated by the company. The court of appeals held that the patients had no right to be resupplied with the drug.

Special Rules for Vulnerable Populations

The Common Rule treats children, pregnant women, fetuses, prisoners, and those with mental disabilities differently from others.[106] With respect to children, prisoners, and those with mental disabilities,[107] there is concern that voluntary informed consent may be neither feasible nor possible, and therefore special rules are required. The rules governing pregnant women and fetuses reflect the tenuous nature of pregnancy and the fickle nature of politics more than any special legal considerations or constraints. This section discusses the special rules pertaining to children and to women who are pregnant. The other special populations (e.g., prisoners and those with mental disabilities) are not discussed here only because the issues are relatively rare.

The rules governing special populations present a unique dialectic. On the one hand, the Common Rule makes it more difficult to accrue women and children into studies. On the other hand, the National Institutes of Health Reauthorization Act of 1993 and various FDA initiatives strive to increase the participation of women and children in clinical trials. Balancing the need for a diverse pool of subjects that permits broad generalizations against the need to protect certain vulnerable populations has proven difficult.

CHILDREN

An informed consent form is a contract between the subject and the researcher. Children, namely those who have not reached the age of majority, which varies from state to state, lack the legal capacity to enter into binding contracts. Take the case of Kim Young, an eighteen-year-old from Tuscaloosa, Alabama. Young, tired of living with her parents, decided to set off on her own. She found a job at a local hardware store and, along with a friend or two, rented an apartment from a Mr. Weaver. In addition to having roommates, Young also brought along her trusty dog; the unnamed dog (also probably a minor) damaged the floors in the apartment. When Young moved out, she neglected to pay Weaver for the damaged floors as well as for some rent. Weaver sued and obtained a $1,300 judgment against her. Young, although young, was tenacious and kept appealing her case and kept losing until she reached the Alabama Court of Civil Appeals, which ruled in her favor. The court found that because Young had not yet reached nineteen, the age of majority in Alabama, she could not be bound by the lease agreement that she had signed unless the apartment was deemed to be a "necessity." Under Alabama law, minors can be held responsible only when purchasing necessities. The court, however, concluded that since Young had not been kicked out of her parents' house and remained free to return anytime she wished, the apartment where she lived with her dog and friends was not a necessity.[108] As Weaver learned the hard way, there are special rules when it comes to leasing property to kids.[109] The same underlying concepts apply when seeking to enlist children as subjects in biomedical research.

The Common Rule treats children differently in fundamental ways. First, it creates classes of "risk" and compels an IRB to find an added benefit for each incremental increase in risk. Second, it establishes a relatively complex system for obtaining consent (i.e., avoiding Mr. Weaver's plight). Third, the exemption in 45 C.F.R. § 46.101(b)(2) for research involving survey or interview procedures, or observation of public behavior, does not apply to research involving children, except for research involving observations of public behavior when the investigators do not participate in the activities being observed.[110]

When it comes to subjects who are minors, the Common Rule creates four categories of risk: (1) research not involving greater than minimal risk;[111] (2) research involving greater than minimal risk, but presenting the prospect of direct benefit to the individual subject; (3) research involving greater than minimal risk and no prospect of direct benefit to the individual subject, but which likely will yield knowledge about the subject's disorder; and (4) research involving greater than minimal risk without direct or indirect benefit to the subject but which presents an opportunity to understand or prevent a serious health problem in children. As the risk increases and the direct benefit to the child decreases, the likelihood that the research will be approved by the IRB decreases.

Research that does not involve greater than minimal risk to the child subject is ordinarily approvable by an IRB. Research that involves procedures that usually present no more than minimal risk to a healthy child include, for example, urinalyses, obtaining small blood samples, electroencephalograms, allergy scratch tests, minor changes in diet or daily routine, and the use of standard psychological or educational tests.

However, research involving greater than minimal risk, but presenting the prospect of direct benefit to the individual subject, can be approved by an IRB only if it determines that the risk is justified by the anticipated benefit to that child, and that the anticipated benefit is at least as favorable as offered by alternative treatments. Research that falls into the third category (i.e., greater than minimal risk, but no direct benefit to the subject) can be approved only if the IRB determines that the risk is only slightly greater than minimal risk, that the intervention is "reasonably commensurate" with those commonly experienced in their medical or educational situations, and that the intervention is likely to yield generalizable knowledge about the child's malady. Research in the fourth category cannot be approved by an IRB alone, but rather requires a special finding by the Secretary of Health and Human Services after consulting with a panel of experts.

Once an IRB decides that the research can be approved, it next has to decide the form of consent. First, it has to decide whether to require the signature of both parents or only one parent on the consent form. By law, it may opt to permit only one parent to provide consent, but only if the research falls into the first two risk categories—not greater than minimal risk or greater than minimal risk but with likelihood of direct benefit to the minor subject. Research that falls into the third and fourth categories can proceed only following the consent of both parents, subject to certain limited exceptions. In addition, the IRB must decide whether to solicit the minor's "assent." Assent differs from consent in one important way: Assent has no legal significance, whereas consent does. Nevertheless, the assent of the minor is normally required, especially if the minor is old enough to understand the nature and risks of the research and, where relevant, its benefits. A child's assent is normally irrelevant if the IRB determines that the research "holds out a prospect of direct benefit that is important to the health or well-being of the child and is available only in the context of the clinical investigation."[112] This consideration is especially important when the parents consent to the research, but the child refuses to participate. In such cases, the research is viewed no differently than any other medical treatment and may proceed against the child's wishes provided the IRB has approved this course of conduct.

The greatest risk for researchers arises when the parents are divorced or separated and share custody. While the law may permit the research to proceed with the consent of only one parent (if the research falls into the first two risk categories), it is normally unwise to proceed without the consent of both parents. I have seen cases in which researchers have inadvertently become fodder in custody battles in which the evidence supporting a parent's lack of fitness is the fact that one parent allowed the child to be a subject in a research study. Animosity between former spouses is a far greater force than reason.

PREGNANT SUBJECTS AND FETUSES

Accruing women of childbearing age into clinical studies, especially when drugs are involved, has raised special issues owing in part to two conflicting principles. On the one hand, NIH is required to promote research into women's health issues and more specifically, the director is required by the NIH Revitalization Act of 1993 "to ensure that (A) women are included as subjects in each project of such research; and (B) members of minority groups are included in such research."[113] On the other hand, women of childbearing age, especially those who are pregnant, face particular risks to themselves and to their fetuses. Certain types of research that would pose little risk for an ordinary subject may pose a substantial risk to a fetus. Attempting to balance these two competing concerns is difficult.[114] Many researchers are reluctant to accrue women of childbearing age into drug studies because of liability concerns. The fact that a woman may not be pregnant is often insufficient to allay concerns about risk because of the potential that a subject may become pregnant during the course of the study and may not be aware of it.

When a researcher proposes enrolling pregnant women into a study, the IRB has special responsibilities to ensure that the informed consent process is informative and free of coercion. IRBs are required to oversee through sampling the actual informed consent process and to actively monitor the research to determine if there are any unanticipated risks.

Although research involving pregnant subjects has primarily focused on the well-being of the mother and the fetus, research involving fetal tissue, although related, has frequently been the focal point of intense ethical debate having little to do with the physical well-being of the mother or the child. Federally funded research involving fetal tissue transplantation has been an "on-again-off-again-on-again" proposition.[115] On April 15, 1988, the last year of the Reagan administration, the Assistant Secretary for Health, without consulting the Secretary, imposed a "moratorium on federally supported research involving the transplantation of fetal tissue" obtained from electively aborted fetuses.[116] The moratorium, which in my view was extralegal, became the focus of acrimonious congressional hearings and a special blue ribbon commission. President Clinton administratively lifted the moratorium soon after taking office. The moratorium was permanently put to rest with the NIH Reauthorization Act of 1993, which expressly authorized federal funding of such research, but subject to extremely tight controls. Fetal tissue transplanation research remains controversial, and the rules that govern it can be tricky; before contemplating this type of research, consult with your institution's attorney.

DONORS OF STEM CELLS AND RESTRICTIONS ON STEM CELL RESEARCH

The saga of fetal tissue research turned out to be the warm-up for the controversy that would emerge at the start of the George W. Bush administration over federal funding for research using human embryonic stem cells (hESC). As with the fetal tissue ban, this was driven by moral fire inflamed by political winds.

It began in 1996, with the enactment of the Dickey-Wicker amendment, which prohibits federal funding to conduct research involving hESC, if a human embryo "is destroyed, discarded, or knowingly subject to risk of injury or death greater than that allowed for research on fetuses in utero."[117] Many argued that Dickey-Wicker had a major loophole: It did not prohibit federally funded research on hESC if the cells were derived through private funding.[118]

No one had an opportunity, though, to test this interpretation. On August 9, 2001, President George W. Bush issued a policy statement and later an executive order precluding NIH from funding research involving hESC unless the cells were from one of a set of twenty-one designated and extant cell lines.[119] Eight years later, on March 9, 2009, President Obama in Executive Order 13,505 rescinded Bush's policy statement and the executive order, and authorized the Secretary of HHS to "support and conduct responsible, scientifically worthy human stem cell research, including human embryonic stem cell research, to the extent permitted by law."[120] The order also required the Secretary to "review existing NIH guidance and other widely recognized guidelines on human stem cell research, including provisions establishing appropriate safeguards, and issue new NIH guidance on such research that is consistent with this order."[121]

The executive order left unresolved a host of issues, including whether the Obama administration would narrowly interpret the Dickey-Wicker amendment so that it has little impact on funding or, instead, whether there would be restrictions on the source of the hESC, whether certain types of research would still be off limits to federal funding, and whether there would be new requirements concerning a donor's informed consent.

On July 7, 2009, after receiving over 49,000 public comments, NIH issued guidelines to govern hESC research.[122] Under the guidelines, which implemented a presidential executive order, NIH funding is limited to hESC meeting certain criteria. Only hESC obtained (1) from embryos created by in vitro fertilization (IVF) for reproductive purposes and are no longer needed for that purpose, (2) with the documented informed consent of the donor, and (3) that meet other requirements, can be used in NIH-funded research. Stem cells that are listed on an NIH registry also may be used, and hESC that were obtained before the effective date of the new guidelines and which meet the new criteria in "spirit," but do not satisfy the strict informed consent requirements of the new guidelines, may qualify if approved by a special NIH advisory committee. In keeping with prior policy, if the donor's identity is stripped from the specimen before it is given over to the researchers, then the in vitro research is not "human subjects research"; according to HHS, "using hESC from which the identity of the donor(s) cannot readily be ascertained by the investigator [is] not considered human subject research" and not subject to IRB review.[123] However, even if the donors do not fall within the Common Rule, the research institution that receives the hESC would be responsible for ensuring that the IVF clinic complied with the requirements of the NIH guidelines. Finally, the guidelines also make clear that the administration is narrowly construing the Dickey-Wicker amendment so that its prescriptive language only prohibits using federal funds to collect the specimens; the federal funds can be used to conduct the research on those "privately" obtained hESC specimens.

IRB-Conducted Investigations

Not infrequently, an IRB learns that an investigator may have been less than thorough in discharging his or her responsibilities under the Common Rule or other governing norms (e.g., university rules, FDA rules, or funding agency rules). For example, what if an IRB suspects that an investigator has not been obtaining informed consent, or has not been documenting informed consent, or has been modifying the approved protocol willy-nilly without IRB approval, or has not been keeping the IRB apprised of adverse events? Does an IRB have the responsibility and jurisdiction to investigate and sanction the researcher? The Common Rule expressly vests each IRB with the authority to terminate or suspend a study that is not being conducted in accordance with the Common Rule.[124] When an IRB does this, it must provide the researcher with reasons why it is terminating or suspending the research. The IRB also is required to inform the institution (e.g., vice president for research) and the head of the funding agency (e.g., Secretary of HHS when the study is funded by NIH). If the study is within FDA jurisdiction, the IRB also must provide this information to FDA.

I have just given you the basic legal answer, but that answer does not take into account a number of factors. First, if the researcher is charged with having violated the Common Rule over an extended period, where was the IRB? Was it asleep at the wheel? Often, when an investigator's compliance is being called into question, so is the IRB's compliance. The IRB may have an inherent conflict of interest that would prevent it from conducting a full and balanced investigation. When this occurs, universities usually appoint an ad hoc committee to conduct the investigation.

Second, even if there is no conflict, relatively few IRBs have the expertise necessary to conduct an investigation. This lack of expertise is only exacerbated by the lack of policies and procedures for conducting these types of investigations. Remember, unlike a scientific misconduct case, in which intent is the primary focus, an IRB investigation focuses exclusively on the conduct. It does not matter as much why the researcher is violating the Common Rule as it does that he or she is violating the Rule.

Rules for Research Conducted Overseas

Conducting research involving human subjects—especially clinical trials—overseas can be complex and perilous, as Pfizer, Inc., the New York-based pharmaceutical giant, recently learned. In 1996, Pfizer conducted in Nigeria a clinical trial of its experimental new antibiotic, Trovan. At the time, there was a widespread outbreak of bacterial meningitis in northern Nigeria, and Pfizer sought to conduct its clinical trial on about two hundred infected children who were patients in Nigeria's Infectious Disease Hospital. The children were divided into two groups; those in one group received Trovan while those in the other group received Ceftriaxone, an FDA-approved antibiotic. After the clinical trial had concluded, a group representing the children claimed that Pfizer, working in complicity with the Nigerian government, had violated international legal norms by not obtaining informed consent, not advising the children or their guardians of the risks associated with Trovan, not providing information in their native language, Hausa, and not advising them that those in the control group would be receiving only a half-dose of Ceftriaxone. The representatives also alleged that the ethics committee approval provided by Infectious Disease Hospital had been backdated. They claimed that between eleven and thirty-four children died, and many others were severely injured and disabled (e.g., blindness, deafness) as a result of the clinical trial. Pfizer denied all the allegations, claiming that meningitis and not Trovan killed and disabled the children.

Not surprisingly, the children, through their parents and guardians, filed suit against Pfizer in both Nigeria and the United States. Also, the Nigerian federal government and the Kano (Nigeria) state government filed both civil claims and criminal charges against Pfizer. Early in 2009, the U.S. court of appeals, sitting in New York, held that a U.S. federal court could entertain the suit. On July 30, 2009, Pfizer announced that it had settled the civil and criminal claims brought by the Kano government for $75 million.[125] This settlement does not affect the suit instituted by the Nigerian government in Nigeria, and it is unclear how the settlement will affect the various private suits still pending in the United States.[126]

Research involving human subjects is complicated enough when conducted in the United States; when conducted abroad with federal funding, it can become a bureaucratic nightmare. For example, let's suppose that Northsouthern University has received an NIH grant to conduct research involving human subjects in Lower Slobovia. The research would have to comply with both the Common Rule and Lower Slobovia's analogous ethical rule. Serious problems can arise if there is a conflict between the U.S. requirements and the foreign requirements. For example, what happens if Slobovian law requires research subjects to waive their right to sue? Under the Common Rule, an informed consent form cannot contain exculpatory language. There are no clear guidelines for how a U.S. researcher or his or her institution should proceed when U.S. and foreign laws are mutually exclusive, making foreign research involving human subjects tricky at best.[127]

PRIVACY AND THE HEALTH INSURANCE PORTABILITY AND ACCOUNTABILITY ACT OF 1996

The Concept of Privacy

Aspects of the right to personal privacy are deeply ingrained in the common law. Battery, libel, and slander all contain components of the right to privacy. Battery, after all, is the ultimate invasion of personal privacy. However, it was not until the 1890s that legal scholars first argued that personal privacy ought to be viewed as a right on its own, not as an aspect of another broader right. In a seminal law review article in 1890, Louis Brandeis and Samuel Warren argued that state laws ought to protect personal privacy.[128] The Warren and Brandeis thesis was not enthusiastically embraced by the courts or state legislatures. Nearly forty years after penning his law review article, Brandeis observed that the Constitution

> sought to protect Americans in their beliefs, their thoughts, their emotions and their sensations. They conferred, as against the Government, the right to be let alone—the most comprehensive of rights and the right most valued by civilized man.

Brandeis wrote these words not as a law professor seeking to publish another erudite article, but as a Justice of the Supreme Court in *Olmstead v. United States*.[129] Brandeis's words, though, were written in dissent. It took another generation before privacy was to become a right, albeit limited by First Amendment considerations.[130] Today, nearly every state has laws that protect certain aspects of one's privacy from intruding eyes. As we shall see, the Health Insurance Portability and Accountability Act of 1996 (HIPAA) is layered on top of these state laws.[131]

HIPAA—Good Statute, Bad Regulations

The Common Rule is designed to help protect a subject's privacy, as well as his or her physical and emotional well-being. In fact, as discussed above, certain experiments that do not involve any physical or emotional risks are covered by the Common Rule and require IRB approval because of the risks to subjects' privacy. However, privacy is secondary to the Common Rule. There are other regulatory regimes, such as HIPAA, that are far more concerned with privacy.

Most folks who have visited a doctor's or dentist's office have been asked to read about that office's so-called HIPAA policy. Amusingly, most physicians and most dentists whom I have questioned have only the vaguest idea what HIPAA is and how it works. Most cannot even tell you what the acronym means. HIPAA, though, has its greatest impact on researchers, especially those who use tissue specimens or databases with patient-specific health information.

The Health Insurance Portability and Accountability Act of 1996 has actually very little to do with privacy. The word "privacy" appears nowhere in the legislation's name. HIPAA does three things. First, it makes it easier for employees to move from one job to another without losing their ability to acquire health insurance coverage through their new employer because of a preexisting condition. This is the "insurance portability" aspect of the law. Second, HIPAA makes it easier for the federal government to prosecute healthcare fraud. This is the "accountability" aspect of the law. And third, HIPAA seeks to make it easier for health insurers and healthcare providers to communicate with each other and to make or receive payments electronically. This is the "administrative simplification" aspect of the law. However, administrative simplification was far down the pecking order in terms of importance to those who crafted the legislation, primarily Senators Edward Kennedy (D-MA) and Nancy Kassebaum (R-KS), so that no hint of it appears in the bill's title. The administrative simplification provisions require the Secretary of HHS to develop standard code sets for medical procedures and diagnoses and standard formats for electronic communications. It also requires the Secretary to develop security standards for those using computers to transmit health information. Finally, almost as afterthought to an afterthought, HIPAA requires the Secretary to develop standards to ensure the privacy of health information that is being electronically transmitted.

In 2000, the Secretary issued an encyclopedia of regulations implementing the privacy provisions of HIPAA.[132] HIPAA and the regulations created two worlds—the world of "covered entities" and the world of "noncovered entities." HIPAA regulates covered entities.[133] A covered entity is any health insurer, any health insurance clearinghouse, and certain providers (e.g., hospitals, physicians, clinical laboratories). All insurers and all clearinghouses are "covered entities," but not all providers are. A provider is a covered entity (and governed by HIPAA) only if the provider bills electronically using specified code sets. Physicians who do not accept insurance payments and only charge their patients as they leave the office and do not bill electronically are usually not covered entities and do not have to bother with HIPAA.

Covered entities are permitted to use and share a person's private health information (called protected health information, or PHI) for treatment purposes, payment purposes, or operational purposes (e.g., quality control, peer review) without a person's consent or authorization. A physician who is not a "covered entity" is permitted to receive PHI for treatment purposes. If a physician is a covered entity, then the physician can use the information only for treatment, payment, or operational purposes. If the physician is not a covered entity, he or she is not governed by HIPAA and can use the information for any purpose he or she wishes consistent with state law.

HIPAA not only regulates the movement of information (e.g., a covered entity can transmit PHI only to another covered entity or provider), but also the use of that information (e.g., covered entities can use PHI only for treatment, payment, and operations, such as peer review). Research is not permitted unless the patient authorizes it. The "authorization" is to HIPAA as the "informed consent" is to the Common Rule. Originally, the authorization had to be on a separate sheet of paper and had to be approved by a special body within the university or hospital, called a Privacy Board. After much hue and cry from the research community, that was changed so that the authorization can be part of the informed consent form; it usually comes at the end, and most institutions require a separate signature. The HIPAA rules also permit an IRB to function as a Privacy Board.

Although HHS claims that its HIPAA rule "was not intended to impede research using records within databases and repositories that include individuals' health information," it has had precisely that effect: HIPAA dramatically restricts what a researcher can do with an existing database or with data collected anew. One study at the University of Michigan found that HIPAA requirements have led to a "significant drop in the number of patients who agree to participate in outcomes research for heart care."[134] HIPAA's untoward effects on research are a result of a series of requirements, none of which address any documented problem.[135] One should always question the wisdom of a regulation in search of a problem.

Under HIPAA, researchers are no longer permitted (1) to use specimens from a database if those specimens are linked to identifiable patients or (2) to use information from a healthcare database with PHI, unless the IRB has granted a waiver or the patient has authorized the researcher to use his or her identified specimen or medical information in specific research. An IRB may grant a waiver (meaning that patient's authorization is not necessary) if the IRB concludes that the research poses a nonsignificant risk to the patient's privacy, obtaining authorization would be impractical (more about that in a moment), the researcher promises not to disclose the PHI to another person, and the researcher will strip away as many identifiers as possible and retain only those necessary for the research.

What does all this really mean? Suppose that you have a tissue specimen collection consisting of renal carcinoma cells that you obtained from hundreds of patients over the course of years. You have been using the collection in your research for the past ten years. Each specimen carries the patient's name, age, date of birth, date of biopsy, and other personal information. The overwhelming majority of donors are no longer alive. Under the Common Rule, these individuals would not qualify as "human subjects" because they are not breathing. HIPAA, though, is not constrained by mortality, and it covers those on both sides of the Styx. It is not practical (or possible) to obtain authorization from these folks, and it will be impossible to maintain and use the specimen collection without a waiver. The waiver process was designed to deal with this type of situation. However, it is important to note that once a waiver has been granted, the researcher is no longer free to transfer the PHI that he or she maintains to other researchers at other institutions.

Another troubling aspect of the privacy rule is that you can maintain and use a patient's information only for a set period and for specified purposes. Moreover, the patient has the right to request that you cease using his or her PHI in research even before the agreed-to period ends. For example, suppose that you have conducted a clinical trial on the safety and effectiveness of a new drug. The subjects in the trial all agree to permit you to retain and use their medical records for ten years to support an FDA application and ancillary publications. You have analyzed all of the data and are on the verge of submitting those data as part of a $150 million New Drug Application (NDA). Suddenly, you receive a letter from one of the subjects requesting that you return his data and not use those data for any purpose, even though there is plenty of time left in the ten-year agreement. This is where things get tricky. The regulations permit the researcher to retain the data and to continue to use them for the authorized purpose, if and only if the researcher "has taken action in reliance" on the authorization.[136] In our example, since FDA requires a sponsor to submit all clinical data with its NDA, removing one patient's information from the submission could undermine the entire submission. In other contexts, however, it is not altogether clear when there is sufficient "reliance" to enable a researcher to continue using the patient's information. Traditionally, if a researcher takes certain action based on his reasonable belief that he will be able to use the data, then the subject is precluded from undermining the researcher's reasonable reliance. However, under the law the subject always has the right to demand that her personal data not be used further. Therefore, how can one ever reasonably rely on a subject's authorization? The regulations created, but do not adequately address, this legal conundrum.

HIPAA not only imposes constraints on research, but also undermines a fundamental policy of NIH and other funding agencies—data sharing. HIPAA effectively precludes data sharing for research purposes if the data include PHI. There are two ways of getting around this limitation, and neither is fully satisfactory. First, a researcher can "de-identify" his or her data set by removing parameters that can be used to identify patients or subjects. The HIPAA regulations list eighteen parameters that must be removed to qualify as de-identified, including the usual suspects (e.g., name, address, Social Security number or other patient numbers). Included in the list of banned parameters, though, are zip code, city, and county if the code, city, or county has fewer than 20,000 residents; dates relating to an individual (e.g., date of birth, date of admission, discharge, death); and anyone's age if that person is older than eighty-nine years of age.[137] Alternatively, you can retain, use, and transmit data associated with otherwise prohibited parameters as if it were a de-identified data set if a statistician opines that the information cannot be used to identify an individual.[138]

If de-identification is not feasible, you may still be able to transmit data by transforming it into a "limited data set." A limited data set is one that excludes what we would normally think of as direct identifiers (e.g., name, address, telephone number, Social Security number, and full-face picture), but does not exclude zip codes, dates, ages, and the like. To transmit a limited data set for research purposes, the two institutions (the transmitting and receiving institutions) must enter into a written agreement that, among other things, precludes retransmission of the data set.[139]

OFFICE OF BIOTECHNOLOGY ACTIVITIES AND THE RECOMBINANT DNA ADVISORY COMMITTEE

Within less than one year after Herbert Boyer and Stanley Cohen first spliced DNA, NIH established the Recombinant DNA Advisory Committee (RAC) "in response to public concerns regarding the safety of manipulating genetic material through the use of recombinant DNA techniques."[140] The original concern was that researchers, in the course of gene splicing, might inadvertently create a drug-resistant pathogen that would escape into the environment.[141] Since its creation in 1974, the RAC's charter and NIH rules have expanded to address the risks to human subjects posed by experimental gene therapy and issues associated with nonhuman cloning. The RAC is one of only three entities designed to regulate recombinant research both at the bench and the bedside: the RAC, the Office of Biotechnology Activities

(OBA) within NIH, and an Institutional Biosafety Committee (IBC), the recombinant analog to an IRB. The RAC, while receiving the most attention, is only an advisory committee; the real decision maker at the federal level is OBA, and at universities and other awardee institutions, it is the institution's Institutional Biosafety Committee. The details governing the operation of OBA, IBC, and the RAC are beyond the scope of this guidebook. What follows provides an overview, and a very short one at that.

Research Regulated by Office of Biotechnology Activities

The Common Rule, the Scientific Misconduct Rule, and the Conflict of Interest Rule (discussed in the next chapter) apply only to the precise research that is federally funded. Universities and other grantees may voluntarily extend those rules to nonfederally funded research, but they have no obligation to do so; and the government has no authority to compel a university to "voluntarily" extend these rules to nonfederally funded research. That is not the case with OBA and the OBA guidelines, which extend to any recombinant research, whether federally funded or not, so long as it is conducted at an institution that receives Public Health Service funding for recombinant research. If Dr. John McTag receives NIH funding for recombinant DNA research at Indiana Jones University, then any recombinant research conducted by anyone at Jones University is governed by the OBA, the RAC, and their "rules."

The NIH Guidelines for Research Involving Recombinant DNA Molecules, as the rules are known, apply to both nonclinical laboratory research and to clinical research, with the degree of scrutiny varying as a function of the overall risk and whether clinical trials are involved. For regulatory purposes, OBA has created six categories of research, each with varying degrees of control and approvals. For example, one class of research involves transferring a drug-resistant trait to microorganisms that are not known to acquire the trait naturally, if the transfer could compromise the ability to control disease agents in humans, animals, or agricultural plants. This type of research can proceed only with RAC review and the approval of the NIH director and the local Institutional Biosafety Committee. By contrast, experiments involving the cloning of toxin molecules with an LD50 of less than 100 ng/kg of body weight can proceed only with OBA and the Institutional Biosafety Committee approval. The oversight decreases as the risk decreases.

Gene therapy trials require RAC review and OBA approval along with IRB and IBC approvals at the institution. There are special adverse event reporting requirements and follow-up reviews.

The Relationship between the RAC and OBA

The RAC is an advisory committee composed of nongovernment employees, which means that all the RAC can legally do is provide advice; it cannot bind OBA, NIH, or the Secretary of HHS, nor can research be conditioned on RAC approval. OBA actually runs the show; it determines policy and decides whether research ought to be approved. The RAC, though, is the public presence of recombinant regulation. Aside from reviewing and making recommendations about certain research and helping develop NIH's guidelines on recombinant research, the RAC investigates mishaps.

For example, on September 17, 2007, the RAC convened a public hearing concerning the death of Jolee Mohr, a thirty-six-year-old mother from Illinois, who enrolled in a phase 1/2 study of an investigational therapy called tgAAC94, an adeno-associated virus (AAV) vector. Her attending physician told the RAC that she started to feel nauseous the evening of July 2, 2007, shortly after receiving her second injection, and her condition continued to worsen until she died on July 24. FDA suspended the study on the same day, but in November 2007, it authorized the sponsor to resume the clinical trial. The RAC issued its report thereafter concluding that while the "possible role of the gene transfer in this clinical course cannot definitively be excluded due to the lack of data[,]" it appeared unlikely that the gene transfer was the cause of death.[142]

The Source of OBA and RAC Authority

The RAC has been in business since the mid-1970s; the "rules" that govern recombinant research have been around for over a decade. Interestingly, there is no statutory or even regulatory authority for the requirements imposed on researchers by NIH or OBA. Indeed, it is likely that the entire "regulatory fabric" lacks legal authority. This is not to say that regulation of this type is not wise. Rather, when one imposes requirements on those outside government or limits the discretion of an agency, the agency must discharge certain procedural formalities, and that was not done here. As was discussed in the materials above, OBA restrictions and requirements constitute what are called substantive rules. Substantive rules can be issued only through notice-and-comment rulemaking or rulemaking procedures that are even more formal. There is an interesting aside. On the one hand, the normal rulemaking strictures do not apply to rules governing grants. Therefore, agencies are free to impose requirements on grantees without going through the laborious rulemaking process. On the other hand, in 1972, Elliot Richardson, the then-Secretary of Health, Education, and Welfare

(HHS's predecessor department), executed what has become known as the Richardson waiver, in which he voluntarily agreed that the department's rules for grants and contracts would be issued only following notice-and-comment rulemaking, even though the law did not require him to do so. The Richardson waiver remains in effect, and courts have held the department to its word. Therefore, NIH cannot regulate recombinant research without going through rulemaking, which it has not done.

CASE STUDIES AND PROBLEMS

Case 1: The Case of the Compromised Collaboration

(Use applicable PHS and HIPAA rules.)

Kevin Motely entered Greater Ascension Treatment Center and Hospital (GATC) as an outpatient for a routine screening colonoscopy. During the procedure, Kevin's physician detected a polyp, which he removed and sent to pathology. The polyp was benign. However, unbeknownst to Motely or his physician, the chief pathologist, Gregor Van Husen, stored the unused portion of all polyps along with each patient's basic information—name, address, date of admission, date of birth, zip code, age, race, religion, and insurer.

A few months later, Van Husen, who is an amateur epidemiologist, obtained IRB approval to conduct a prospective study of colonoscopy patients. Under the approved protocol, he contacted each patient who had had a benign polyp removed to find out whether the patient would be interested in participating in a prospective study of eating habits and the development of future polyps, both benign and otherwise. If a subject was interested, he or she would be given a lengthy informed consent form that described the study. Specifically, the IRB-approved informed consent form contained the following:

> This study poses no medical risks to you. You will be asked to complete a questionnaire containing questions about your eating and lifestyle habits, medical history, and ancestry. Each year, someone from the study will contact you and ask if you have had a colonoscopy during the year, and if so, the results. The person contacting you will also ask about your eating and lifestyle habits during the year. The information that we obtain will be confidential, and the results will be published only in the aggregate.

After about two years, Van Husen found some interesting trends, but nothing really worth publishing. He was somewhat disappointed.

One Saturday, while mulling over what to do with all of his data, he decided that it would be interesting to run some basic genetic screens on the polyp samples that he collected. Van Husen, however, was not a bench scientist. He called his good friend, Arthur Crickson, who had done a fair amount of gene screening. Crickson was a professor at nearby North Eastern Research and Development Center, part of Tech U. During the next nine months, Crickson and Van Husen ran various screens on the tissue samples. The results were, in Crickson's view, earth shattering. Crickson had located a gene in those with Mediterranean ancestry that correlated remarkably well with the development of polyps; the results were even more startling because they were relatively insensitive to diet or lifestyle.

They eagerly set about writing up their results for submission to a journal. The article was accepted almost immediately, was published, and made the national news. The day following their numerous press conferences, both are contacted by the chairs of their respective IRBs. What problems do either or both gentlemen have, and what should GATC and Tech U. do, if anything?

Case 2: The Case of the Random Regime

(Use the Common Rule and FDA rules.)

Howard L. Vegas is a rising star in the Department of Medicine at OxCam University in New Columbia. Vegas hates mandatory clinic duty where he is required to be the attending physician; it interferes with his research. Virtually all of the patients that he sees at the clinic are indigent, most don't speak much English, and few have interesting or challenging cases. To make clinic duty a little more bearable, Vegas decides to compare two drug treatments for Bollix Disease (BD), an otherwise rare autoimmune disease that he has been seeing with greater frequency in his clinic patients, especially among the immigrants. One drug, Defenestratia, has been approved by FDA to treat BD. Another drug, OmniAll, has been approved by FDA to treat a related condition, but not BD. Vegas is interested in seeing how the two drugs actually compare, especially since Defenestratia has serious side effects and OmniAll has none. Each time he sees a patient with BD, he flips a coin. If the coin shows heads, he prescribes Defenestratia; if the coin shows tails, he prescribes OmniAll.

He never tells the patients or anyone else about his little experiment. A few of the patients that he sees are under eighteen years of age. The patients who received OmniAll are totally cured within two weeks. Most, but not all, of the patients who received Defenestratia are cured, but it takes much longer.

One of the Defenestratia patients, a sixteen-year-old boy, dies of a drug-related event. His parents sue the drug manufacturer; they also sue OxCam and Vegas for malpractice. During discovery, the attorney for the parents learns that the sixteen-year-old was actually part of an "informal" experiment. The attorney amends his lawsuit to allege battery and violations of the Common Rule and FDA rules. What should the university do? What should FDA do?

Case 3: The Case of the Naïve Nephrologist

(Use applicable FDA rules.)

Nancy Newcomb is professor of medicine at North East West University (NEW U); she is chairman of the Department of Medicine within the Medical College and is a nephrologist. In addition to her administrative duties, Newcomb is an active bench and clinical researcher. She is currently the principal investigator on a National Cancer Institute grant to test a new genetically engineered vaccine (i.e., biologic) designed to treat those with early-stage kidney cancer. The treatment, if successful and approved by FDA, could replace nephrectomy as the treatment of choice for this ailment.

NCI scientists invented the vaccine, and NCI is sponsoring the multicenter clinical trials. NCI submitted the necessary Form 1571 to FDA; each principal investigator, including Dr. Newcomb, submitted an FDA Form 1572 to the sponsor, which in turn forwarded them to FDA. In the Form 1572, each investigator of an investigational new drug, such as the one being tested by Newcomb, certifies as follows:

- I agree to conduct the study(ies) in accordance with the relevant, current protocol(s) and will only make changes in a protocol after notifying the sponsor, except when necessary to protect the safety, rights, or welfare of subjects.
- I agree to personally conduct or supervise the described investigation(s).
- I agree to inform any patients, or any persons used as controls, that the drugs are being used for investigational purposes, and I will ensure that the requirements relating to obtaining informed consent in 21 CFR Part 50 and institutional review board (IRB) review and approval in 21 CFR Part 56 are met.
- I agree to report to the sponsor adverse experiences that occur in the course of the investigation(s) in accordance with 21 CFR 312.64.
- I have read and understand the information in the investigator's brochure, including the potential risks and side effects of the drug.
- I agree to ensure that all associates, colleagues, and employees assisting

in the conduct of the study(ies) are informed about their obligations in meeting the above commitments.

- I agree to maintain adequate and accurate records in accordance with 21 CFR 312.62 and to make those records available for inspection in accordance with 21 CFR 312.68.
- I will ensure that an IRB that complies with the requirements of 21 CFR Part 56 will be responsible for the initial and continuing review and approval of the clinical investigation. I also agree to promptly report to the IRB all changes in the research activity and all unanticipated problems involving risks to human subjects or others. Additionally, I will not make any changes in the research without IRB approval, except where necessary to eliminate apparent immediate hazards to human subjects.
- I agree to comply with all other requirements regarding the obligations of clinical investigators and all other pertinent requirements in 21 CFR Part 312.

Under the protocol, patients who are diagnosed with early-stage kidney cancer, are not suffering from other ailments, and meet a variety of other criteria are eligible to participate in the study. The precise location of the tumor is documented by a special Doppler sonogram. The vaccine is made from the patient's own cancer cells and is supposed to induce a massive immune response to the cancer. Patients are hospitalized, their cancer surgically removed, and the cancer cells are then manipulated in the laboratory by Newcomb to create the vaccine. Under the protocol, the vaccine is administered over a one-week period while the patient is still hospitalized. During hospitalization, a variety of tests are run on the patient and recorded in the Case Report Form (CRF), which contains the patient's signed informed consent form, hundreds of data points, and the signature of the physician attesting to the fact that he or she reviewed all data points and that they are accurate. The data recorded in the CRF not only support each patient's eligibility (e.g., negative HIV, negative hepatitis B virus, negative metatases) but also document the course of the experimental treatment and the results of various tests run during and after each treatment session. Immediately before discharge, each patient receives a second Doppler sonogram.

Following discharge and for two months thereafter, each patient is asked to come weekly for some follow-up tests, which are also recorded in the CRF by the investigator. After two months, each patient is asked to provide monthly urine and blood samples. Special diagnostic tests are run on each urine and blood sample to see if there is any evidence of cancer. At six months, each patient is asked by the study's clinical nurse to either return to the hospital or go to a facility near them for a Doppler sonogram, their third since being accrued into the study. The sonograms are not read by the investigator, but sent immediately by the technician to the sponsor, a group at the National Cancer Institute.

NEW U is one of ten centers involved in the clinical trial, and during the two-year study, over fifty patients have been accrued into the study at NEW U. On the NIH grant and the FDA Form 1572, the following individuals are listed as participating researchers: Newcomb; Katz, a surgeon; Kringle, a surgeon; Moon, a resident; Wagner, a clinical research nurse; and Holmes, a clinical research nurse. All patients who are admitted to NEW U with suspected kidney cancer are subjected to a normal battery of tests that would be given anywhere; in addition, all patients are also given an HIV test and various other screening tests to ascertain eligibility to participate in the study. Those who meet the eligibility criteria are then asked if they wish to participate and, if so, are asked to sign the informed consent form.

Karen Wagner, one of the clinical research nurses, is relatively new to the hospital. Although she appears to be energetic and is very good with patients, her organizational skills are somewhat wanting, and she is not detail-oriented. Each week, Newcomb reviews all the CRFs for patients and checks each CRF entry against the patient's hospital records to ensure that the CRFs are accurate. All sonogram records, except the six-month sonogram, which is sent directly to NIH, are included in the CRF. After each CRF is completed for each patient, Dr. Newcomb corrects errors, signs at the end certifying that the data are accurate, and sends it off to NIH. Wagner makes an unusually large number of recording errors, a fact that Newcomb notices and corrects, but does nothing else about.

At an investigators' meeting sponsored by NIH, one of the NIH staffers responsible for overseeing and monitoring data collection walks up to Newcomb, who is enjoying a nice refreshing gin and tonic, and states: "We've had some concerns with your six-month sonograms. Something seems to be wrong; in one case, we received a six-month sonogram for a patient who had died in an auto accident four months earlier." Newcomb nearly chokes on her drink. She races back to the hotel, checks out, and catches the first flight to NEW U. She starts reviewing all copies of all the six-month sonograms and immediately notices that many of them appear to be nothing more than copies of earlier sonograms with the dates changed. Newcomb confronts both Wagner and Holmes. Wagner breaks down and admits having falsified at least fifteen of the sonograms and also to having made up blood pressures and body temperatures for many of the patients. NEW U immediately fires Wagner, institutes an investigation of Wagner, and advises both NIH and the HHS Office of Research Integrity.

A few months later, an FDA inspector visits Newcomb to review the data from the study. He asks if anything unusual occurred. Newcomb is aghast. She asks, "Hasn't anyone at NIH or HHS told you that we had a major problem with data collection?" The inspector, equally surprised, replies "no," whereupon Newcomb fills him in. Thereafter, the inspector conducts his normal inspection during which he notices that patients were screened for eligibility before they signed their informed consents.

Yesterday, Newcomb received a letter from FDA proposing to disqualify her from receiving investigational new drugs because, in FDA's view, she "repeatedly or deliberately" (i) submitted false data to a sponsor, (ii) failed to adequately supervise a clinical trial, (iii) failed to conduct a study in accordance with the protocol, (iv) failed to maintain accurate case histories, and (v) failed to obtain informed consent prior to enrolling patients in the study.

Discuss the merits of the FDA case.

NOTES

1. I use the word "involving" rather than the more direct word "on" because research does not have to be "on" humans to be considered regulated "human subjects research."

2. There are also state laws that may affect human research in that state. A few states actively regulate all research on humans. For example, Maryland applies the Common Rule to all research conducted in the state irrespective of the funding source. *See* note 56, *infra*. Significantly, many states preclude certain types of research on moral grounds. For example, many state laws prohibit human cloning and the use of fetal tissue. *See* Forbes v. Napolitano, 236 F.3d 1009 (2000), *amended at* 240 F.3d 903 *and* 260 F.3d 1159 (9th Cir. 2001) (invalidating Arizona's ban on fetal tissue research as void for vagueness); Jane L. v. Bangerter, 61 F.3d 1493, 1499 (10th Cir. 1995), *rev'd and remanded on other grounds sub nom.* Leavitt v. Jane L., 518 U.S. 137 (1996) (striking down on vagueness grounds a Utah law that provided that "[l]ive unborn children may not be used for experimentation, but when advisable, in the best medical judgment of the physician, may be tested for genetic defects."); Margaret S. v. Edwards, 794 F.2d 994, 998–99 (5th Cir. 1986) (striking down on vagueness grounds a Louisiana law almost identical to the Utah law); Lifchez v. Hartigan, 735 F. Supp. 1361, 1363–76 (N.D. Ill.), *aff'd mem.*, 914 F.3d 260 (7th Cir. 1990) (striking down an Illinois law that banned research on fetal tissue from an aborted fetus).

3. *See* Grimshaw v. Ford Motor Co., 174 Cal. Rptr. 348 (Cal. Ct. App. 1981).

4. *See* Mohr v. Williams, 104 N.W. 12 (Minn. 1905).

5. *See, e.g.*, Schloendorff v. N.Y. Hosp., 105 N.E. 92, 93 (N.Y. 1914) (holding that a surgeon who performs an operation without his patient's consent commits an assault, for which he is liable in damages).

6. In some cases, the courts have been willing to "imply" consent. For example, in *O'Brien v. Cunard Steamship Co.*, 28 N.E. 266 (Mass. 1891), an unvaccinated immigrant stood in line to leave a ship on her way to being interviewed and examined by immigration officials. While in line, she held up her arm and was vaccinated; she later sued, arguing that she had not consented. The court held that since vaccination was a precondition to entry into the United States and

since she was standing in line to enter the United States and held up her arm as if she wanted to be vaccinated, she impliedly consented. At the very least, she learned quickly the American way—sue first, ask questions later.

7. *See* Pedesky v. Bleiberg, 59 Cal. Rptr. 294, 298 (Cal. Ct. App. 1967); Bradford v. Winter, 30 Cal. Rptr. 243 (Cal. Ct. App. 1963).

8. Canterbury v. Spence, 464 F.2d 772 (D.C. Cir. 1972).

9. *Id.* at 776.

10. The irony of the legislation is that the Tuskegee study, unlike most government-supported research, was actually conducted by Public Health Service researchers working in cooperation with various state and local health departments. *See* National Archives and Records Administration Southeast Region (Record Group 442—Centers for Disease Control and Prevention).

11. National Research Service Award Act of 1974, Pub. L. No. 93-348, § 212, 88 Stat. 342, 352–53; *see* S. Rep. No. 381 (1974). A year before congressional action, the then-Department of Health, Education, and Welfare (HEW) issued a proposed rule (dubbed "proposed policy") that sought to codify, for the first time, prior HEW policy for protecting human subjects in HEW-funded research. *See* 38 Fed. Reg. 27,882 (Oct. 9, 1973). The HEW rule, which was to be codified at 45 C.F.R. pt. 46, was to govern all research programs funded by HEW. A final rule was issued the following year. *See* 39 Fed. Reg. 18,914 (May 30, 1974).

12. *See* Community Mental Health Centers Extension Act of 1978, Pub. L. No. 95-622, 92 Stat. 3412.

13. *See* 47 Fed. Reg. 13,272 (Mar. 29, 1982).

14. *See* 56 Fed. Reg. 28,002 (June 18, 1991); 51 Fed. Reg. 20,204 (June 3, 1986) (proposed rule).

15. *See* Model Federal Policy § ___.116; 45 C.F.R. § 46.116.

16. The most widely referenced codification of the Common Rule is the one issued by HHS, the largest funder of biomedical research. *See* 45 C.F.R. pt. 46, which is reproduced, as revised through October 2008, at http://www.access.gpo.gov/nara/cfr/waisidx_08/45cfr46_08.html.

17. There used to be an extraordinarily convoluted system by which a putative grantee assured NIH that it was complying or would comply with the Common Rule. Entities could file any one of a number of different types of assurances. That system has been replaced by a single Federalwide Assurance (FWA) that provides grantees with significantly more flexibility than the Byzantine system it replaces. *See* note 84, *infra*, for a more complete discussion of this past system.

18. *See* Model Federal Policy § ___.107; 45 C.F.R. § 46.107.

19. *See* 45 C.F.R. § 46.110.

20. *See id.* § 46.116(b).

21. OHRP replaced the Office for Protection Against Research Risks (OPRR), which had been part of the National Institutes of Health. *See* 65 Fed. Reg. 37,136 (June 13, 2000). OHRP, as was the case for OPRR, is burdened with an overly bureaucratic, noun-string name that is difficult to remember and silly. How can someone protect another from risk associated with research, especially when it is impossible to protect people from the risk of crossing the street?

22. 45 C.F.R. § 46.123(a).

23. *See* Gina Kolata, *Johns Hopkins Death Brings Halt to U.S.-Financed Human Studies*, N.Y. TIMES, July 20, 2001.

24. *See* OFFICE FOR HUMAN RESEARCH PROTECTIONS, INSTITUTIONAL REVIEW BOARD GUIDEBOOK ch. 1, at 10 (n.d.); 45 C.F.R. pt. 76.

25. *See* Robert Charrow, *Protection of Human Subjects: Is Expansive Regulation Counterproductive*? 101 Nw. U. L. Rev. 707 (2007); Robert Charrow, *Wheat, Guns, and Science: The Commerce Clause and Human Subjects*, 9 J. NIH Res. 55 (1997); Robert Charrow, *Whose Tissue Is It Anyway*? 6 J. NIH Res. 79 (1994); Robert Charrow, *Informed Consent: From Canterbury Tales to* Canterbury v. Spence, 5 J. NIH Res. 75 (1993).

26. It is recommended that if an IRB, based on the advice of counsel, determines that specific activities either do not involve human subjects or do not qualify as research, then that determination should not be part of the normal IRB records. To preserve the attorney-client privilege, those determinations, along with counsel's advice, should be stored in a secure location. They should not be divulged to outside inspectors without consulting with counsel.

27. *Compare* Immanuel Kant, Critique of Pure Reason (1781) (J. M. D. Meiklejohn trans., 1855), Karl R. Popper, The Logic of Scientific Discovery (rev. ed. 1972), and Francis Bacon, The New Organon or True Directions Concerning the Interpretation of Nature (1620) (James Spedding et al. trans., 1863), *with* the Common Rule. The Supreme Court has adopted a Popperian view of science—one must be able to articulate a testable hypothesis, one that can be falsified. *See* Daubert v. Merrill Dow Pharms., Inc., 509 U.S. 579 (1993).

28. Under California law, a "medical experiment" requiring informed consent is defined as "(a) [Use] . . . in or upon a human subject in the practice or research of medicine in a manner not reasonably related to maintaining or improving the health of the subject or otherwise directly benefiting the subject. (b) The investigational use of a drug or device [or] . . . (c) Withholding medical treatment from a human subject for any purpose other than maintenance or improvement of the health of the subject." Cal. Health & Safety Code § 24174.

29. It would be difficult to argue that OHRP, or for that matter, the Secretary of Health and Human Services, has authority to interfere with the practice of medicine. This is especially so when the Medicare program expressly provides that "[n]othing in this title [Medicare] shall be construed to authorize any Federal officer or employee to exercise any supervision or control over the practice of medicine or the manner in which medical services are provided, or over the selection, tenure, or compensation of any officer or employee of any institution, agency, or person providing health services; or to exercise any supervision or control over the administration or operation of any such institution, agency, or person." Social Security Act § 1801, 42 U.S.C. § 1395.

30. Skeptical Inquirer, Guide for Authors, http://www.csicop.org/si/guide-for-authors.html (last visited July 31, 2009).

31. Taus v. Loftus, 151 P.3d 1185, 1193 (Cal. 2007).

32. *Id.* at 1194.

33. *Id.*

34. *See* James Lindgren, Dennis Murashko & Matthew R. Ford, *Foreword: Symposium on Censorship and Institutional Review Boards*, 101 Nw. U. L. Rev. 399 (2007).

35. *See* Rust v. Sullivan, 500 U.S. 173, 200 (1991) (holding that the university "is a traditional sphere of free expression so fundamental to the functioning of our society that the Government's ability to control speech within that sphere" is extraordinarily limited).

36. California has an unusual statute called an Anti-SLAPP (Strategic Lawsuit Against Public Participation) that is triggered when a lawsuit may chill free speech. Once that statute is triggered, a plaintiff must present evidence early in the proceedings that the case is meritorious. The California Supreme Court held that the Anti-SLAPP statute applied and that the plaintiff had failed to present enough evidence that her case was meritorious.

37. Office for Human Research Protections, Guidance on Research Involving Coded Private Information or Biological Specimens (Oct. 16, 2008), *available at* http://www.hhs.gov/ohrp/humansubjects/guidance/cdebiol.htm.

38. *See* U.S. Food & Drug Admin., Information Sheets, Guidance for Institutional Review Boards and Clinical Investigators: 1998 Update, at 9 (1998) ("[I]nformed consent must be obtained prior to initiation of any clinical procedures [including screening tests] that are performed *solely* for the purpose of determining eligibility for research. . . .") (emphasis added).

39. 45 C.F.R. § 46.102(f); *see* 56 Fed. Reg. 28,002, 28,013 (June 18, 1991) (emphasis supplied).

40. Many legal scholars have suggested that applying the Common Rule to non-federally funded social science research may raise significant First Amendment concerns. *See* Reneé Lettow Lerner, *Unconstitutional Conditions, Germaneness, and Institutional Review Boards*, 101 Nw. U. L. Rev. 775 (2007); Philip Hamburger, *The New Censorship: Institutional Review Boards*, 2004 Sup. Ct. Rev. 271, 290. This is a view that I share. *See* Charrow, *Protection of Human Subjects, supra* note 25.

41. *See* 18 U.S.C. § 1708.

42. *See* Letter from Michael A. Carome, Director, Division of Compliance Oversight, Office for Human Research Protections, to Eugene P. Trani, President, Virginia Commonwealth University (Dec. 15, 2000), *available at* http://www.dhhs.gov/ohrp/detrm_letrs/dec00i.pdf; *id.* to Eugene P. Trani, President, & Roy Pickens, Associate Vice President for Research, Virginia Commonwealth University (Sept. 22, 2000), *available at* http://www.dhhs.gov/ohrp/detrm_letrs/sep00d.pdf.

43. *See* National Human Research Protections Advisory Committee, Clarification of the Status of Third Parties When Referenced by Human Subjects in Research (Jan. 28–29, 2002), http://www.hhs.gov/ohrp/nhrpac/documents/third.pdf.

44. Statement of Bernard Schwetz, then-director, OHRP, presented at the National Council on Ethics in Human Research National Conference, Ottawa, Ontario (March 5–6, 2005), as reported in 13 Communiqué 13, *available* at http://www.ncehr-cnerh.org/english/English%20Communique%20March2005.pdf (last visited Sept. 14, 2009) (noting there has been a "hard time reaching agreement . . . between the federal agencies within DHHS" on how to treat third parties).

45. 45 C.F.R. § 46.102(f).

46. *See* Minnesota v. Carter, 525 U.S. 83 (1998) (holding that open window shades obviate any expectation of privacy).

47. The American Heritage Dictionary 960 (4th ed. 2002).

48. *See* Ethical Principles and Guidelines for the Protection of Humans Subjects of Research (Apr. 18, 1979), *reprinted in* 44 Fed. Reg. 23,192 (Apr. 18, 1979); OHRP Institutional Review Board Guidebook, *supra* note 24 (extensively discussing the Belmont Report and its significance).

49. *See* 44 Fed. Reg. 23,193 (col. c) (Apr. 18, 1979).

50. To obviate some of the adverse consequences associated with dissemination of genetic information, Congress enacted the Genetic Information Nondiscrimination Act of 2008 (GINA), Pub. L. No. 110–233, 122 Stat. 881 (2008) (codified in various titles of the U.S.C.). GINA prohibits discrimination in health coverage and employment based on genetic information; it also prohibits discrimination in other areas, such as life insurance. GINA goes into effect for health coverage sometime between May 22, 2009, and May 21, 2010, and for employ-

ment on November 21, 2009. *See* 122 Stat. 896. The extent to which one discusses GINA in an informed consent form is an open question. On the one hand, if subjects know about GINA, that knowledge may give some subjects a false sense of security, especially given GINA's limited scope. On the other hand, mentioning GINA in a consent form and discussing its limitations may distract subjects from other, more concrete, risks associated with the research. *See* OFFICE FOR HUMAN RESEARCH PROTECTIONS, GUIDANCE ON THE GENETIC INFORMATION NONDISCRIMINATION ACT: IMPLICATIONS FOR INVESTIGATORS AND INSTITUTIONAL REVIEW BOARDS (Mar. 24, 2009), *available at* http://www.hhs.gov/ohrp/humansubjects/guidance/gina.html.

51. David Bazelon, *Probing Privacy*, 12 GONZ. L. REV. 587, 588 (1977), and Charles Fried, *Privacy*, 77 YALE L.J. 475, 482, 483 (1968), respectively. David Bazelon was for many years a judge on the United States Court of Appeals for the D.C. Circuit (1949–79). Charles Fried is a Harvard law professor, former associate justice of the Supreme Judicial Court of Massachusetts (1995–99), and former solicitor general of the United States (1985–89).

52. 45 C.F.R. § 46.101(b)(2).

53. *Id.* § 46.101(b)(3)(ii).

54. The authors of the Common Rule may have been thinking about Certificates of Confidentiality. A researcher may apply to the Secretary of Health and Human Services for a Certificate of Confidentiality. If the Secretary issues the certificate, the researchers "may not be compelled in any Federal, State, or local civil, criminal, administrative, legislative, or other proceedings to identify [any research subjects]." Public Health Service Act § 301(d), 42 U.S.C. § 241(d). However, a Certificate of Confidentiality does not prevent the researcher from revealing confidential information about subjects; it merely prevents third parties from forcing the researcher to reveal private information. Other, more focused, types of certificates are also available to those conducting research on drug addiction, mental illness, and the like. *See* 42 C.F.R. pts. 2 and 2a. Even those more focused certificates are not absolute. The Privacy Act, which applies only to systems of records maintained by the federal government, contains numerous exceptions, all of which permit federal agencies to share confidential information under certain circumstances. *See* 5 U.S.C. § 552a(b).

55. 45 C.F.R. § 46.101(b)(4).

56. *See* OFFICE FOR HUMAN RESEARCH PROTECTIONS, GUIDANCE ON RESEARCH INVOLVING CODED PRIVATE INFORMATION OR BIOLOGICAL SPECIMENS (Aug. 10, 2004), *available at* http://www.dhhs.gov/ohrp/humansubjects/guidance/cdebiol.htm.

57. Three states—Maryland, New York, and Virginia—regulate research on humans conducted in their respective states irrespective of the funding source. *See* MD. CODE ANN., HEALTH-GEN. § 13-2001 *et seq.*; N.Y. PUB. HEALTH LAW § 2441(2); VA. CODE ANN. §§ 32.1–162.16; California, as noted earlier, has a well-developed set of rules concerning informed consent. *See* CAL. HEALTH & SAFETY CODE § 24173 *et seq.* Many states also have laws regulating or prohibiting certain types of research (e.g., human cloning) or research on certain classes of individuals (e.g., prisoners).

58. Most medical devices are actually "cleared," rather than approved, by FDA. A relatively small number (about 1 percent) of medical devices, namely those involving high risk or used to sustain life, are affirmatively approved by FDA. The clearance process, variously referred to as the premarket notification process or the 510(k) process, normally takes ninety days. In contrast, the approval process for devices, referred to as the premarket approval or PMA process, can take over a year of review (this excludes the time to conduct the clinical trials).

59. *See* FDCA § 501 *et seq.*, 21 U.S.C. § 351 *et seq.* (drugs); FDCA § 513, 21 U.S.C. § 360c *et seq.* (devices); and 42 U.S.C. § 262 (biologics). The FDA also regulates food, cosmetics, and virtually anything that emits radiation or even uncoupled electrical and magnetic fields. *See* FDCA § 401 *et seq.*, 21 U.S.C. § 341 *et seq.* (food); FDCA § 601 *et seq.*, 21 U.S.C. § 361 *et seq.* (cosmetics); and FDCA § 531 *et seq.*, 21 U.S.C. § 360hh *et seq.* (radiation-emitting devices). FDA regulates about 25 percent of the nation's economy.

60. *See* FDCA § 201(g), 21 U.S.C. § 321(g).

61. *See id.* at § 201(h), 21 U.S.C. § 321(h).

62. Certain types of low-risk devices, such a bed pans, thermometers, and the like can be distributed without clinical evidence of safety or effectiveness. Interestingly, most devices (except those that are truly risky) can be marketed without overt FDA approval. Rather, as mentioned in note 58, FDA "clears" the device if the manufacturer can show that its device is substantially equivalent to a device that was lawfully marketed prior to May 28, 1976, or to a device that is being marketed on or after that date, but which itself is substantially equivalent to a pre-May 28, 1976, device. Even for many of the "substantially equivalent" devices, FDA will demand some clinical data.

63. *See* 45 C.F.R. § 46.116; 21 C.F.R. § 50.25.

64. *See* Michael Paasche-Orlow & Frederick Brancati, *Assessment of Medical School Institutional Review Board Policies Regarding Compensation of Subjects for Research-Related Injury*, 118 AM. J. MED. 175 (2005).

65. Constructing a reliable and valid multiple-choice test requires expertise beyond that of most biomedical researchers and normally needs to be pilot tested. A "paraphrase task" provides the most valid and reliable way of testing comprehensibility, but the results usually can be assessed only by a psycholinguist, of which there are relatively few. *See* Robert Charrow & Veda Charrow, *Assessing the Comprehensibility of Standard Jury Instructions: A Psycholinguistic Approach*, 79 COLUM. L. REV. 1305 (1979).

66. 45 C.F.R. § 46.116(a)(2); *see id.* §§ 46.116(a)(4), (b)(1).

67. Lenahan v. Univ. of Chicago, 808 N.E.2d 1078 (Ill. App. Ct. 2004).

68. Stewart v. Cleveland Clinic Found., 736 N.E.2d 491 (Ohio Ct. App. 1999).

69. *See* Letter from Patrick J. McNeilly, Division of Compliance Oversight, Office for Human Research Protections, to Ronald Newbower, Senior Vice President for Research & Technology, Massachusetts General Hospital (Apr. 3, 2002), *available at* http://www.hhs.gov/ohrp/detrm_letrs/YR02/apr02i.pdf; *see* Letter from Michael Carome, Director, Division of Compliance Oversight, Office for Human Research Protections, to Arthur S. Levine, Senior Vice Chancellor for Health Sciences & Dean, School of Medicine, University of Pittsburgh (Mar. 25, 2002), *available at* http://www.hhs.gov/ohrp/detrm_letrs/YR02/mar02i.pdf.

70. Moore v. The Regents of the Univ. of Cal., 793 P.2d 479 (Cal. 1990), *cert. denied*, 499 U.S. 936 (1991).

71. *See* Letter from Rina Hakimian, Compliance Oversight Coordinator, Division of Compliance Oversight, Office for Human Research Protections, to Jane E. Henney, Senior Vice President & Provost for Health Affairs, University of Cincinnati (Apr. 5, 2004), *available at* http://www.hhs.gov/ohrp/detrm_letrs/YR04/apr04a.pdf.

72. *See* W. Levinson, D. L. Roter, J. P. Mullooly, V. T. Dull & R. M. Frankel, *Physician-Patient Communication: The Relationship with Malpractice Claims among Primary Care Physicians and Surgeons*, 277 J. AM. MED. ASS'N 553 (1997) (interestingly, the authors found that while communication differences affected the likelihood of a claim against a primary care physician, it had no effect on claims against surgeons).

73. Lett v. Sahenk, No. 05-3452 (6th Cir. Apr. 4, 2006).

74. 21 C.F.R. § 50.27(a).

75. *See id.*

76. *See id.* pt. 56.

77. *See* 45 C.F.R. § 46.107.

78. *See* Office for Human Research Protections, Guidance on Written IRB Procedures (Jan. 15, 2007), *available at* http://www.hhs.gov/ohrp/humansubjects/guidance/irbgd107.htm.

79. The IRB can require that the protocol be renewed more frequently than once per year.

80. Office for Human Research Protections, Guidance on Continuing Review (Jan. 15, 2007), *available at* http://www.hhs.gov/ohrp/humansubjects/guidance/contrev0107.htm (numbering added).

81. There used to be a relatively complex and arcane system of providing "assurance." That system recognized two types of assurances—a single project assurance that was good for only one grant or project, and a multiple project assurance (MPA) that covered all research at the institution. However, as a condition of obtaining an MPA, an institution had to promise that all of its research, whether federally funded or not, would hew to the Common Rule. Since, as a practical matter, all research institutions had to obtain an MPA, OHRP's predecessor hoped to use the MPA as a way of extending the Common Rule to all research, whether federally funded or not. Conditioning an MPA on an assurance that a university would follow the Common Rule for all of its research was likely improper. *See* Charrow, *Protection of Human Subjects, supra* note 25, at 716. The MPA was gradually phased out in favor of the Federalwide Assurance. After December 31, 2005, all then-existing MPAs were deactivated. *See* 45 C.F.R. § 46.103(a).

82. *See* 45 C.F.R. § 46.501 *et seq.*, as added by 74 Fed. Reg. 2399 (Jan. 15, 2009) (modifying and formalizing the IRB registration system) (effective July 14, 2009). FDA also requires IRBs involved in reviewing clinical research under the jurisdiction of FDA to register with FDA. *See* 74 Fed. Reg. 2358 (Jan. 15, 2009).

83. See Association for the Accreditation of Human Research Protection Programs, Inc., http://www.aahrpp.org/www.aspx?PageID=17 (last visited Sept. 14, 2009).

84. The Centers for Disease Control and Prevention (CDC) has awarded a grant to the Association for the Accreditation of Human Research Protection Programs (AAHRPP) to assess whether accreditation enhances "the protection of participants in public health research." *See* Marjorie A. Speers, *AAHRPP's Peer-Reviewed Approach to Protecting Research Participants*, 3 Med. Res. L. & Pol'y Rep. 55 (2004). It would have been preferable for such a study to be conducted by an entity that had no economic interest in the outcome of the research. It also would have been preferable for such a study to have been undertaken before universities and other research institutions rush to pay tens of thousands of dollars to secure accreditation. Designing and running such a study would not be all that difficult. The ideal vehicle would be a multicenter trial where IRBs at some of the sites are not accredited and IRBs at other sites are accredited. My hypothesis is that there would be no discernible difference in patient safety between the two groups, although there might be differences in the quality of recordkeeping. It never ceases to amaze me that scientists are actually more willing than lawyers to adopt major policy initiatives without any empirical support whatsoever.

85. Most FDA studies are conducted at multiple sites. An investigator at site A is obligated to inform his or her IRB of adverse events at site A, but that investigator may not know what has occurred at sites B and C unless the sponsor informs him or her and also the IRBs at sites B and C.

86. *See* 21 C.F.R. § 312.64(a).

87. *Id.* §§ 312.66, 312.64 (emphasis supplied).

88. 45 C.F.R. § 46.103(b)(5)(i).

89. *See* Serna v. Roche Labs., 684 P.2d 1187 (N.M. Ct. App. 1984).

90. *See* U.S. Food & Drug Admin., Adverse Event Reporting—Improving Human Subject Protection (Apr. 2007).

91. 21 C.F.R. § 312.32(c)(1)(A), (c)(2) (emphasis supplied). The rules further define the terms "serious adverse drug experience," "unexpected adverse drug experience," "associated with the use of the drug," and "life-threatening adverse drug experience." *Id.* § 312.32(a).

92. *See id.* § 312.32(c)(1)(B).

93. *See id.* § 312.33(b)(1)–(4).

94. *Id.* § 812.150(a)(1).

95. *Id.* § 812.3(s).

96. *See id.* § 812.46(b)(2).

97. *See, e.g.*, Notice of Initiation of Disqualification Proceedings and Opportunity to Explain, to Stephen D. Rossner, MD, (May 10, 2006), *available at* http://www.fda.gov/foi/nidpoe/default.html/.

98. *See* 45 C.F.R. § 46.110(b).

99. *See* 63 Fed. Reg. 60,364 (Nov. 9, 1998) (this notice sets out examples of types of specimens that may be collected under expedited review and conditions for collecting blood to qualify for expedited review).

100. *Id.*; Office for Human Research Protections, Categories of Research That May Be Reviewed by the Institutional Review Board (IRB) through an Expedited Review Procedure, *available at* http://www.dhhs.gov/ohrp/humansubjects/guidance/expedited98.htm.

101. *See* 21 C.F.R. § 312.36 (emergency use), §§ 312.34, 812.34 (treatment use); 21 C.F.R. § 56.104 (emergency use). *See* U.S. Food & Drug Admin., Guidance Documents (Medical Devices) (1985), *available at* http://www.fda.gov/cdrh/manual/unappr.html.

102. In many instances, FDA will require the sponsor to submit an application for a single-subject IND.

103. 21 C.F.R. § 312.34(b); *see* 21 C.F.R. § 812.36(b).

104. 495 F.3d 695 (D.C. Cir. 2007) (en banc).

105. Abney v. Amgen Inc., 443 F.3d 540 (6th Cir. 2006). *See also* Pollack v. Rosalind Franklin Univ., No. 04-CH-12098 (Ill. Cir. Ct., filed July 28, 2004) (suit to halt termination of a breast cancer vaccine, settled after the university agreed to make all records and remaining vaccine available to the fifty-two plaintiffs, and to finance the patients' pursuit of FDA approval of the vaccine).

106. It is unfortunate that these five groups are treated as members of the same set—the vulnerable—just as it is unfortunate that attorneys are classified in the same federal employment category as those with mental disabilities. *See* 71 Fed. Reg. 42,241 (July 26, 2006); U.S. Office of Personnel Management, Attorneys in the Federal Service, http://www.usajobs.opm.gov/ei24.asp (last visited Sept. 14, 2009).

107. Currently, there are special rules when children, pregnant women, and prisoners are asked to be subjects. There are no special rules, per se, for those who are mentally impaired. On September 5, 2007, HHS published a notice seeking comments on whether additional tailored regulations should be adopted for this vulnerable population. *See* 72 Fed. Reg. 50,966 (Sept. 5, 2007).

108. *See* Young v. Weaver, 883 So. 2d 234 (Ala. Civ. App. 2003).

109. Those rules can be especially complex when the subject is a resident of a state with a relatively high age of majority (e.g., twenty-one years), but the research is being conducted in a neighboring state with a lower age of majority (e.g., eighteen years). Which state's law governs—the state of residence or the state where the research is being conducted? You should adopt procedures so that you never have to confront what lawyers call a "choice of law" question.

110. *See* 45 C.F.R. § 46.401(b).

111. "A risk is minimal where the probability and magnitude of harm or discomfort anticipated in the proposed research are not greater, in and of themselves, than those ordinarily encountered in daily life or during the performance of routine physical or psychological examinations or tests. . . . For example, the risk of drawing a small amount of blood from a healthy individual for research purposes is no greater than the risk of doing so as part of routine physical examination." ROBIN LEVIN PENSLAR, OHRP-INSTITUTIONAL REVIEW BOARD GUIDEBOOK ch. 6 (1993), *available at* http://www.hhs.gov/ohrp/irb/irb_chapter6.htm#g4.

112. 21 C.F.R. § 50.55(c)(2).

113. 42 U.S.C. § 289a-2(a)(1).

114. *See* Otis W. Brawley, *The Study of Accrual to Clinical Trials: Can We Learn from Studying Who Enters Our Studies*? 22 J. CLIN. ONCOLOGY 2039 (2004).

115. The history of regulation of fetal tissue research traces back to shortly after *Roe v. Wade*, when Congress imposed a moratorium on federal funding until the National Commission on the Protection of Human Subjects of Biomedical and Behavioral Research could address the issues. *See* National Research Act of 1974, Pub. L. No. 93-348, § 202(b), 88 Stat. 342, 348 (1974). The Commission and subsequently the Secretary addressed the matter by recommending and then issuing, respectively, the human subjects regulation. *See* 45 C.F.R. pt. 46.

116. *See* Robert Charrow, *Fetal Tissue Transplantation Ban: Illegal Political Solution to Moral Problem*, 3 J. NIH RES. 20 (1991).

117. The Balanced Budget Downpayment Act, Pub. L. No. 104-99, § 128, 110 Stat. 26, 34 (1996) (appropriating funds for the Department of Health and Human Services for fiscal year 1996). The Dickey-Wicker rider has appeared in every HHS appropriations act since 1996, including the one signed by President Obama on March 11, 2009. *See* Omnibus Appropriations Act, Pub. L. No. 111-8, 123 Stat. 524 (2009).

118. *See* Robert E. Wanerman, *Congressional Activity on hESC Research*, 8 MED. RES. L. & POL'Y REP. 253 (2009).

119. See http://www.pbs.org/newshour/bb/health/july-dec01/bushspeech_8-9.html; Exec. Order No. 13,435 (June 20, 2007), 72 Fed. Reg. 34,591 (June 22, 2007). It is likely that Bush's action was an improper attempt to make regulations through executive orders and that Obama's executive order was proper, but only to the extent that it repealed the Bush order. *See* Robert P. Charrow, *The Propriety of Executive Branch Rulemaking Through Executive Orders*, 8 MED. RES. L. & POL'Y REP. 253 (2009).

120. Exec. Order No. 13,505, § 2, 74 Fed. Reg. 10,667 (Mar. 9, 2009).

121. *Id.* § 3.

122. *See* 74 Fed. Reg. 3217 (July 7, 2009).

123. *See* OFFICE FOR HUMAN RESEARCH PROTECTIONS, GUIDANCE FOR INVESTIGATORS AND INSTITUTIONAL REVIEW BOARDS REGARDING RESEARCH INVOLVING HUMAN EMBRYONIC STEM CELLS, GERM CELLS AND STEM CELL-DERIVED TEST ARTICLES 3 (Mar. 19, 2002), *available at* http://www.hhs.gov/ohrp/humansubjects/guidance/stemcell.pdf.

124. 45 C.F.R. § 46.113.

125. *See Pfizer, Kano State Reach Settlement of Trovan Cases*, BUSINESS WIRE, July 30, 2009, http://mediaroom.pfizer.com/portal/site/pfizer/index.jsp?ndmViewId=news_view&newsId=20090730005769&newsLang=en.

126. *Litigation Pfizer, Nigeria Settlement in Trovan Trial Stalled by Connecticut Injunction Hearing*, 8 MED. RES. L. & POL'Y REP. 362 (2009).

127. For an excellent discussion of the various issues that one confronts when conducting research abroad, see William F. Ferreira, *Conducting Research and Sponsored Programs Overseas Part I: Setting Up Operations*, 7 MED. RES. L. & POL'Y REP. 441 (2008), and Maria A. Garner, *Conducting Research and Sponsored Programs Overseas Part II: Laws and Policies Affecting Research Projects*, 7 MED. RES. L. & POL'Y REP. 487 (2008).

128. Samuel Warren & Louis Brandeis, *The Right to Privacy*, 4 HARV. L. REV. 193 (1890).

129. Olmstead v. United States, 277 U.S. 438, 478 (1928) (Brandeis, J., dissenting).

130. *See* Time, Inc. v. Hill, 385 U.S. 374 (1967).

131. Pub. L. No. 104-191, 110 Stat. 1936 (1996). The privacy provisions of HIPAA are found at section 262, 110 Stat. 2021–2031.

132. *See* 45 C.F.R. pts. 160 & 164. *See* 65 Fed. Reg. 82,462 (Dec. 28, 2000), as amended at 68 Fed. Reg. 8334 (Feb. 20, 2003).

133. As a result of recent amendments, HIPAA also requires the Secretary to regulate directly those who perform services for "covered entities," such as software consultants and others who might see or come into contact with patient information. *See* American Recovery and Reinvestment Act of 2009, tit. XIII, Pub. L. No. 111-5, 123 Stat. 115, 258.

134. *HIPAA Privacy Rule Mandates Cut Participation in Heart Attack Research by Two-Thirds*, 4 MED. RES. L. & POL'Y REP. 440 (2005).

135. The HIPAA Rules are a prime example of what can happen when a regulation is aimed at a nonexistent problem and is drafted by individuals who have no experience in the sector that they are seeking to regulate. There is no example in any of the preambles to any of the various rulemakings of any researcher leaking or disclosing private healthcare information. A regulation in search of a problem is dangerous.

136. 45 C.F.R. § 164.508(b)(5).

137. *See id.* § 164.514(b)(2)(i).

138. *Id.* § 164.514(b)(1).

139. *See id.* § 164.514(e).

140. Office of Biotechnology Activities, About Recombinant DNA Advisory Committee (RAC), http://oba.od.nih.gov/rdna_rac/rac_about.html (last visited on Sept. 14, 2009).

141. Various localities have adopted ordinances regulating recombinant DNA research within their town or city limits. *See* Cambridge, Mass., Rev. Ord. ch. 11, art. II, § 11-7 (1977); Princeton, N.J., Rev. Ord. ch. 26A, §§ 1–13 (1978); Amherst, Mass., Rev. Ord. ch. 22, §§ 22-1, 22-2 (1981); Berkeley, Cal., Ord. 5010-N.S. (Oct. 21, 1977); Emeryville, Cal., Resolution 77-39 (Apr. 1977).

142. Recombinant DNA Advisory Committee, Minutes of Meeting of December 3–5, 2007, at 6.

CHAPTER 5

Financial Conflicts of Interest: The Legacy of Shoeless Joe Jackson

Professional codes and federal laws designed to monitor, regulate, or prohibit conflicts of interest are not new. In one way or another, they affect virtually every calling. Undertakers, architects, engineers, and lawyers have codes of professional responsibility. The rules governing state and federal employees, irrespective of station, are particularly well developed and deeply rooted in our nation's consciousness, tracing their origin to the First Congress. On September 2, 1789, President Washington signed into law "an Act to establish the Treasury Department." It was the twelfth law enacted by the United States, and one section prohibited any officer in the new department from directly or indirectly carrying on any trade or business.[1] The new Congress was so concerned that matters of personal wealth not affect matters of public policy that the law rewarded anyone who unearthed a conflict. Over the years, this twelfth law, which is actually our first whistleblower statute, has mushroomed, and now actions of federal employees are governed by a complex web of criminal conflict of interest statutes and administrative regulations.[2] Even the Congress has gotten into the act; each time there is a scandal, one chamber or the other ratchets up its conflict of interest rules as a way of demonstrating concern. In 2007, for instance, the House of Representatives and Senate modified their gift rules constricting some of the loopholes that allowed congressional members and their staffs to accept all sorts of nifty gifts, like lunches, tickets to sporting events, plane rides, and the like.[3] The loopholes still exist; they're just tighter to fit through than in the past.

Frequently, codes are adopted in response to some scandal or perceived need. For instance, the Code of Judicial Conduct was a byproduct of the famous 1919 Chicago Black Sox scandal. After five White Sox players, including Shoeless Joe Jackson, had been accused of accepting money to throw the 1919 World Series, major league baseball hired Judge Kenesaw Mountain Landis as its first commissioner to investigate the matter and then to oversee the two major leagues. Landis, however, was a sitting federal judge, and he had no intention of resigning; he indicated that he could and would hold both positions. Landis's actions led the American Bar Association to adopt the Code of Judicial Conduct, one provision of which expressly barred a sitting judge from holding other employment; second jobs created too much opportunity for hidden conflicts to arise. Soon after the code was adopted, Landis resigned his judgeship.

There is no universally accepted definition of the term "conflict of interest." Usually, the code, rule, or law regulating the specific conflict at issue will provide either a general definition or, in some cases, an operational definition, one that is keyed to specific facts. One definition, however, that is useful for analytical, although not necessarily legal, purposes is as follows:

> A conflict of interest exists if a personal interest or duty to a third party interferes with or is inconsistent with an existing duty to another.

This definition encompasses a broad range of potential conflicts, from effete intellectual conflicts to base financial ones. While philosophers and ethicists devote considerable attention to intellectual conflicts, the law is more pragmatic and tends to focus almost exclusively on financial conflicts. Most financial conflict of interest laws consist of two elements: disclosure and prohibition. For example, the Ethics in Government Act of 1978,[4] enacted in response to the Watergate scandal, requires members of Congress, judges, certain high-ranking officials within the executive branch, and certain candidates to disclose annually on (unreadable) forms their assets, earnings, and liabilities. In addition, other provisions of federal law prohibit a federal employee from taking action in any matter in which the employee or any member of the employee's family has a financial interest.

The research community has always been concerned that those who hold the purse strings, whether private or public, could affect research agendas. More recently, though, the relationship between science and money has given rise to a more fundamental question—can money influence, either subjectively or otherwise, the way in which a scientist designs or performs an experiment or analyzes its results? These concerns are driven by changing tides. First, in 1962, federal law changed to require drug companies to prove the efficacy of their products as a condition of approval by FDA.[5] Proof of efficacy required large clinical trials, and what better place to perform them than at large academic medical centers? As clinical trial money made its way into university medical centers, some were concerned that researchers might believe—correctly or incorrectly—that continued access to clinical trials might be conditioned on positive clinical results. And second, starting in the early 1980s, federal law changed to permit scientists and their universities to retain the patent rights on inventions developed with federal funding.[6] In some instances, scientists would set up companies to further develop the inventions that they had conceived while at the university. Suddenly, scientists had an opportunity to own significant equity in new companies. The fact that faculty could see the influx of private funding (drug company money for clinical testing and venture capital for applied research) left many with an uneasy feeling and stimulated others to wonder what effect this would have on basic research. After all, the mission of a university and the mission of a corporation are inherently different. The university is supposed to provide a protected environment where inquiring minds are free to ponder the workings of nature. The corporation, on the other hand, is driven by the more mundane quest for maximum return on investment; it provides protection for none. Can corporations "purchase" scientific results, much as they can purchase buildings or machinery? If so, what risk to the ethical reputation and nurturing environment of the university would private funding by corporations pose? Are scientists immune from the normal human drive for financial well-being? Would "purchased research" undermine the trust that the public has in scientists? These questions were valid when they were first asked more than twenty-five years ago, and they remain so. Some early evidence suggested that corporations could seek to improperly influence science.

The case of Betty J. Dong, a pharmacologist at the University of California at San Francisco (UCSF), gave credence to those who feared the impact of corporate money on university research. Dong had published a limited study that suggested that a drug, Synthroid, manufactured by Knoll Pharmaceutical Company of Mt. Olive, New Jersey, was superior to its competitors for the treatment of hypothyroidism. When Knoll learned of the study, it approached Dong and offered to pay the university $250,000 to conduct a clinical trial to replicate her earlier, more limited study. UCSF accepted the funding, and Dong conducted the trial. The only problem was that Dong's results showed that Synthroid was no better than its competitors, even though it cost significantly more. Dong wrote up her results and, in 1990, submitted her paper to the *Journal of the American Medical Association*. However, Knoll, claiming that its contract with UCSF gave it the right to approve all publications of data that it paid to collect, sought to stop publication and successfully did so for years. When Dong's results came to the attention of the Food and Drug Administration (FDA), Knoll sought to discredit the study in meetings with FDA.[7] Dong's paper was finally published in 1997 in *JAMA*.[8]

Most corporations are not as crass as Knoll had been in this case, and most recognize that good science is in everyone's best interest. Still, the early lessons linger, and concerns are now being magnified as many researchers are finding it necessary, in light of reductions in federal research funding, to look to private profit-making corporations to fund their basic research.

It is against this backdrop that this chapter focuses on the three sorts of financial conflict of interest laws that affect the scientific research community. The first section examines personal financial conflicts pertaining to those who are conducting federally funded or FDA-regulated research. In that regard, we examine the Public Health Service (PHS) and FDA rules, the National Science Foundation (NSF) policy, and the disclosures that prudence dictates ought to be included in an informed consent. The second section explores institutional financial conflicts likely to arise when a university or nonprofit research institute enters into a long-term agreement with a profit-making entity, under which the entity provides research funding in exchange for an option to license a significant portion of the inventions in a given field. The third section reviews the complex web of federal criminal statutes and regulations that govern the activities of federal employees, including those scientists who serve on advisory committees and study sections. The chapter then ends with some problems.

CONFLICT OF INTEREST LAWS GOVERNING THOSE RECEIVING FEDERAL RESEARCH FUNDING OR CONDUCTING CLINICAL TRIALS REGULATED BY FDA

The conflict of interest laws governing scientific research are unique: They were not issued in response to any documented problem. Instead, the move toward regulating the financial interests of researchers was largely an outgrowth of two independent events. First, there was concern that various laws such the Bayh-Dole Act of 1980[9] and the Federal Technology Transfer Act of 1986[10] provided scientists with new opportunities to financially exploit and profit from their research: Bayh-Dole gave universities and their faculty the right to patent inventions developed with federal grants, and the Technology Transfer Act gave federal employees a share of any royalties that the government might realize by licensing out their inventions. As a result of these laws, scientists, whether employed by the federal government or by state or private universities, were suddenly encouraged to do things, such as patent inventions and earn royalties from those inventions, which previously had been either discouraged or unlawful. And second, at about the same time, there was the general push to regulate science by treating researchers no differently than bankers, lawyers, or politicians.[11]

In this section, we first examine in detail the conflict rules and policies governing PHS- and NSF-funded research. An overview of the way these directives operate is as follows.[12] First, before applying for NSF funding or spending National Institutes of Health (NIH) funding, the grantee (e.g., a university) must establish written procedures for collecting and reviewing the financial disclosures of its investigators (e.g., a listing of "significant financial interests" that "would reasonably appear to be affected by the research"). Second, the institution must appoint a "designated official" to review each financial disclosure to determine if a conflict of interest exists as a result of the significant financial interest (e.g., whether the interest could significantly affect the research). Third, if a conflict exists, the official must determine what action should be taken to manage, reduce, or eliminate the conflict (e.g., public disclosure of the interest, monitoring of the research, divestiture of the interest). Fourth, before spending any of the funds from a new grant, the institution must report to the funding agency the existence of the conflict and certify that the conflict has been managed. The institution, however, need not report the exact nature of the conflict or precisely how it was managed, although it must retain the disclosure and documentation for three years. And fifth, as a condition of receiving the funding, the institution must certify that it has complied with the PHS or NSF directives, as the case may be. The directives leave it up to the institution to decide which investigators are obligated to file a financial disclosure, which significant financial interests are reportable to the institution's officials, and which of those interests create a conflict. Furthermore, the directives also leave it to the institution to decide the nature of the action that ought to be taken to manage a conflict.

In the second subsection, we examine the FDA rules that govern clinical research regulated by FDA.[13] Under those rules, an applicant seeking approval to market a new drug, device, or biologic (e.g., vaccine, blood product) must either certify to FDA that no researcher who conducted clinical trials that are relied upon by the applicant has any significant financial conflict, or disclose each investigators' significant financial interests, as defined in the rules.

Research Funded by PHS or NSF

HISTORICAL PERSPECTIVE

In June 1989, NIH sponsored an open forum to discuss various issues associated with regulating conflicts of interest in PHS-funded research. Following the forum, on September 15, 1989, the NIH Office of Extramural Affairs published for public comment in the *NIH Guide for Grants and Contracts* a document entitled "Proposed Guidelines for Policies on Conflict of Interest."[14]

The proposed guidelines were poorly conceived and demonstrated the perils of attempting to regulate hypothetical problems. Among other things, the proposed guidelines would have prohibited any researcher receiving PHS funding from owning any stock or options in "any company that would be affected by the outcome of the research or that produces a product or equipment being evaluated in the research project." In essence, the proposal would have banned certain funding that Congress had expressly authorized. For instance, pharmaceutical companies would effectively have been precluded from seeking PHS funding because employees of most such companies own stock or stock options in their employer. The proposal generated a firestorm: Universities, private research institutions, and commercial entities bombarded NIH and the Office of the Secretary with adverse comments. Many questioned the legality of imposing such sweeping requirements on the research community without going through notice-and-comment rulemaking.[15] Others questioned the wisdom of or need for such draconian requirements, especially given the fact that NIH "had difficulty conjuring up more than a handful of cases in which the objectivity of a researcher was arguably threatened by pecuniary interests."[16] Ultimately, the Secretary of HHS ordered the proposal withdrawn.

Congress was not to be dissuaded by the research community's opposition to conflict of interest laws. After all, while members of Congress might not understand the nuances of scientific research, they were experts on financial improprieties. In 1993, as part of the National Institutes of Health Revitalization Act of 1993,[17] Congress enacted a provision requiring the Secretary of HHS to develop regulations to manage financial interests "that will, or may be reasonably expected to, create a bias in favor of obtaining results in such project that are consistent with such financial interest."[18] However, unlike NIH's abortive attempt to address conflicts in all research, the Revitalization Act addressed only "clinical research whose purpose is to evaluate the safety or effectiveness of a drug, medical device, or treatment. . . ."[19]

PHS RULE AND NSF POLICY

Despite the act's modest reach, the HHS regulation that ultimately emerged was broad and covered all research, whether clinical or other. The PHS rule is divided into two parallel sections—one governing projects funded by grants or cooperative agreements, excluding Small Business Innovation Research (SBIR) Program grants, Phase I,[20] and the other governing projects funded under procurement contracts.[21] On the same day that HHS issued its rule, NSF revised its policy to be "more consistent with the provisions of the final" HHS rule.[22] The NSF policy and PHS rule are virtually identical. They both require that recipients of research funding have in place a system for collecting and evaluating financial information and for managing potential conflicts of interest with respect to research funded by that agency. The main difference between the two, as noted below, relates to the nature of the institutions subject to the rule or policy, as the case may be, and whether reporting requirements extend to a subgrantee.[23] Given the identity between the NSF policy and the PHS rule, the discussion below is keyed to the PHS rule.[24]

Scope of Rule: Who Is Covered by the PHS Rule and NSF Policy?

The PHS rule covers all biomedical and social science research funded under the Public Health Service Act. The size or nature of the grantee is immaterial, except that the rule does not apply to applications for phase I, Small Business Innovation Research Program grants, or Small Business Technology Transfer Program grants.[25] The NSF rule covers all research funded by NSF, but excludes grantees with fifty or fewer employees. Institutions receiving PHS or NSF research funding must have in place a system for reviewing potential financial conflicts of interest of all "investigators" (and their families, i.e., spouse and dependent children). NSF will not process a grant application unless the institution has filed an assurance with NSF indicating that it has such a system in place. In contrast, PHS will review an application without an assurance, but will not fund the grant until an assurance has been filed.[26] In addition and as discussed later, most Institutional Review Boards (IRBs) will independently review any financial conflicts and may require the researcher to disclose his or her relevant financial interests to each research subject as part of the informed consent form. (See chart 2 for how to navigate the conflict maze.)

For the purpose of the conflict rules, an investigator is the principal investigator (or co-principal investigator, in the case of NSF) or any other person who is "responsible for the design, conduct, or reporting" of such research.[27] The rule makes no attempt to define the degree of responsibility necessary to trigger a financial disclosure. In the preamble, PHS notes that the

> degree to which individuals are responsible for the design, conduct, or reporting of the PHS-funded research will vary. In some circumstances students, technical personnel and administrators may not be "responsible," but in other circumstances, they may be. . . . [W]e believe the institutions are in the best position to determine who is responsible. . . . [28]

Given the rule's uncertainty, most large institutions with significant PHS or NSF funding have found it legally safer and administratively simpler to require all those involved in federally funded research to file a financial disclosure with the university rather than attempting to make a case-by-case determination. It should be noted that the rule applies to everyone (and their immediate families, i.e., spouse and dependent children) who is "responsible" for some aspect of the research, even if they receive no funding under the grant.

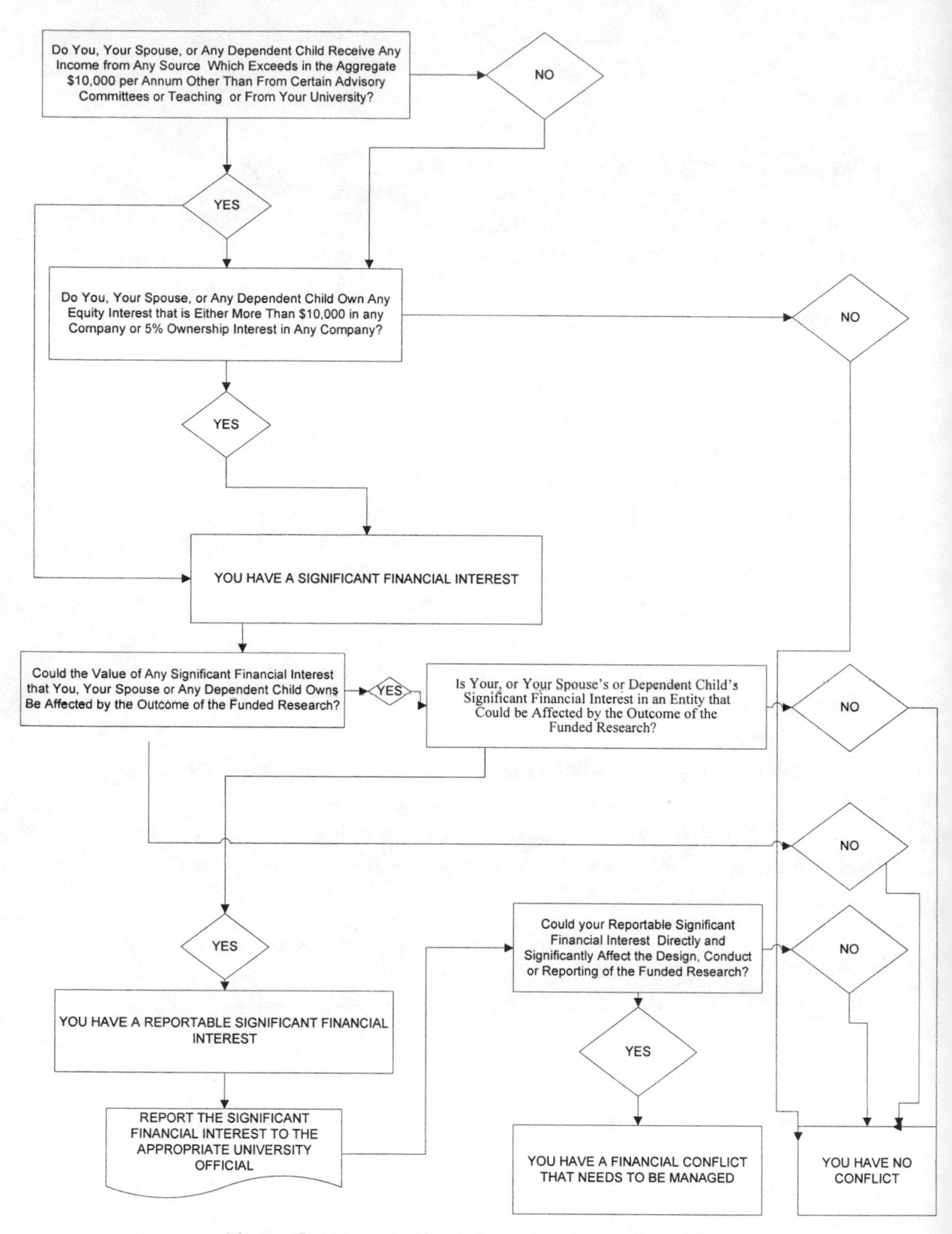

CHART 2. The Conflict Maze: Deciding Whether You Have a Financial Conflict under the PHS Rules

Conflict of Interest: When Does a Financial Conflict Exist?

To ascertain whether a financial interest creates a conflict, an institution must ask and answer the following three questions:

1. Is the financial interest a "significant financial interest," as that term is defined in the regulation?
2. If so, is the significant financial interest reportable to the institution?
3. If so, does the reportable significant financial interest create a conflict of interest?

The rule requires a researcher to report to his or her institution only financial holdings that are both "significant" and related to the research involved. Such interests are called "reportable significant financial interests." For example, $20,000 earned by a clinical researcher moonlighting in an emergency room is a "significant financial interest," but it would likely not need to be reported because the extra money that the researcher earns by practicing medicine would not be affected by his or her research, nor would the research have any financial effect on the practice group's income or fiscal viability. Chart 2 contains a "conflict of interest" decision tree.

Significant Financial Interest. The linchpin of both the PHS rule and NSF policy is the "significant financial interest." The rule defines the term as anything of "monetary value" including all earnings (e.g., salary, consulting fees, honoraria), equity interests (e.g., stocks, stock options), and intellectual property (e.g., patents, copyrights, and the royalties from them). However, the rule excludes from this broad definition any interest that falls into one of six exempt categories:

1. Remuneration of any sort from the applicant institution
2. Ownership interest in an entity that is an applicant under the SBIR program
3. Income from seminars, teaching, lectures sponsored by public (i.e., governmental) or nonprofit entities
4. Income from service on advisory committees or review panels for public or nonprofit entities
5. An equity interest that when aggregated with the interests held by family members (i) does not exceed $10,000 and (ii) does not represent more than 5 percent ownership interest in any single entity
6. Salaries, royalties, and other payment that when aggregated for the investigator and his or her immediate family are not expected to exceed $10,000 over the next twelve months[29]

The definition of "significant financial interest" is not a model of clarity and fails to address many issues. For example, the rule probably requires one to aggregate all equity interests in a single entity. Thus, a $15,000 interest in a company would be considered significant, even if that interest consisted of $5,000 worth of class A common stock, $5,000 worth of class B common stock, and $5,000 worth of preferred stock. It is unclear, though, how one would treat a $100 corporate bond. In theory, one could argue that the bond constitutes a significant financial interest since it is something of "monetary value" and does not fall within any of the exemptions. Because it is not an "equity interest," that exemption would not apply. Therefore, a $100 corporate bond would be a significant financial interest, whereas $5,000 worth of stock in the same company would not be.

Let's compare a husband and wife. Suppose the wife works for a federal agency that has nothing to with science (e.g., Federal Deposit Insurance Corporation). The husband is a professor at George Masonite University where he developed a new drug, "scritoase," an enzyme that everyone believes can treat that unpleasant skin disease, the "dry scritos." In keeping with Masonite's patent policy, the husband, Richard Wyethe, assigns his patent rights to the university, which in turn licenses the patent on the drug to American Chemical and Nutrition Enterprise (ACNE), a large international pharmaceutical company. The university is paid a signing fee of $60,000 and an annual maintenance fee of $100,000; in addition, the university is entitled to receive a 6 percent royalty on sales once the drug is approved by FDA. Under Masonite's patent policy, Wyethe receives from Masonite 25 percent of whatever it receives from ACNE, i.e., $15,000 at the time Wyethe signed the license agreement, and $25,000 per year in maintenance fees. ACNE is about to conduct a clinical trial of the drug and selects Masonite as one of the trial sites and Wyethe as the principal investigator (PI) at that site. NIH intends to fund a portion of the trial at Masonite. Under the PHS rule, Wyethe's wife has a significant financial interest because of her government salary. It is a payment that is greater than $10,000 per year and does not fall into one of the exemption categories. In contrast, Wyethe's royalty income from the drug that he is about to test on humans is not a significant financial interest because that income fits within one of the exceptions, namely it is "remuneration of any sort from the applicant institution" (i.e., Masonite).

It is also unclear when aggregating income across the family whether one aggregates by source or across sources. For example, suppose that the investigator's spouse earns $8,000 working part-time in a bookstore and the investigator's dependent child earns $2,100 working during the summer selling shoes. Does one add across sources to yield a financial interest of $10,100, or does one add across the family only when the income source for each family member is the same (e.g., they both received wages from the bookstore)?

The definition of "significant financial interest" becomes more complicated when the researcher's equity interest is an option to purchase stock.[30] How does a researcher determine whether his or her options are worth more than $10,000? This is no easy undertaking and may require the researcher (or, more likely, someone in the economics department of the researcher's institution) to apply the Black-Scholes option pricing formula.[31]

Most economists would argue that using discrete dollar thresholds as a way of separating the significant from the insignificant makes no sense. This is especially so when the purpose of the threshold is to separate those interests that are so small that they would be unlikely to lead one to fudge data or take other liberties with an experiment. However, an ownership interest of $100,000 is likely to have no impact on the thinking of a scientist with $100 million, whereas a $5,000 interest could conceivably influence an impoverished graduate student. Although a more "personalized" approach would make more sense to an economist, it would be difficult to implement. Universities would be unable to determine whether an interest was significant without reviewing everyone's entire stock portfolio and other assets. The administrative burden and the intrusion would be too great. A one size (e.g., $10,000) fits all approach is much simpler.

Both NSF and HHS have taken the position that ownership of diversified mutual funds likely would not create a reportable significant financial interest. Both agencies have indicated that the regulations that govern federal employees' ownership of mutual funds should provide "guidance on how [grantees] might wish to treat [such interests]," although in a more recent issuance, NIH seems to suggest that diversified holdings may be too remote to create a conflict, but that the researcher should "refer to your Institution's policy."[32] However, both agencies treat a $15,000 stake in a highly diversified mutual fund as a significant financial interest. Whether it is reportable is a separate question. Thus, your retirement interest in TIAA-CREF, for example, would likely constitute a significant financial interest, but probably not be reportable.

It is likely that many universities will resolve any ambiguities in favor of requiring their researchers to treat a given financial interest as if it were a significant one. It should be remembered that nothing in the rule precludes a university from having more stringent reporting requirements than those imposed by federal law.

Reportable Significant Financial Interests. Under the rule, not all significant financial interests are required to be reported to the institution and not all reportable significant financial interests will create a conflict. Only those significant financial interests held by the investigator (or his or her immediate family)

> (i) [t]hat would reasonably appear to be affected by the research for which the PHS funding is sought; *and*
> (ii) [i]n entities whose financial interests would reasonably appear to be affected by the research

need to be reported.[33] Each institution is required, under the rule, to appoint an individual to receive and to review financial disclosures by researchers and others.

Literally read, the rule is remarkably limited and fails to cover interests that the drafters probably intended to cover and the organic legislation required the rule to cover. As written, the reporting requirement applies only if both the significant financial interest can be affected by the research being conducted by the investigator and the entity can be affected by the research. It would appear, however, that the agencies inadvertently used the conjunction "and" in the rule when they likely intended to use the disjunctive "or."[34] Such an error is significant because it in theory allows researchers not to report significant financial interests that could affect their research. Thus, for example, under a literal reading of the rule, a researcher who receives an honorarium of $5 million from a drug company at the start of a nonclinical efficacy study that he or she is performing would not have to report that honorarium to the university because the "interest" (i.e., the $5 million honorarium), having been received in cash, cannot, by definition, be affected by the research. The researcher keeps the $5 million whether the research generates good results or bad results. This is so even though the research could have a significant effect on the value of the company's stock and even though it would appear to most disinterested observers that a $5 million honorarium could affect the way the research was performed or reported.

As a practical matter, however, this apparent drafting error is likely to have little impact on most researchers. This is because many universities require investigators to report all of their significant financial interests whether the interest is affected by their research or not. In fact, some universities require faculty to divulge all of their financial interests whether significant or otherwise. For example, Stanford University School of Medicine, which traditionally has been among the most "entrepreneurially friendly" institutions in the country, nevertheless requires that "all personal financial interests related to Stanford activities must be reported, regardless of dollar amount."[35] On the other hand, Harvard School of Medicine's policy, which is considered to be among the most restrictive in the country, is so poorly written that it is difficult to ascertain, at least from the Web, what is reportable and what is not.[36] To further complicate matters, Harvard has two sets of monetary thresholds, one for NIH- and NSF-funded research that corresponds to the thresholds in the PHS rule and NSF policy, and another set of higher thresholds for non-NIH, non-NSF-funded research. Yale Medical School requires faculty to disclose their ownership interests of any size in "related entities," which is defined to be any non-Yale, nongovernmental entity that "conducts activities that could relate to" the faculty member's research.[37] The one thing that becomes quickly apparent is that every school imposes different reporting requirements on their faculty, from a "report all" to a "report only that which is required by federal law." The variation across universities in the way financial interests are reported, evaluated, and managed is well documented in the literature.[38]

An additional complication is that many peer-reviewed journals require authors to disclose their conflicts or to certify that they have none. The International Committee of Medical Journal Editors (ICMJE) is one of only a few organizations recognizing that lucre is not the only source of conflicts and "conflicts can occur for other reasons, such as personal relationships, academic competition, and intellectual passion."[39] However, it is extraordinarily difficult, bordering on the impossible, to regulate conflicts attributable to intellectual passion or academic competition.

Managing, Reducing, or Eliminating Reportable Significant Financial Interests that Create a Conflict of Interest. Years ago, Harry Houdini wagered that he could break out of a cell in a New York City jail; the city was quick to take him up on his bet. Houdini was stripped and searched just to make sure that he had not secreted any tools or other implements. After much fanfare, the jailers and others shook his hand and then he was locked in the cell and they left. Within hours, Houdini had escaped. It seems that the last person to shake Houdini's hand palmed him the key to the cell. Like Houdini's trick, regulation of conflicts depends more on the last step, namely "managing conflict," than it does on all the front-end fanfare, and it is in the managing of conflicts where we see the greatest variation from university to university.

Under the PHS rule, the designated official at each institution must review the completed financial disclosure forms and determine whether a given reportable significant financial interest creates a conflict or potential conflict of interest. A conflict is deemed to exist when the designated official "reasonably determines that the Significant Financial Interest *could directly and significantly* affect the design, conduct, or reporting of the PHS [or NSF] funded research."[40] Once this determination is made, the official is required to "manage, reduce, or eliminate" that conflict and to report the existence of the conflict (although not its details) to the funding agency.[41] Examples of conditions or restrictions that might be imposed to manage, reduce, or eliminate conflicts of interest include, but are not limited to, the following:

1. Public disclosure of significant financial interests
2. Monitoring of research by independent reviewers
3. Modification of the research plan
4. Disqualification from participation in all or a portion of the research funded by PHS
5. Divestiture of significant financial interests
6. Severance of relationships that create actual or potential conflicts

Suppose that Adam Smith, a senior member of the faculty at a medical school's Institute for Genetic Research, Engineering, Evaluation, and Development (GREED) owns $50,000 of stock in OncoCure, a small drug company. He acquired the stock over the years in exchange for providing consulting services to the cash-strapped company. Recently, owing to an extremely successful phase 2 clinical trial, OncoCure has managed to raise significant capital and is preparing to begin a pivotal phase 3 clinical trial of the drug. Smith is one of many researchers who have been asked by the company to participate as investigators in the phase 3 multicenter clinical trial of that drug, and Smith's federal funding agency has agreed to permit him to use some federal funds for aspects of the study. If the clinical trial goes well, the likelihood of FDA approval increases, and the value of Smith's stock should dramatically increase as well. Smith dutifully reports his interest in OncoCure as a reportable significant financial interest. The vice dean of the medical school, who also happens to be the conflict officer, isn't sure whether a conflict exists and, if so, how to manage it. She could, of course, ban Smith from participating as an investigator in the FDA-approved trial. However, that might not be in the best interests of the patients participating in the trial because Smith happens to know more about the new drug than just about anyone in the country. She sits down with Smith and tries to determine the aspects of the research that would involve Smith's subjective assessment. They conclude that although a clinical trial is really "cookie cutter" science, many aspects involve medical judgment. For example, while many of the inclusion and exclusion criteria are black and white (e.g., age, various blood counts), many are not. However, since the experiment is blinded, involves two arms (the new drug and the standard treatment), with randomization to occur after a patient is accrued into the study, the likelihood that Smith's financial interest will have any impact on the initial aspect of the study is slight. They go through each aspect of the study and ultimately decide that there is a slight probability that Smith could really influence the study's outcome. However, the test, at least with respect to whether a conflict exists, is whether the interest could directly and significantly affect some aspect of the study. The word "could" implies that there is a possibility, albeit small. Accordingly, the dean reluctantly labels Smith's interest as a conflicting financial interest and "manages the conflict" by requiring that all end points and all adverse events be reviewed by another senior faculty member who will also act as an investigator in the study. Once the dean labels the interest as a conflict, she must advise the funding agency of the general nature of the conflict and the steps that she has taken to mitigate its impact on the study.

In addition to managing the conflict, the administration and the IRB have to decide whether the informed consent form provided to each patient should include some statement indicating that one of the investigators, Dr. Adam Smith, has a significant financial interest in OncoCure. They decide that this issue is best left to the IRB.

Smith, though, indicates that he has no objection to disclosing his ownership to putative patients in the clinical trial, but decides to discuss the matter with GREED's general counsel, Harley Whipple. Whipple explains that many years earlier, the California Supreme Court, in Moore v. The Board of Regents of the Univ. of Cal., held that researchers had an obligation to inform subjects of their financial interest in any aspect of the clinical trial.[42] Moore was a patient at UCLA Medical Center suffering from hairy-cell leukemia. The physicians treating Moore recommended that his spleen be removed as a way of slowing the progress of the disease. Moore agreed, and over the next seven years, the physicians took blood and tissue from Moore. Moore's cells were then used by the UCLA physicians to develop a cell line, which they patented and commercialized. Moore sued the physicians and UCLA, alleging that the defendants had stolen his cells. Although the court rejected Moore's interest in the patent, it went on to hold that "(1) a physician must disclose personal interests unrelated to the patient's health, whether research or economic, that may affect the physician's professional judgment; and (2) a physician's failure to disclose such interests may give rise to a cause of action for performing medical procedures without informed consent or breach of fiduciary duty."[43] *Moore* is required reading in most law school torts classes; however, it has gained surprisingly little traction outside of California and has been expressly rejected by a number of courts.[44] In fact, Pennsylvania has gone so far as to state affirmatively in legislation that a physician has no duty to reveal his or her financial interests to patients.[45] Whipple cautions that merely because the law does not require this form of disclosure, it would probably be a good idea to include the information in the consent form; that way no one can come back and sue later.

At many universities, most notably Harvard, managing the type of conflict that Adam Smith had is remarkably simple: The researcher is not permitted to participate in the research. This "black and white" approach, while simple to administer, has many drawbacks, such as precluding those with the most knowledge about a particular drug or condition from participating in the study. It is also scientifically infirm: There is no evidence to indicate that a pecuniary conflict is so much worse than other forms of conflict that it ought it be singled out for special treatment. In that regard, there is certainly no evidence to suggest that ego-based conflicts are significantly more benign than pecuniary ones, yet a Harvard-type rule necessarily assumes that money (and only money) is the root of all evil.

What is fascinating is that FDA, which relies heavily on the outcome of clinical research in formulating public policy (e.g., should a given drug be approved for a specific condition), has never contemplated such an extreme approach. As discussed below, FDA's approach is one of the more reasoned approaches, but one that would be unacceptable to funding agencies and to academia. FDA, unlike universities and the funding agencies, does not have to rely on inferences, hunches, or guesses to ascertain whether a conflict is influencing a clinical trial: FDA randomly audits clinical trials, and the likelihood of being audited increases if any investigator has a financial conflict. If the data are incorrect, if the protocol was not followed, or if informed consent was not appropriately documented, FDA can take administrative action against the investigator. The reason why the errors or deviations occurred is not as important to FDA as is the simple fact that they did occur. FDA punishes sloppiness whether the cause is a financial interest or the lack of any interest in the study.

FDA's Conflict of Interest Rules

FDA, like NIH and NSF, regulates economic conflicts of interest and requires each researcher to assess whether he or she has a financial interest in the research. FDA, like NIH and NSF, recognizes that

> one potential source of bias in clinical studies is a financial interest of the clinical investigator in the outcome of the study because of the way payment is arranged (e.g., a royalty) or because the investigator has a proprietary interest in the product (e.g., a patent) or because the investigator has an equity interest in the sponsor of the covered study.[46]

However, there are a number of salient differences between FDA's rules and the rules of the others. First, FDA uses different economic thresholds to determine whether a financial interest is significant or needs to be reported. Second, while NIH and NSF require the institution to "manage" conflicts, FDA has no such requirement. And third, while NIH and NSF require an awardee institution to decide up front whether the researcher has a reportable significant financial interest that creates a conflict that needs to be managed, FDA does everything at the end of the research. Although FDA expects that the university or sponsor will manage any conflict, FDA does not require management. But unlike NIH and NSF, which have no interest in the results, FDA does. It uses the results to decide to approve a drug, device, or biologic. Accordingly, FDA randomly audits the results of clinical trials to ensure that they have been properly conducted and that the data on which FDA is about to rely are valid and reliable. A clinical trial conducted by an investigator with a significant financial stake in the outcome may receive significantly more scrutiny from FDA than a trial conducted by someone without any stake in its outcome. How then does one reconcile the FDA, PHS, and university rules, especially when they appear to conflict? As a general rule, the most restrictive set of rules must be applied. Thus, for example, suppose a researcher with a significant financial interest, as judged by his or her university's standards, is conducting a clinical trial, and the university bars researchers with financial interests in a drug company from conducting a clinical trial. That researcher is barred from conducting the trial, not by virtue of FDA rules, but rather by virtue of the university's rules.

SCOPE OF FDA RULES

The FDA rules apply only to clinical trials of new drugs, devices, and biologics if a marketing application is actually filed. The FDA rules do not apply to basic research or to clinical trials that are not conducted under the jurisdiction of FDA. The FDA rules do not apply to research that is funded by FDA. The PHS rule applies to that research. The FDA rules also do not apply to clinical trials that are not intended to affect the labeling of the drug, biologic, or device.[47] For example, if you are studying an "off-label" use of a drug and have no intention of submitting the data to FDA, but your university nevertheless requires that you submit an Investigational New Drug application to FDA, the FDA conflict rules do not apply because you will not be submitting those data to FDA to gain approval of the off-label use. If the FDA rule applies, but the research is also being funded by NIH, then both the FDA and PHS rules apply.

The FDA rules apply to investigators and subinvestigators and their spouses and dependent children. A subinvestigator is viewed as anyone who is involved in the treatment of the subjects or the collection or evaluation of the data. The rule does not apply to full-time employees of the sponsor (e.g., a drug company) because it is assumed that the sponsor and its employees have a conflict; as a result, the agency already gives any data the sponsor submits close scrutiny.

FINANCIAL INTERESTS

FDA is concerned with financial interests that fall into one of four categories: (1) compensation that is affected by the outcome of the study, (2) significant equity interest in the sponsor, (3) proprietary interest in the article being tested, or (4) significant payments.[48] If an investigator's or subinvestigator's financial interest falls into one of these four categories, then FDA views it as a potential conflict and requires the sponsor to inform it of the interest. Here, I refer to any of these four as "reportable interests."

Compensation Affected by Study Outcome

Any compensation in any form received by an investigator or subinvestigator is reportable if it will either increase with favorable results or be worth more with favorable results. Compensation falls into this category if it is tied to a favorable outcome or is in the form of an equity interest in the sponsor or is tied to sales of the product.[49] Thus, a straight payment of $5,000 would not fall into this category, but $5,000 worth of the common stock in the non-publicly traded pharmaceutical company whose drug you are testing would.

Significant Equity Interest in Sponsor

FDA draws a distinction between securities whose value can be objectively and easily determined through publicly available sources (e.g., publicly traded securities in the sponsor) and those that cannot be objectively and easily determined (e.g., nonpublicly traded securities in the sponsor). When the sponsor is not a publicly traded company, any equity interest, whether through ownership of stock or options, is deemed "significant." When the sponsor is publicly traded, the ownership interest is significant if it exceeds $50,000 at any time while the investigator is conducting the study and for one year following completion of the study. If the value hovers around $45,000 while the clinical trial is being conducted, spikes up to $51,000 for one day about a month after the trial is completed, but then drops down again to $45,000, the interest is "significant" because it was above $50,000 at some point during the relevant temporal window.[50]

Proprietary Interest in the Tested Product

A proprietary interest in the tested drug, biologic, or device means "property or other financial interest in the product including, but not limited to, a patent, trademark, copyright or licensing agreement."[51] Therefore, if an investigator is one of the inventors of the drug, then he or she has a reportable interest irrespective of value of the ownership interest. In our discussion of Dr. Wyethe, ACNE, and George Masonite University, we concluded that Dr. Wyethe did not have a significant financial interest under the PHS rule, even though he was receiving $25,000 as a result of licensing the drug he was about test and would receive much more if the drug were approved and commercialized. Under the FDA rule, though, Wyethe's interest would be a reportable interest because it is compensation that could be affected by the results of his research.

Significant Payments to Investigator or Institution

FDA divides payments into two categories:

1. Payments made by the sponsor to the institution or the investigator for the purpose of carrying out the clinical trial. These payments are not counted in determining whether a conflict exists.
2. Any other payments to the institution or the investigator.

Payments to an investigator or subinvestigator that exceed $25,000 in the aggregate during the time the investigation is being conducted by that investigator, and for one year thereafter, are reportable. Thus, grants, honoraria, consulting fees, and the like count toward the $25,000. Grants or payments to the institution that are then made available by the institution to an investigator for research are also reportable.[52] Thus, for example, if an institution receives $7,500 per patient enrolled in a clinical trial and the investigator receives a portion of that from the institution, those payments are not reportable so long as they are payment for conducting the clinical trial. If, however, the sponsor also makes a $500,000 grant to the institution with the understanding that the investigator will receive that money, less overhead, for his or her research, then that payment is reportable.

The interests of FDA are somewhat different from those of NIH, NSF, or most universities. FDA is concerned only with financial interests in or payments from the sponsor of the trial, or an interest in the product being tested. FDA's rules do not apply to indirect ownership interests that might be affected by the research; the PHS rule and NSF policy arguably apply to any interest that could be affected by the research, irrespective of whether that interest is in the sponsor of the research or another entity. Thus, for example, if a researcher owns stock in a company that in turn owns an interest in the sponsor, then that indirect ownership interest does not fall within the FDA rule, but it does fall within the rules of the PHS and NSF.

REPORTABLE FINANCIAL INTERESTS

FDA's reporting requirements are also different from those of the funding agencies. Rather than seeking financial information before a study is funded or the funds are actually used, FDA requires the sponsor to make certain disclosures to it only once the sponsor submits its clinical data in support of an application to market a drug (i.e., New Drug Application [NDA]), biologic (i.e., Biologics License Application [BLA]), or device (i.e., Premarket Approval application [PMA] or 510(k)).[53] As part of the application, for each investigator and subinvestigator, the sponsor must certify either that that person did not have a reportable financial interest, or, if the person had one, the sponsor must reveal to FDA the precise nature of the financial interest and any steps taken by the sponsor or the institution to manage that potential conflict.[54] Given the responsibility placed on sponsors to report their investigators' financial conflicts accurately, sponsors in turn require clinical investigators to make full disclosures to them about their financial interests; investigators are required to update that information promptly in the event of changes.[55] When an investigator has a reportable financial interest, FDA may elect to audit that investigator's data more carefully than might otherwise have been the case.

There is one additional difference between FDA and the funding agencies: FDA regulates "intellectual conflicts" among its advisory committees. These committees, which consist of nongovernment scientists and others, recommend whether a new product ought to be approved. The criteria that FDA uses to screen for intellectual conflicts are murky at best and have been the subject of congressional scrutiny.[56] Should scientists who have expressed views about a new drug be precluded from participating in the committee's deliberations? Should scientists who have conducted studies concerning a new drug be precluded? At what point do one's knowledge and opinions about a product cross the line? These are not easy questions to answer because intellectual conflicts, unlike financial ones, cannot be easily measured. One has to wonder, though, whether Congress is equipped to weigh intellectual conflicts, especially in the sciences.

INSTITUTIONAL CONFLICTS OF INTEREST

Many have raised concerns that financial interests potentially affect not only the objectivity of researchers, but also institutional decision making in a number of ways. First, many universities, especially through their endowments, are extraordinarily large equity holders, and those equity positions could somehow affect research agendas and the like. Second, university ownership of certain patent rights, especially in new technology, may affect research agendas and even the way the university regulates the research. Third, a quasi-institutional conflict can arise when a university official sits on the board of a company that is either dealing with the university on research matters or could be affected by university policy.[57] The fourth type of institutional conflict can arise when the institution has been retained by both sides to a scientific controversy and fails either to inform the sides of the conflict or to create an ethics wall between the two teams of researchers.

In most cases, institutional conflicts are more potential than real. True, university endowments can be significant. At the end of 2008, the top three were Harvard ($36.5 billion), Yale ($22.9 billion), and Stanford ($17.2 billion).[58] The recent dramatic downturn in the economy has eroded these figures by about 25 percent. Even so, the endowments for these three institutions are so large that a fraction of the income (one-sixth in Harvard's case) would pay for the tuition, room, and board of the every undergraduate student enrolled at each institution. Even though endowments can be large, it is unlikely that any single holding in the portfolio would affect university behavior. University endowments tend to be so actively managed that equity owned in one company in one month may be replaced by equity in a totally different entity in the subsequent month. Also, there tends to be a firewall between those who manage the endowment and those who run the university. Portfolio managers for the endowments tend to have little to do with the rest of the university. At major research universities, they tend to be the highest paid university employees, higher than even the football coach. Thus, Harvard's portfolio manager received over $7 million in 2004, considerably more than Ohio State's football coach.[59] The situation might, of course, be different at a newer institution with a much smaller or more focused endowment. The Association of American Medical Colleges has attempted to address this form of institutional conflict by highlighting, in its model policy, the limited types of institutional holdings that *might* trigger a conflict. For example, under the AAMC guidelines, a conflict might occur when a university acquires, as result of licensing its technology, an equity interest in a publicly traded company that exceeds $100,000, and human trials of the technology are to take place at the university.[60] Under the AAMC guidelines, mere ownership of equity in a publicly traded company is insufficient to create a conflict.

University ownership of patents and technology can, in unusual circumstances, raise significant and real conflicts of interest. These conflicts have arisen in two ways. Most notably, a university may wish to enter into an exclusive licensing arrangement with a single drug company. Under the arrangement, the drug company would pay the university a fixed fee, let's say $200 million, and in exchange, the university would agree to give the drug company a right of first refusal to license all of the technology developed at the university for a set period. This type of arrangement, when it is university-wide, raises a number of problems. First, it effectively transforms a not-for-profit research institute or university that receives federal grant funds into a conduit for a profit-making entity. Second, it restricts the ability of researchers at the university to work with other drug companies or potentially with other researchers at other institutions, thereby potentially stifling creativity. In 1992, the Scripps Research Institute (Scripps) announced that it and Sandoz, a Swiss pharmaceutical company, had entered into a long-term arrangement giving Sandoz rights of first refusal to any invention developed by Scripps researchers and effectively precluding Scripps researchers from sharing their results with scientists at other research institutions. Dr. Bernadine Healy, the then-director of NIH, objected to the arrangement as contrary to the spirit of science and possibly illegal. As a result of Healy's objections and congressional hearings into the arrangement, the agreement was scaled back to encompass slightly less than 50 percent of the technology developed at Scripps. The agreement was recently renewed with Sandoz's successor in interest, Novartis.

It is also theoretically possible that a university's ownership of certain intellectual property could influence how it makes various decisions that could conceivably affect the value of that property. For example, suppose Dr. Jones, a researcher at North-South University, invents a new drug and obtains a patent on that technology, which is then assigned to the university. Suppose further that the university has licensed the patent to NewMole Pharmaceutical Company, which wants to conduct a phase 1 safety trial at North-South and hopes that Jones will be the investigator on that trial. Both Jones and the university stand to make a significant amount of money if the drug proves successful. Does the university have too much of a conflict to even decide whether Jones has a conflict and, if so, to manage that conflict? Some undoubtedly would argue that the university is in no better position to make these decisions than Jones is to act as an investigator on the trial. Despite the fact that these types of conflicts can and do arise, there is no evidence so far that university decision making has been affected by such conflicts. After all, a university is hardly a monolith. The AAMC in its model policy, though, strongly suggests that this scenario may well raise inappropriate institutional conflicts.

Institutional conflicts can also arise when a university wants to amend a preexisting licensing arrangement with a company and finds that the only leverage it has is over a faculty member who is on the board of the company. One major institution, for instance, sought to charge one of its senior faculty members with violating certain unstated conflict rules unless the company on whose board the faculty member served agreed to increase royalty payments to the university above the amounts previously negotiated and memorialized in an agreement. The company bowed to the university's pressure. However, unbeknownst to the university, the company let its peers in the pharmaceutical industry know in no uncertain terms that this university could not be trusted. As a result, drug companies have treated the university as a leper; the university now has one of the lowest royalty return rates per dollar of grant funding in the nation. It has spent a fortune on consultants in an effort to ascertain why drug companies have avoided it like the plague (so to speak).

The final type of institutional conflict involves government contractors and so-called beltway bandits that may end up on both sides of the same controversy. It is akin to each side in litigation hiring the same law firm to represent it. Under the ethics rules governing lawyers, this is referred to as a nonwaivable conflict. The same law firm cannot represent the plaintiff and the defendant in a single litigation. There are many other potential conflicts that can be waived by the parties, but the parties have to be put on notice of the conflict. The rules that govern lawyers, however, do not necessarily pertain in other callings. Recently, for instance, NIH terminated a contractor that it had retained to review the safety of certain chemicals after it learned that the consulting company was simultaneously working for companies manufacturing the chemicals under scrutiny. The contractor, Sciences International, was hired by NIH to run the federal Center for the Evaluation of Risks to Human Reproduction. According to news reports, at the same time it was running the Center, it was also consulting with "Dow Company and BASF, 3M, and other companies that produce some of the chemicals under scrutiny" at the Center.[61]

FEDERAL GOVERNMENT EMPLOYEES HAVE SPECIAL RULES

The headline on the Department of Justice press release said it all: "NIH Senior Scientist Charged with Conflict of Interest."[62] The release went on to recount that the government had filed criminal charges against Pearson "Trey" Sunderland III, chief of the Geriatric Psychiatry Branch at the National Institute of Mental Health (NIMH), for accepting "$285,000 in consulting fees and additional travel expenses from a drug company without the required approval of and disclosure to National Institutes of Health (NIH) officials." According to the charges, Sunderland arranged for NIMH to provide Pfizer with "approximately 600 tubes of cerebrospinal fluid samples that had been previously collected from NIMH research subjects . . . and . . . approximately 200 additional samples of cerebrospinal fluid, blood serum and related clinical data were provided." On December 22, 2006, Sunderland pleaded guilty and was sentenced to two years' probation, and was ordered to forfeit the money he had been paid by Pfizer and to perform four hundred hours of community service. The Sunderland case was one of a handful that came to light as a result of investigative reporting and congressional hearings.[63]

On December 7, 2003, the *Los Angeles Times* carried a lengthy article detailing the financial relationships between various researchers at NIH and drug companies, and, in one case, readers were invited to infer that the financial relationship may have affected the NIH researcher's decision making. The relationships reported in the article, unlike the one involving Sunderland, were all legal and approved by NIH, and the income was appropriately reported in each employee's annual financial disclosure form. Nonetheless, the article and congressional interest triggered a firestorm that forced the NIH director to revise the agency's rules concerning outside activities and financial interests—but it took a few tries. The first attempt ended with a rule that was so draconian that secretaries, custodians, and others would have had to choose between staying at NIH or seeking to alter their spouse's 401(k) plan so that it held no interest in any healthcare entity. The initial set of rules was overly broad, pushed more by political fear than anything else, a fact that was acknowledged in the rulemaking itself:

> Moreover, in light of recent Congressional oversight and media reports, HHS has determined that the existing rules governing outside activities have not prevented reasonable public questioning of the integrity of NIH employees and the impartiality and objectivity with which agency programs are administered.[64]

Eight months later, the rules were refined. The second go-around made more sense and focused on those at NIH who could actually influence policy or research. The provisions that would have applied to all investments held by all employees were refocused and moderated so that investment rules would apply only to senior officials and would permit a senior official to hold an investment not exceeding $15,000 in a given healthcare-related entity and $50,000 in a sector-specific mutual fund.[65]

In this section, I attempt to put the NIH-specific financial restrictions in context by first discussing the criminal statutes and regulatory provisions that govern all federal employees irrespective of their rank or seniority. This section also highlights the special rules that govern those who are members of study sections or other government advisory committees. The section is not intended to provide an in-depth treatment of federal conflict of interest and related laws; there are treatises on the subject.

Felonies Are Not Good

OVERVIEW

The conflict of interest regulations that apply to federal employees spring from a small set of statutory provisions aimed at prohibiting bribery (§ 201),[66] banning a federal employee from representing private parties before government agencies (§ 203),[67] precluding a federal employee from participating as an official in any matter that can affect his or her financial interests (§ 208),[68] and preventing a federal employee from accepting any supplements to his or her salary for activities as a federal employee (§ 209).[69] These provisions are found in Title 18 of the United States Code, also known as the federal criminal code. Every year, a few federal employees are indicted and convicted for violating one or more of these prohibitions, each of which is a felony. Thus, based on the news accounts, Trey Sunderland could have been indicted for any one of these four crimes. Nor is Sunderland the only NIH researcher to have run into the Title 18 quartet. In 1992, Prim Sarin, a senior researcher at the National Cancer Institute, agreed to perform certain trials on an anti-human immunodeficiency virus (HIV) drug for a German pharmaceutical company. The company agreed to pay NCI $50,000 in two payments. Apparently, the checks were made payable to both NCI and Sarin. Sarin deposited one of those $25,000 payments from the company into two personal bank accounts. Sarin was convicted of embezzlement; he was sentenced to three years in prison, all but two months of which were suspended. The conviction was upheld on appeal.[70]

Researchers are normally not the target of choice for those investigating conflict of interest or bribery. Most criminal investigators focus on where the action and money are, namely procurement officials. Take the case of Darleen A. Druyun, former deputy assistant secretary of the Air Force for acquisition and management. Druyun was the Air Force's top career procurement official and, as a result, wielded tremendous power and the ability to influence multibillion dollar acquisitions. In 2000, Boeing, at Druyun's request, hired her future son-in-law. In 2002, Boeing's then-chief financial officer recruited Druyun for an executive position with Boeing following her retirement. However, during the two-year period immediately prior to her retirement, Druyun oversaw many Boeing contracts. More critically, though, she helped award a $20+ billion procurement to lease from the Boeing a fleet of aerial tankers. Druyun was indicted for violating various conflict of interest laws and eventually pleaded guilty in April 2004 "to negotiating employment with Boeing while she was participating personally and substantially as an Air Force official overseeing the negotiation of the proposed multi-billion dollar lease of Boeing KC 767A tanker aircraft."[71]

Interestingly, the Department of Defense canceled the award to Boeing because of these improprieties and conducted a new round of competition, which ended in early 2008 with a European-American consortium (EADS, the parent of Airbus, and Northrop Grumman) being awarded the $35 billion tanker contract. The award to EADS became an issue during the 2008 elections, with each company taking out ads in national newspapers emphasizing the virtues of its tanker. Shortly before the election, the Department of Defense canceled the award for a second time, leaving the decision to the next administration.[72]

While procurement officials certainly have the opportunity to cross the line, the champion line crossers, at least when adjusted for cohort size, are those responsible for drawing the lines, namely members of Congress. Where else can a former federal judge impeached by the House, convicted by the Senate, and removed from the bench for accepting bribes, turn around and be elected to the very body that impeached him and then rise to become vice chair of the House Permanent Select Committee on Intelligence? Nor was there anything subtle about former congressman Duke Cunningham's schemes. Cunningham accepted millions of dollars in gifts from government contractors in exchange for steering large defense contracts in their direction. In 2005, Cunningham resigned his seat in the House and pleaded guilty to conspiracy to commit bribery, mail fraud, wire fraud, and tax evasion. He was sentenced to eight years and four months in prison and ordered to pay $1.8 million in restitution. It does not take a lawyer to recognize that what Druyun and Cunningham did was illegal; there was nothing subtle about their conduct, and in fact, that is the way it is with most criminal violations of sections 201, 203, 208, and 209 of Title 18.

Given the extremely broad scope, though, of these sections, how is it possible for a person to serve on a federal advisory committee? After all, when you serve on an advisory committee, including a study section, you become a "special government employee" and become subject to section 201, 203, 208, and 209 restrictions. A special government employee is anyone who is employed outside the normal civil service system either for a short term, usually no more than 130 days during a year, or for intermittent work.[73]

CRIMINAL CONFLICT OF INTEREST LAWS

The criminal conflict of interest laws proscribe conduct that clearly creates a conflict of interest. One law prohibits a federal employee from representing a person (e.g., private company, state, university) for compensation in a proceeding or other matter in which the United States is a party or has a direct and substantial interest. For example, an NIH employee may not represent a government contractor in negotiations with the Department of Defense concerning the award of a contract.[74] In contrast, a special government employee, such as a study section member, is permitted to represent his or her university (or anyone else for that matter) before any federal agency, provided the matter is not one in which the person participated in personally and substantially while a special government employee.[75]

These laws, while simple on their face, are not simple to apply, as illustrated by the case of David Baird, an engineer who served in the Coast Guard and who resigned his commission to go into the private sector. After being laid off from his private sector job, he joined the Coast Guard Reserve, effectively as a full-time employee but serving only through a series of short-term appointments; the first was 139 days and the second 69 days. While serving on one those appointments, he learned that a private engineering firm was negotiating a contract with the Coast Guard and was looking for a consultant to assist it. Baird was hired for the job and represented the company in negotiations with the Coast Guard. He received less than $1,000 for the consultancy. Shortly thereafter, he was indicted for representing private parties before government agencies while serving as a special government employee, in violation of 18 U.S.C. § 203. Following his conviction, he appealed, arguing that he was not a special government employee. The government argued that it did not matter because if he was not a special government employee, then he was a regular government employee and still subject to section 203. The court of appeals agreed that he was either one or the other, but reversed the conviction for other reasons.[76]

Section 208 is particularly broad and prohibits a federal employee from participating personally and substantially in a matter that can affect the employee's financial interests. Thus, for example, a federal employee, including a special government employee, is prohibited from simultaneously negotiating with a company for prospective employment and participating personally and substantially in any matter in which that company has a financial interest.[77] When there is such a conflict, the employee is supposed to recuse him- or herself, and in that way, will have no influence on any government decision. Thus, a study section member is not permitted to participate in considering a grant application submitted by that person's employing university.

An interesting issue involving section 208 has arisen under the Federal Technology Transfer Act of 1986. One provision of that act permits a federal laboratory to retain the domestic patent rights to employee inventions but cedes the foreign rights to the employees. In one case, three HHS employee-inventors shared the rights to obtain certain foreign patents. The United States owned the domestic patent. The employees licensed their foreign rights to a specific company. At the same time, the agency (probably NIH) employing the three inventors awarded the domestic rights to the same licensee. The agency intended to enter into a Cooperative Research and Development Agreement (CRADA) with the licensee under which that firm would undertake the clinical trials necessary to test and evaluate the invention for FDA approval and commercialization. Two of the three employee-inventors were supposed "to be directly involved, as part of their official duties, with work related to the invention through the CRADA."[78] It is "typical for the inventor and the Government to enter into licensing agreements with the same firm," and "it is often in the Government's best interest to allow inventors who hold foreign rights to continue to develop their work."[79]

The Office of Government Ethics concluded, and the Office of Legal Counsel agreed, that the employee-inventors had a section 208 "financial interest" in their inventions "because they own the foreign patent rights from which they receive royalties," and that they could not "officially act on any matter involving the private firm to which they assigned their patent rights. This prohibition would include work by the employee-inventors on the research and development agreement with the private firm."[80]

There are certain exceptions to section 208, and individuals can receive "waivers." Indeed, it would be virtually impossible to populate advisory committees with knowledgeable members if section 208 were to apply to those individuals.[81] For example, many members of the Secretary's Advisory Committee on Organ Transplantation are transplant surgeons who could be financially affected by the committee's recommendations. Correspondingly, if one views the allocation of research grants as a zero-sum game, then members of study sections and of each institute's advisory committee have vested interests in seeing that grant applications submitted by entities other than their employing institutions receive unfavorable scores or recommendations. As a result, the Secretary routinely grants so-called 208 waivers to those serving on study sections and other advisory committees, permitting them to participate in one or more particular matters in which they have a financial interest or an interest that is imputed to them because their regular employers have a financial interest.[82]

CRIMINAL SALARY AUGMENTATION STATUTE

How It Applies to Federal Employees and Special Government Employees

As noted above, federal law also prohibits a federal employee from accepting payment from a nongovernment source for activities that relate to his or her government duties.[83] A government employee, such as Dr. Sunderland, is not permitted to act as a paid consultant to an outside entity where there is overlap between the employee's official duties and the consultancy. A government employee is also not permitted to give speeches for compensation concerning his or her official duties or which relate to his or her official duties. A government employee, though, is permitted to give speeches and write articles about subjects outside the scope of his or her employment. This was not always the case. The Ethics in Government Act of 1978 prohibited all federal employees from writing articles or giving speeches for compensation, even when those articles or speeches had nothing to do with their federal jobs. Shortly thereafter, a group of federal employees filed a suit challenging the statute as an unconstitutional abridgment of their freedom of speech. The speeches and articles for which the employees had received honoraria in the past concerned matters such as religion, history, dance, and the environment; with few exceptions, neither their topics nor the persons or groups paying them had any connection with their official duties. The Supreme Court struck down the law as overly broad; when Congress regulates speech, it has to impose the least burdensome alternative, and the Ethics in Government Act failed to do this. The Court noted that the "ban imposes a far more significant burden on [ordinary federal employees] than on the relatively small group of lawmakers [i.e., members of Congress] whose past receipt of honoraria motivated its enactment."[84] Although federal employees may engage in First Amendment activities for compensation, they may nonetheless be required to obtain approval for such activity. The approval provides the employee with an assurance that the agency does not consider the outside activity to be "job related."

The prohibition on outside activities that relate to one's government position applies to special government employees as well.[85] Researchers become "special government employees" if they are paid to serve on an advisory committee, for a few days or even weeks per year.[86] However, special government employees "may accept compensation for speaking on subjects within their discipline or inherent area of expertise based on their educational background or experience. They cannot accept compensation for the outside activity if the speaking concerns matters they participate in due to their position on the Federal advisory committee. The outside speaking cannot relate to the work the [special government employee] is providing to the Government."[87] Special government employees, such as study section members, do not need prior approval to engage in paid outside activities.[88]

Federal Employees Receiving Royalties under a Patent

The Federal Technology Transfer Act of 1986 required the federal government to share with inventors the royalties that it received on an invention developed at federal laboratory. Some were concerned that royalty payments to federal employees under the FTTA constituted "supplementation" of their salaries, which is not permitted by section 209. Both the Office of Government Ethics and the assistant attorney general for the Office of Legal Counsel have concluded that these payments are government compensation and, therefore, are not subject to section 209.[89]

The Rules Provide Some Guidance and Some Leeway

The statutes are fairly bleak, appearing to ban most everything with even the slightest odor. The regulations are somewhat more forgiving, however, and allow federal employees a whiff of the private sector. Generally, executive branch employees, including high-level officials, may not accept gifts that are given because of their official position or that come from certain interested parties ("prohibited sources"). These prohibited sources include persons (or an organization made up of such persons) who (1) are seeking official action by the employee's agency, (2) are doing or seeking to do business with the employee's agency, (3) are regulated by the employee's agency, or (4) have interests that may be substantially affected by the performance or nonperformance of the employee's official duties.[90] Entities that have or are seeking government contracts with a specified agency are considered prohibited sources with respect to that agency's employees. The prohibition on gifts is broad and includes everything from airfare to lunches.

DE MINIMIS EXCEPTION

There are certain exceptions, however, to this blanket prohibition. First, a federal employee is permitted to accept gifts of *de minimis* value (i.e., worth $20 or less per occasion and not exceeding $50 per year).[91] However, the *de minimis* exception does not apply to small gifts that are made so frequently that a reasonable person could infer that the employee is using his office for private gain. The Office of Government Ethics (OGE) gives as an example a procurement official within the Department of Veterans Affairs who meets with pharmaceutical representatives during lunch hour on a regular basis. The representatives provide the employee with a sandwich and a soft drink worth less than $6. OGE concluded that acceptance of these modest gifts on a recurring basis would be improper, even though each is well within the $20 *de minimis* exception and the aggregate value is less than $50 per year.[92] Furthermore, even though the gift rules might permit these *de minimis* gifts, many corporations prohibit their employees from making any gift of any value to any federal employee; some federal agencies have adopted a "zero gift" policy as well.

As a rule, federal employees should not accept even a gift of *de minimis* value from a prohibited source, even though the rules may allow them to do so. It is easier to accept no gifts than to risk accepting one that is of greater value than is permitted.

WIDELY ATTENDED EVENTS EXCEPTION

Second, an employee, under two sets of circumstances, is permitted to attend a widely attended event paid for by another. In one instance, an employee is permitted to accept free attendance at a widely attended event from the sponsor of the event if "there has been a determination that his attendance is in the interest of the agency because it will further agency programs and operations."[93] This determination may be made orally or in writing.[94] In a second instance, if the entity that extends the invitation has "interests that may be substantially affected by the performance or nonperformance of an employee's official duties," then the employee may accept the invitation "where there is a written finding by the agency designee that the agency's interest in the employee's participation in the event outweighs the concern that the acceptance of the gift of free attendance may or may appear to improperly influence the employee. . . ."[95]

The rule divides "widely attended events" into two categories for gift purposes: events where (1) the sponsor is providing the free attendance, and (2) someone other than the sponsor is providing the free attendance.[96] The rules dealing with the latter are more rigid than the rules dealing with the former.[97] With respect to free attendance provided by a nonsponsor, the rule defines a "widely attended event" as one in which "more than 100 persons are expected to attend."[98] There is no similar numerical restriction when the free attendance is provided by the sponsor. The fact that the event is primarily social in nature does not undermine the exception.[99] In Advisory Letter 99 × 2, the OGE pointedly noted that "receptions [as opposed to fundraising events], may be more clearly conducive to the exchange of ideas or otherwise more readily support a finding of agency interest."

If an entity is deemed to be the sponsor of a widely attended event and if an invitee's employing agency makes the determination required by 5 C.F.R. § 2635.204(g)(3)(i), then there is no limit to the value of the invitation to either the employee or his or her spouse, as the case may be.[100] If, on the other hand, the entity is not deemed to be the sponsor, then there would be a $250 limit on the aggregate value of the invitations to the employee and his or her spouse.

The gift rule itself can be violated only by government employees; however, there are criminal statutes related to the gift rule that can be violated by private sector entities.[101] Thus, to avoid difficulties, any entity that is sponsoring a widely attended event and intends to invite federal officials should advise those agencies' ethics officials of the pending invitations so as to better enable them to promptly address any questions that employees within their respective departments might have. I believe that this conservative approach is warranted because if a federal employee fails to obtain the appropriate waiver, then the gift exception no longer applies. At that point, the gift becomes an illegal gift, which can adversely affect both the donor and the recipient. As noted above, making an improper gift can create potential criminal liability under 18 U.S.C. §§ 201, 203, and 209, for example.

Special Rules for Those on Study Sections

NIH has special rules that apply to those conducting peer reviews of applications for NIH funding (e.g., sitting on NIH study sections). Interestingly, these along with those issued by FDA are the only conflict rules that apply to both financial and nonfinancial conflicts (e.g., ego conflicts and academic disagreements).[102] The rules define two types of conflicts: an appearance of a conflict and a real conflict.

> Appearance of a conflict of interest means that a reviewer or close relative or professional associate of the reviewer has a financial or other interest in an application or proposal that is known to the reviewer or the government official managing the review and would cause a reasonable person to question the reviewer's impartiality if he or she were to participate in the review; the government official managing the review (the Scientific Review Administrator or equivalent) will evaluate the appearance of a conflict of interest and determine, in accordance with this subpart, whether or not the interest would likely bias the reviewer's evaluation of the application or proposal.[103]

A reviewer has a real conflict of interest if he or she or a close relative or professional associate of the reviewer:

> (1) Has received or could receive a direct financial benefit of any amount deriving from an application or proposal under review;
> (2) Apart from any direct financial benefit deriving from an application or proposal under review, has received or could receive a financial benefit from the applicant institution, offeror or principal investigator that in the aggregate exceeds $10,000 per year; this amount includes honoraria, fees, stock or other financial benefit, and additionally includes the current value of the reviewer's already existing stock holdings. The Director, NIH, may amend the dollar threshold periodically, as appropriate, after public notice and comment; or
> (3) Has any other interest in the application or proposal that is likely to bias the reviewer's evaluation of that application or proposal. Regardless of the level of financial involvement or other interest, if the reviewer feels unable to provide objective advice, he/she must recuse him/herself from the review of the application or proposal at issue.

> The peer review system relies on the professionalism of each reviewer to identify to the designated government official any real or apparent conflicts of interest that are likely to bias the reviewer's evaluation of an application or proposal.[104]

Under the rule, a study section reviewer must recuse him- or herself from any involvement in reviewing a given application if the individual has (1) a real conflict of interest, and (2) is employed by the organization submitting the proposal, unless the director of NIH determines that the "components of a large or multicomponent organization are sufficiently independent to constitute, in effect, separate organizations, and provided that the reviewer has no responsibilities at the institution that would significantly affect the other component."[105] The director may also waive certain financial conflicts. An individual with an appearance of a conflict must also recuse him- or herself, except that too is subject to waiver by the director under limited circumstances.

CASE STUDIES AND PROBLEMS

Case 1: The Case of the Curious Conflict

(Use applicable PHS rules.)

In 1974, Carl Schute and David Ladder conceptualized ways of sequencing genes. For this work, they were awarded a Nobel Prize. They were also awarded numerous patents on the underlying basic technology, and, not surprisingly, various business opportunities came their way. Their employing institution, the famed Colorado Technology Institute (CTI), licensed most of Schute and Ladder's technology to companies. Indeed, Schute and Ladder became major investors in one of these companies, MicroMolecular Machinery (M^3).

Schute and Ladder have a large laboratory at CTI where they do NIH- and NSF-funded research. They are working on a new technology that would enable ordinary clinical labs to sequence a significant portion of a chromosome in virtually no time. More significantly, the new technology would sell for about one-tenth the price of the highly specialized instruments sold by M^3. To test out the viability of their technology, Schute and Ladder draw their own blood and sequence their entire genomes. They then recruit everyone in their lab—scientists, technicians, and staff—to participate. In the end, Schute and Ladder present to each person that person's fully sequenced genome. Schute and Ladder had reviewed the genomes first and determined who had risk factors for which diseases, but they have kept this information to themselves. The scientists in the lab begin the process of putting all the data together to write up some papers for publication.

Shortly thereafter, Schute and Ladder discuss their new findings with Ted Tight, the CEO of M^3. Given the low costs of the new technology, Tight is concerned that it will undermine M^3's viability and Schute and Ladder's significant investment. He suggests that perhaps they ought to "shelve" the research in the interest of business.

At the next laboratory meeting, they discuss their concocted "concerns" about the quality of the data that the new technology is generating. Specifically, Schute tells everyone gathered that "I have serious reservations about the quality of the data and what would happen if we published prematurely. Accordingly, Dave and I have decided that until we can confirm the accuracy of the technology, we should go no further. No one will publish anything about the technology until we are satisfied that it really works. However, there are more pressing issues that need to be addressed. Dave and I are therefore redirecting our efforts." Schute goes on to set out new scientific issues and experiments unrelated to the technology.

Laboratory morale takes a nosedive. One of the more senior researchers, Charles Wise, secretly continues to perfect the new technology. If the technology were to pan out, he would be on the patent and entitled to large royalties. As it turns out, Wise is desperate. He has a wife, three young children, and little money. When viewing his genome, he realizes that he has all the risk factors for serious coronary artery disease (CAD); his father and grandfather each died of CAD at thirty-five; Wise is now nearly forty. Wise is also suspicious because he believes his genomic sequence is probably very accurate. After working for months, he concludes that the original sequences and data were all good and that Schute and Ladder were probably lying to the lab when they "put the kibosh" on further research. Wise confronts them with his data. They become furious that he disobeyed and they immediately fire him, noting that he probably doesn't have long to live anyway.

Wise goes to the dean. What should the dean do? Carefully read the materials on human subjects (chapter 4) and conflicts of interest (this chapter). Use the PHS definitions to analyze this hypothetical.

Case 2: The Case of the Painful Panacea

Dr. V. Learned McCash is a noted microbiologist and expert in pain receptors at Spartan University in California. McCash was recently appointed a member of an FDA advisory committee that makes recommendations to the commissioner about various analgesic drugs. McCash is not a great believer in analgesics for three reasons. First, he believes that there are too many pain receptors, making any analgesic less than fully effective. Second, he believes that pain is natural and that people should grin and bear it. And third, he believes that all analgesics carry significant risk not fully appreciated by the medical community or the patients.

McCash has put his money where his mouth is and has invested heavily in No-Pain Centers, Ltd. (NPCL), a chain of holistic clinics that use herbs and psychotherapy, not drugs, to treat pain. The value of NPCL stock rose dramatically following the revelations about Vioxx and similar drugs. McCash has disclosed this investment to FDA and to Spartan.

Three months ago, FDA approved a new prescription analgesic, OpMorph, to replace Vioxx. OpMorph is a "miracle drug"; it is more effective than morphine, but carries none of that drug's side effects. McCash believes that OpMorph is a sham and immediately undertakes laboratory work to test its safety. His initial set of experiments using a special cell line reveals that the drug appears (1) to damage a cell's repair mechanism making it more vulnerable to becoming cancerous and (2) to promote the formation of plaque within blood vessels. He predicts that over the short run, those who use the drug will have elevated risk of stroke; over the long run, those who use the drug will have an elevated risk of developing various types of cancer. He quickly summarizes his results in a letter and sends it to *Nature's Science*, a top-line medical journal, which immediately publishes the letter. He intends to follow that up with a full-blown article in the next few months.

OpMorph's manufacturer, Pandemic, Inc., is disturbed by the McCash letter, but is more concerned with the very preliminary results of its phase 4 study (postmarketing study) that seems to confirm McCash's prediction with respect to strokes. Pandemic informs FDA, which convenes an advisory committee meeting for the following week.

In the interim, McCash is unable to replicate his earlier results and soon discovers that his cell line was contaminated. Using a noncontaminated line, the effects that he originally noticed evaporate. McCash says nothing. He attends the advisory committee meeting and, based on his published letter in *Nature's Science* and Pandemic's own preliminary phase 4 results, urges the committee to recommend that the drug be withdrawn. By a closely divided vote, five to four, the committee recommends that the commissioner suspend approval to market OpMorph. On that news, NPCL stock skyrockets.

The Commissioner of Food and Drugs has no intention of suspending marketing. He believes that the drug is safe. The day before he is scheduled to make his announcement, the chairman of a House committee with FDA oversight responsibility dies of a massive stroke. He had been taking OpMorph to control pain associated with his arthritis. The commissioner does an about-face and orders an immediate suspension of marketing.

Two months later, all of the data from the phase 4 results are analyzed, and the original safety concerns disappear. Pandemic has also unsuccessfully attempted to replicate McCash's study. Pandemic contacts the inspector general, the FDA commissioner, and the VP for research at Spartan. What should each do?

Case 3: The Case of the Stock-Swapping Scientist

Dr. Ivan Pabst is a well-respected clinical nephrologist at the North Iowa Central Eastern University's College of Medicine. In addition to his teaching and clinical responsibilities, Pabst acts as an investigator on clinical drug trials, usually sponsored by large pharmaceutical companies. During his thirty-year tenure at NICE U., Pabst has been the principal investigator on more than fifty clinical drug and biologics trials involving thousands of patients. Most of the trials were phase 3 multicenter clinical trials involving scores of academic medical centers.

On November 22, 2009, Dr. Pabst is approached by Dr. Lars Gold, a friend and fellow nephrologist at Halava University. It seems that Halava has entered into a contract with Rex Drug Company to coordinate a large, multicenter clinical trial of Rex's new drug YourX, which is designed to retard certain types of kidney disease. Under its contract with Rex, Halava University would recruit one hundred academic medical centers, distribute protocols and case report forms, monitor data collection, and review data as they are being collected for any safety problems. Gold asks Pabst if he would be interested in participating as a principal investigator. Gold indicates that NICE U. would be paid $1,000 per patient and that Pabst would be paid $1,000 per patient either in Rex stock or in cash. If the stock payment is selected, the value of the stock will be the trading price at the close of the trading day on which Pabst signs the agreement with Rex. Pabst agrees to participate in the study and is confident that he will be able to easily recruit one hundred patients who meet the eligibility criteria. He opts to be paid in stock. He does not reveal the payment to his IRB, although NICE U. has a conflict policy identical to HHS policy that applies to all research irrespective of the funding source.

About a quarter of the way through the clinical trial, Pabst notices that one weekend he has a fair number of adverse event reports on his desk; some are serious. That Monday morning, he sells the Rex stock that he has been paid.

About midway through the clinical trial, Gold contacts Pabst with some disturbing news. The Data Review and Safety Committee (DRSC) at Halava has discovered that a significant number of subjects have developed liver failure after receiving YourX. Moreover, the drug does not appear to be working that well. Pabst is not surprised, but he immediately sells the rest of the stock in Rex, which he had accumulated since his last sell-off. What, if anything, should NICE U. do?

Case 4: The Case of the Consulting Theoretician

Sam Swann is a theoretical physicist specializing in string theory. He fervently believes in the laws of thermodynamics. He further believes that people who excuse their obesity because of their "slow metabolism" are only fooling themselves. Swann has NSF and Defense Department grants to study string theory. He also appears on radio and television programs where he rails against fad diets and the like. He earns more than $10,000 per annum from these programs. He has a diet of his own: "Take in fewer calories than you expend and you will lose weight, guaranteed." Swann owns about $100,000 worth of stock in Calorie Counters, a weight-loss program that relies exclusively on portion control.

He also does medical education programs for Panpill Pharmaceuticals, Inc., the largest drug company in the world, and is paid about $200,000 per year to teach physicians about the laws of thermodynamics and that fad diets just don't work. Among its thousands of products, Panpill makes some drugs that speed up metabolism. It markets these diet pill products by trashing fad diets. Swann does not inform his university, NSF, or the Defense Department about his stock ownership or income from Panpill. He does not tell the radio or television stations either. Does Swann have problems? If yes, what kind exactly?

NOTES

1. *See* Act of Sept. 2, 1789, ch. 12, § 8, 1 Stat. 65, 67.

2. *See, e.g.*, 18 U.S.C. §§ 207, 208; 5 C.F.R. pt. 2600 *et seq.*

3. *See* H.R. Res. 6, 110th Cong. (2007) (changing the House gift rule by resolution); Honest Leadership and Open Government Act of 2007, Pub. L. No. 110-81, 121 Stat. 735 (changing the Senate gift rule through legislation and affecting other reforms government-wide).

4. Pub. L. No. 95-521, 92 Stat. 1824.

5. *See* Kefauver-Harris Drug Amendments of 1962, Pub. L. No. 87-781, 76 Stat. 781 (codified at 21 U.S.C. § 355 *et seq.*).

6. *See* Amendments to the Patent and Trademark Laws § 6 (Bayh-Dole Act), Pub. L. No. 96-517, 94 Stat. 3015, 3018 *et seq.* (1980) (codified at 35 U.S.C. §§ 200–211).

7. *See* Lawrence K. Altman, *Drug Firm, Relenting, Allows Unflattering Study to Appear*, N.Y. TIMES, Apr. 16, 1997, at A-1.

8. *See id.*

9. *See* note 6, *supra.*

10. Pub. L. No. 99-502, 100 Stat. 1785 (codified at 15 U.S.C. § 3710a *et seq.*). The Federal

Technology Transfer Act of 1986 amended the Stevenson-Wydler Technology Innovation Act of 1980, Pub. L. No. 96-480, 94 Stat. 2311 (codified at 15 U.S.C. § 3701 *et seq.*).

11. *See* DANIEL KEVLES, THE BALTIMORE CASE 11 (1998) (noting the "tensions emergent in the late twentieth century between biomedical sciences and American political culture").

12. The PHS rule was issued as a "rule," as that term is used in the Administrative Procedure Act, 5 U.S.C. § 551. It covers research funded by any Public Health Service agency including (1) NIH, (2) Centers for Disease Control and Prevention, (3) Health Resources and Services Administration, (4) Indian Health Service, (5) Agency for Healthcare Research and Quality, (6) Agency for Toxic Substances and Disease Registry, (7) Substance Abuse and Mental Health Services Administration, and (8) FDA. In contrast, the NSF conflict of interest requirements were issued as a "notice" and not a rule, even though the notice has the same effect. For ease of reference, when referring to the PHS rule and NSF policy, I use either the term "PHS and NSF rules" or the more neutral term "directives."

13. *See* 21 C.F.R. pt. 54.

14. *See* 18 NIH GUIDE FOR GRANTS AND CONTRACTS no. 32 (1989).

15. *See* Comment of the Association of American Universities (Nov. 27, 1989).

16. Robert Charrow, *Weighing Conflicting Solutions to Conflict of Interest*, 1 J. NIH RES. 135 (1989).

17. Pub. L. No. 103-43, 107 Stat. 122.

18. Section 164 of the NIH Revitalization Act added a new section 493A to the Public Health Service Act, codified at 42 U.S.C. § 289b-1, which provides, in part:

> (a) The Secretary shall by regulation define the specific circumstances that constitute the existence of a financial interest in a project on the part of an entity or individual that will, or may be reasonably expected to, create a bias in favor of obtaining results in such project that are consistent with such financial interest. Such definition shall apply uniformly to each entity or individual conducting a research project under this Act. In the case of any entity or individual receiving assistance from the Secretary for a [clinical] project of research described in subsection (b), the Secretary shall by regulation establish standards for responding to, including managing, reducing or eliminating, the existence of such a financial interest. The entity may adopt individualized procedures for implementing the standards.

19. 42 U.S.C. § 289b-1(b).

20. *See* 42 C.F.R. § 50.601 *et seq.*

21. *See* 45 C.F.R. § 94.1 *et seq.* and 60 Fed. Reg. 35,810 (July 11, 1995). Technical corrections were made to the PHS rules on July 31, 1994. *See* 60 Fed. Reg. 39,076 (July 31, 1995).

22. 60 Fed. Reg. 35,820 (July 11, 1995) (NSF Notice).

23. The PHS rule applies to subgrants, whereas the NSF policy does not.

24. The acting director of NIH recently published an Advance Notice of Proposed Rulemaking soliciting suggestions for amending the current rules to address a variety of issues, including the disclosure thresholds, institutional enforcement, and institutional conflicts. *See* 74 Fed. Reg. 21,610 (May 8, 2009). *See also Conflict of Interest: Revisions to NIH COI Regulations Likely, Medical College Association's Counsel Says*, 8 MED. RES. L. & POL'Y REP. 97 (2009). It could take a few years before HHS adopts final regulatory changes.

25. *See* 42 C.F.R. § 50.602.

26. *See* 61 Fed. Reg. 34,839 (July 3, 1996) (answering frequently asked questions about the

PHS rule and NSF policy); *see* NIH, Frequently Asked Questions (Oct. 28, 2008), http://grants.nih.gov/grants/policy/coifaq.htm#a5.

27. 42 C.F.R. § 50.603. PHS generally does not recognize a co-principal investigator; each PHS Act grant has only one principal investigator. *See* 42 C.F.R. § 52.2. In contrast, NSF recognizes co-principal investigators.

28. 60 Fed. Reg. 35,809, 35,812 (col. b) (July 11, 1995).

29. *See* 42 C.F.R. § 50.603 (PHS grants); 45 C.F.R. § 94.3 (PHS contracts); NSF Grant Policy Manual 510(b) (July 2002).

30. *See* 42 C.F.R. § 50.603 (noting that an equity interest includes stock and stock options).

31. The Black-Scholes models were developed by Fischer Black, Myron Scholes, and Robert Merton. Merton and Scholes received the 1997 Nobel Prize in Economics for this and related work; Black had died two years earlier and was therefore ineligible to receive the prize.

32. NIH, *supra* note 26; *see* 5 C.F.R. § 2634.310(c).

33. *See* 42 C.F.R. § 50.604(c)(1); 45 C.F.R. § 94.4(c)(1) (emphasis supplied).

34. The use of the conjunction can be traced to the Notice of Proposed Rulemaking in which the rule was organized in such a way that the conjunction was entirely appropriate. When the rule was issued in final form, though, the conjunction was retained, even though it was no longer appropriate. *See* 42 C.F.R. § 50.605(a), as proposed in 59 Fed. Reg. 33,242, 33,248 (July 28, 1994). It should be noted that in a guidance document, both agencies treat the two reporting triggers as if satisfying either is sufficient to mandate reporting. *See* Answer to Question 19, 61 Fed. Reg. 34,839 (July 3, 1996).

35. Stanford School of Medicine, Disclosure of Financial Interests, http://med.stanford.edu/coi/disclosure.html (last visited Aug. 3, 2009).

36. *See* Harvard Medical School, Policy on Conflicts of Interest and Commitment, http://www.hms.harvard.edu/integrity/conf.html (last visited Aug. 3, 2009).

37. Yale University, Conflict of Interest/Commitment Disclosure Form (2007) http://www.yale.edu/provost/html/disclosure_combined_07.pdf.

38. *See* S. Van McCrary, Cheryl B. Anderson, Jelena Jakovljevic, Tonya Khan, Laurence B. McCullough, Nelda P. Wray & Baruch A. Brody, *A National Survey of Policies on Disclosure of Conflicts of Interest in Biomedical Research*, 343 N. Eng. J. Med. 1621 (2000); Mildred K. Cho, Ryo Shohara, Anna Schissel & Drummond Rennie, *Policies on Faculty Conflicts of Interest at US Universities*, 284 J. Am. Med. Ass'n 2203 (2000); David Korn, *Conflicts of Interest in Biomedical Research*, 284 J. Am. Med. Ass'n 2234 (2000). On February 28, 2008, the Association of American Medical Colleges and the American Association of Universities, recognizing the great disparity in conflicts policies from institution to institution, issued a single set of guidelines that it hoped its members would adopt within the next two years. *See* Press Release, Ass'n of American Medical Colleges, AAMC, AAU Issue New Guidelines on Managing Conflicts of Interest (Feb. 28, 2008), *available at* http://www.aamc.org/newsroom/pressrel/2008/080228.htm.

39. International Committee of Medical Journal Editors, Uniform Requirements for Manuscripts Submitted to Biomedical Journals: Writing and Editing for Biomedical Publication (Oct. 2008), *available at* http://www.icmje.org/#conflicts/.

40. 42 C.F.R. § 50.605.

41. *Id.*

42. *See* Moore v. The Regents of the University of California, 793 P.2d 479 (Cal. 1990), *cert. denied*, 499 U.S. 936 (1991).

43. *Id.* at 483.

44. *See, e.g.*, Greenberg v. Miami Children's Hosp. Research Inst., 264 F. Supp. 2d 1064 (S.D. Fla. 2003); Duttry v. Patterson, 771 A.2d 1255, 1259 n.2 (Pa. 2001). *But see* S. 301, Physician Payments Sunshine Act of 2009, 111th Cong. (2009).

45. *See* 40 PA. CONS. STAT. § 1303.504(b).

46. 21 C.F.R. § 54.1(b).

47. *See id.* § 54.1(a).

48. *See id.* §§ 54.2(a), (b), (c), (f).

49. *See id.*§ 54.2(a).

50. *See id.*§ 54.2(b).

51. *Id.* § 54.2(c).

52. *Id.* § 54.2(f).

53. Drugs are marketed pursuant to an approved "New Drug Application," biologics pursuant to an approved "Biologic License Application," and medical devices pursuant to either an approved "Premarket Approval" application or a clearance pursuant to a premarket notification, referred to as a "510(k)." *See* 21 C.F.R. pt. 314 (NDA), pt. 601 (BLA), pt. 807 (510(k)), pt. 814 (PMA). Some medical devices are so common and pose so little risk (e.g., beds, bed pans, tongue depressors) that they can be marketed without any premarket review by FDA.

54. *See* 21 C.F.R. § 54.4(a).

55. *See id.* § 54.4(b).

56. *See Prasugrel Intellectual Bias: FDA Makes Changes, Congress Demands Answers*, PINK SHEET Mar. 9, 2009, at 33.

57. I do not view this type as a true "institutional conflict," but rather as an individual conflict involving two competing fiduciary obligations—the official's obligation to the university on the one hand and his or her obligation, as a board member, to the company and its stockholders on the other hand.

58. National Ass'n of College and University Business Officers, NACUBO Endowment Study, http://www.nacubo.org/documents/research/NES2008PublicTable-AllInstitutionsByFY08MarketValue.pdf.

59. *See* Rachel Zimmerman, *Harvard Dropouts: Endowment's Chief to Leave with Others*, WALL ST. J., Jan. 12, 2005, at C1, *available at* http://www.williamstrauss.com/Articles/harvard_dropouts.htm/.

60. ASS'N OF AMERICAN MEDICAL COLLEGES, PROTECTING PATIENTS, PRESERVING INTEGRITY, ADVANCING HEALTH: ACCELERATING THE IMPLEMENTATION OF COI POLICIES IN HUMAN SUBJECTS RESEARCH 37 (Feb. 2008).

61. Lyndsey Layton, *NIH Drops Contractor for Conflict of Interest*, WASH. POST, Apr. 14, 2007, at A8.

62. Press Release, U.S. Attorney's Office for the District of Md., NIH Senior Scientist Charged with Conflict of Interest (Dec. 4, 2006), *available at* http://www.usdoj.gov/usao/md/Public-Affairs/press_releases/press06a.htm.

63. http://www.usdoj.gov/usao/md/Public-Affairs/press_releases/press06/NIH%20Senior%20Scientist%20Pearson%20Sunderland%20Sentenced%200n%20Conflict%200f%20Interest%20Charge.html (last viewed July 4, 2007).

64. 70 Fed. Reg. 5543, 5546 (Feb. 3, 2005).

65. *See* 70 Fed. Reg. 51,559 (Aug. 31, 2005).

66. *See* 18 U.S.C. § 201.

67. *See id.* § 203.

68. *See id.* § 208.

69. *See id.* § 209.

70. *See* United States v. Sarin, 10 F.3d 224 (4th Cir. 1993).

71. *Hearings before the S. Comm. on Armed Services* (Aug. 1, 2006) (statement of Paul J. McNulty, Deputy Attorney General), *available at* http://www.usdoj.gov/dag/testimony/2006/080106dagmcnultystatementsenate.htm; *see* Jerry Markon & Renae Merle, *Hearing Set for Boeing Ex-CFO Recruitment of Air Force Official Added to Tanker Controversy*, WASH. POST, July 29, 2004, at A7.

72. *See* CNN, *Pentagon Puts Hold on Tanker Contract Process*, CNN.COM, Sept. 10, 2008, http://www.cnn.com/2008/US/09/10/pentagon.tankers/index.html.

73. *See* 18 U.S.C. § 202.

74. *See id.* § 203.

75. If the special government employee worked more than 60 days for the government during the immediately preceding 365 days, then the individual is not permitted to appear before his or her employing agency on behalf of a private entity.

76. *See* United States v. Baird, 29 F.3d 647 (D.C. Cir. 1994).

77. *See* 18 U.S.C. § 208.

78. *See* U.S. Dep't of Justice, Ethics Issues Related to the Federal Technology Transfer Act of 1986 (Sept. 13, 1993), http://www.usdoj.gov/olc/208.htm.

79. Draft Opinion of Office of Government Ethics 4 (1993).

80. *Id.* at 5.

81. In FY 2003, more than 62,000 individuals served on federal advisory committees. *See* GOVERNMENT ACCOUNTABILITY OFFICE, FEDERAL ADVISORY COMMITTEES 65 (Rpt. No. GAO-04–328) (Apr. 2004), *available at* http://www.gao.gov/new.items/d04328.pdf.

82. *See* 18 U.S.C. § 208(b)(3).

83. *See id.* § 209 (anti-supplementation law).

84. *See* United States v. National Treasury Employees Union, 513 U.S. 454, 469 (1995). Although the federal statute at issue in *National Treasury Employees Union* applied to all federal employees, the Supreme Court's decision involved only employees in the executive branch at the GS/GM grade 15 or lower. Employees in the other branches of government (i.e., legislative and judicial) and those in the senior executive service were not parties to the case.

85. *See* 5 C.F.R. § 2635.807.

86. One does not become a special government employee by accepting grant funding.

87. *See* http://www1.0d.nih.gov/cmo/faqs/main.html#Answer13/ (last viewed April 26, 2009).

88. *See* 61 Fed. Reg. 39,756, 39,758 (col. c) (July 30, 1996).

89. *See* U.S. Dep't of Justice, *supra* note 78.

90. *See* 5 C.F.R. pt. 2635; 18 U.S.C. §§ 201, 203, 209.

91. *See* 5 C.F.R. § 2635.204(a).

92. *See id.* § 2635.202(c)(3) (Example 1).

93. *Id.* § 2635.204(g)(2).

94. *See id.* § 2635.204(g)(3).

95. *Id.* § 2635.204(g)(3)(i).

96. *See id.* § 2635.204(g)(2).

97. *Id.*

98. *Id.*

99. *See* Office of Government Ethics, Informal Advisory Letter 99 × 2; *id.*, Informal Advisory Letter 93 × 15.

100. *See* 5 C.F.R. § 2635.204(g)(6) (dealing with authorization of spouse to attend).

101. *See* 18 U.S.C. § 201 (bribery); *id.* § 203 (providing compensation to government employees and members of Congress with intent to influence them); *id.* § 209 (supplementing a federal employee's salary).

102. *See* 42 C.F.R. pt. 52h.

103. *Id.* § 52h.2(b).

104. *Id.* § 52h.2(q).

105. *Id.* § 52h.5(b)(1).

CHAPTER 6

Who Can Gain Access to Your Data?

We like to believe that our government is, in the vernacular of inside the Beltway, "transparent," and that that transparency is an inherent feature of a republican form of democracy, one that existed from the outset. After all, the House has conducted the people's business in an open forum since the First Congress; the Senate has done the same since December 9, 1795;[1] and, for the past twenty-five years or so, the proceedings in both Chambers have been aired on television daily.[2] Correspondingly, the courts have a long tradition, antedating the Norman Conquest, of conducting their business in public. "The right to an open public trial is a shared right of the accused and the public, the common concern being the assurance of fairness."[3]

What goes for the Congress and the courts, though, does not necessarily go for the executive branch, which thrives on opacity; it is much easier to make decisions out of the harsh glare of public scrutiny, protected by closed doors. Those doors, though, started being pried open in the mid-1960s with the enactment of the Freedom of Information Act (FOIA) and later with enactment of other "open government laws," such as the Privacy Act, Government in the Sunshine Act, the Federal Advisory Committee Act, the Information Quality Act, and the Shelby Amendment.

The benefits of FOIA and all open government laws are relative. Those seeking to learn what the government is doing find them beneficial; those seeking to protect information find them intrusive, at best, and embarrassing at worst. Open government is a two-way street: Documents that you submit to the government, including grant applications, might be subject to public disclosure. In addition, the move toward more open government has increased pressure on funding agencies to ensure that data, articles, and reagents developed with federal funding are shared with other scientists and with the public. This chapter examines these open-government laws and initiatives and how they might affect your ability to control your grant applications, your research data, your scientific publications, and your reagents and specimens.

THE FREEDOM OF INFORMATION ACT

The Freedom of Information Act[4] is the grand-daddy of federal disclosure laws. Enacted on July 4, 1966, it has become, in a single generation, such an accepted feature of American government that its acronym has been transformed into a verb, as in "I just FOIAed Henry's grant application; he will not be a happy camper when he finds out." Understanding FOIA's structure is important to understanding what it does and how it really works.

FOIA requires government agencies to disclose, upon request, all documents in their possession (called "agency records") unless any one of nine exemptions applies. If an exemption applies, the government can refuse to disclose the records. However, merely because an exemption applies does not mean that the government is precluded from disclosing those documents. Even if an exemption applies, the government, in most but not all instances, has the option of either disclosing or not disclosing, as the case may be.

When Does FOIA Apply?

At first blush it might appear that FOIA's application is relatively broad, opening up everything related to federal spending, decision making, and the like. It is not. It applies only to "agency records." But what are "agency records"? Are data developed by a principal investigator under a federal grant "agency records," because the government funded the research and may compel a grantee to provide the data to the funding agency? In other words, are your data subject to a FOIA request? Let's take the case of the University Group Diabetes Program (UGDP), a consortium of academic medical centers coordinated by the University of Maryland, which studied the effectiveness of various treatment regimens for diabetes. The study was funded solely by $15 million in grant funds from the National Institutes of Health (NIH) and generated more than 55 million records documenting the treatment of over one thousand diabetic patients who were monitored for a five- to eight-year period. In 1970, UGDP presented its initial findings, which suggested that treatment of type II diabetes with the hypoglycemic drug tobutamide increased the risk of death from cardiovascular disease as compared to the four other treatments examined by the study. Later, UGDP indicated that its data suggested that another hypoglycemic drug (phenformin hydrochloride) had a similar propensity. These findings generated significant controversy within the medical community.

One group, the Committee on the Care of the Diabetic (Committee on Care), a national association of physicians involved in treating diabetes, was particularly critical and skeptical of UGDP's conclusions and asked UGDP for its raw data so that the Committee on Care could reanalyze them. UGDP refused to share its data with the Committee on Care, whereupon the Committee on Care requested under FOIA that NIH provide it with UGDP's data. The Committee on Care argued that the data were agency records because NIH and its mother agency, the Department of Health, Education, and Welfare (HEW), had the right to demand the data, had actually audited the data, and one its constituent agencies, the Food and Drug Administration (FDA), had used the data to warn physicians about the risks associated with phenformin and tobutamide. Nevertheless, NIH and HEW refused to provide the data, concluding that they neither owned nor possessed the data, and therefore the data were not agency records and not subject to FOIA. The Committee on Care instituted suit.

The case—*Forsham v. Harris*—meandered its way to the Supreme Court, where a seven-member majority concluded that "where a grant was used, there is no dispute that the documents created are the property of the recipient, and not the Federal Government."[5] Accordingly, the Court held that the data were not agency records, were not subject to FOIA, and "that Congress did not intend that grant supervision short of Government control serves as a sufficient basis to make the private records 'agency Records' under [FOIA]."[6] While holding that Congress intended to keep federal grantees free from the direct obligations imposed by FOIA, the Court strongly implied that the result would be different had the data been generated under a procurement contract. (Procurement contracts are discussed in chapter 2.)[7]

The Court's holding in *Forsham* that raw data generated under a grant are not subject to FOIA unless they are in the physical possession of the government disturbed some members of Congress and led, many years later, to two legislative efforts to end-run the Court's decision in *Forsham*—the Shelby Amendment and the Data Quality Act, both of which are examined later in this chapter.

Not only is FOIA limited to records in the possession of the federal government, but it is also limited to "records" in the form of documents, which do not include reagents, specimens, and the like. It does include photographs and disks with computer information or data, but does not include tangible items that are not used to store information. Thus, a specimen contains information, but the information was not put there by a human. Since a specimen does not store information put there by humans, it is not a "record" subject to FOIA.

How Do the Exemptions Work?

Merely because documents are deemed to be agency records doesn't automatically mean that they must be disclosed. The government can invoke any one of nine exemptions,[8] three of which are commonly invoked to justify an agency's refusal to provide research-related documents: (1) "trade secrets and commercial or financial information obtained from a person and privileged or confidential" (Exemption 4); (2) "personnel and medical files and similar files, the disclosure of which would constitute a clearly unwarranted invasion of personal privacy" (Exemption 6); and (3) documents that reveal an agency's deliberative process in reaching a decision (Exemption 5). If documents fall within an exemption, the government is authorized to withhold the information. How does this all play out if you are seeking information from the government? And how does this play out if you seek to prevent some of your information from being disclosed?

TRADE SECRET AND COMMERCIAL CONFIDENTIAL: EXEMPTION 4

This exemption permits the government to withhold those portions of documents that contain either a trade secret or commercial information that is confidential or privileged. In fact, an agency is prohibited from releasing trade secret information, but has more discretion over the release of confidential commercial information. Normally, a trade secret is information in a formula, process, program, compilation, and the like (1) that is valuable, in part, because it is secret; and (2) that the owner seeks to keep secret.[9] The quintessential example of a trade secret is the formula for Coca-Cola. It is not altogether clear, though, that this broad definition should apply to FOIA requests, where the driving force is openness. For FOIA purposes, some courts have taken a narrow view of what constitutes a trade secret.

Confidential commercial information includes customer lists and the like. Even if such information does not rise to the level of trade secret, it deserves protection nonetheless. Pricing information is on the cusp, and most courts would likely afford it trade secret protection. Whether something qualifies as a trade secret, as opposed to confidential commercial information, is important in two interrelated ways. First, federal law makes it a crime for a federal employee to knowingly divulge a trade secret.[10] Second, if information qualifies as a trade secret, the government is affirmatively prohibited from releasing the information. If it is merely commercial confidential information, the government may, under certain circumstances, release the information.

What impact does all this have on your research-related documents in the possession of the government? Do research designs and related information contained in funded grant applications warrant trade secret protection? This question was addressed more than thirty years ago when a group of policy wonks at the Washington Research Project filed a FOIA request with National Institute of Mental Health (NIMH) seeking the initial grant application, site reports, and the summary statements concerning eleven specific projects funded by NIMH, all involving research into the comparative effects of various psychotropic drugs on children with learning disabilities. NIMH refused to provide the information, noting that the requested documents contained research designs that were the researchers' "stock-in-trade," and their premature release could deprive the scientists of career advancement and material rewards in much the same way that the release of trade secrets can deprive one of a competitive advantage.

The court of appeals, in concluding that the information in a grant application does not qualify for trade secret or commercial protection, held that "it is clear enough that a non-commercial scientist's research design is not literally a trade secret or item of commercial information, for it defies common sense to pretend that the scientist is engaged in trade or commerce. This is not to say that the scientist may not have a preference for or an interest in nondisclosure of his research design, but only that it is not a trade or commercial interest."[11] Similar attempts over the years to protect the contents of funded grant applications because they contain trade secrets or commercial confidential information have failed.[12]

A decade after *Washington Research Project*, it was FDA's turn to protect data. FDA, in the course of monitoring the use of the intraocular lens, had amassed a wealth of data on adverse reactions and other problems. An intraocular lens, a small plastic lens, is most commonly used to surgically replace a human lens that has clouded over as a result of cataracts. The Health Research Group, a nonprofit group that wanted to ensure that FDA was appropriately monitoring intraocular lens use, sought two undated summary reports of complications and adverse reactions in intraocular lens studies and related information gathered during the course of clinical trials. FDA refused to provide the information, arguing, among other things, that the information qualified as either trade secrets or commercial confidential information. Again, the case went to the Court of Appeals for the District of Columbia Circuit.

In the course of its discussion of trade secret information, the court held that it would be inappropriate to use the normal definition of what constitutes a trade secret where FOIA is involved because there is strong policy under FOIA that favors the release of information. Accordingly, the court narrowly defined a trade secret for FOIA purposes only as "a secret, commercially valuable plan, formula, process, or device that is used for the making, preparing, compounding, or processing of trade commodities and that can be said to be the end product of either innovation or substantial effort."[13] In other words, to be a trade secret under FOIA the secret has to be used to make something. The court then went on to hold that none of the information at issue was related to making anything and this, of course, would always be the case with any data from a clinical trial. But merely because the information did not qualify for trade secret protection did not end the inquiry.

The court next examined whether the clinical trial data were commercial information intended to be kept confidential. First, the court held that because the health and safety data at issue would be instrumental in gaining market approval from FDA, the intraocular lens manufacturers had a commercial interest in the information. Second, the court examined whether the information was "confidential." To be confidential under FOIA, release of the information either has to impair the government's ability to obtain the information in the future or cause substantial harm to the competitive position of the person who provided the information. The court sent the case back to the lower court for a more careful analysis on that point. The parties likely resolved their differences or were satisfied with the eventual outcome because the case did not generate any further published decisions.

PRIVACY PROTECTION: EXEMPTION 6

Under FOIA, the government is not supposed to release documents containing personnel files, or medical files, or similar records where the disclosure of these documents "would constitute a clearly unwarranted invasion of personal privacy." Courts apply this exemption by balancing the public's interest in knowing against the individual's interest in maintaining privacy. As with any balancing test, sometimes the scales favor release and other times they do not.

In the late 1970s, George Kurzon, a physician and former clinical researcher, wanted to test whether NIH's peer review system was biased against unorthodox proposals. To do the study, he filed a FOIA request for a list of investigators who had applied for National Cancer Institute (NCI) grants but were unsuccessful. NIH denied the request and Kurzon sued. The trial court ruled in the government's favor, finding that to provide the list would be an invasion of the privacy rights of unsuccessful grant applicants. The appeals court, however, disagreed. It found that the public interest in the study was significant; it also found that the information sought did not trigger the privacy exemption because "approximately twice as many applications are rejected as are not." The court concluded there was no stigma associated with not receiving an NIH grant.[14] In the end, the court found that a list of unsuccessful grant applicants does not fit within the exemption for personnel and medical files and similar files, "the disclosure of which would constitute a clearly unwarranted invasion of personal privacy." Two decades later, Kurzon submitted an identical request to NIMH. NIMH, like NCI, refused to provide the information, arguing that it was protected under the privacy exemption. Courts tend to be consistent, and this case was no exception. The trial court ordered NIMH to turn over its list of unsuccessful grant applicants.[15]

Nor does the privacy exemption shield the financial interests of those who serve on NCI advisory committees. In the early 1980s, the *Washington Post*, in the course of investigating possible conflicts of interest at NIH, sought certain rudimentary financial information (e.g., full-time employer and those for whom they consult) about those sitting on NIH study sections and advisory council. NIH refused to provide the information on privacy grounds. The court of appeals rebuffed the agency. It concluded that the information collected by the government and sought by the *Post* did not reveal rates of pay or value of assets and therefore was only minimally intrusive.[16]

Even though there is a strong presumption favoring the release of information, personal privacy can trump this policy. In 1990, Dr. Charles McCutchen, a physicist and an eccentric multimillionaire, filed a FOIA request seeking the release of the names of all scientists who had been the subject of a scientific misconduct investigation by the Office of Research Integrity (ORI) but had been cleared. The government normally releases this list, but with the names of the scientists redacted. McCutchen wanted the names. The trial court ruled in McCutchen's favor; the Department of Health and Human Services (HHS), joined by various university and research organizations, appealed.[17] The court of appeals balanced the privacy interests of the anonymous scientists against the public's right to know. With respect to the scientists, the court concluded that scientists "who have been investigated and exonerated, have a substantial privacy interest in remaining anonymous."[18] The court also held that the public interest did not outweigh the individual interests because there was no public interest in knowing the names of individuals against whom allegations had been levied but who had been exonerated.[19]

DELIBERATIVE PROCESS: EXEMPTION 5

Robert Casad was a dissatisfied grant applicant, but not for the usual reason. Casad's grant application was funded by the National Institute on Aging. However, he wanted the whole truth about what the peer reviewers really thought of his application. At issue was the summary sheet, which contained the study section's recommendation, the priority score, and a summary of the factors considered during peer review. It also recounted the views of the individual peer reviewers, along with a summary of any group discussion. Casad received most of the summary sheet, but those portions that revealed the give-and-take of the review process were redacted. Casad, acting as his own attorney (usually a mistake), sued. The trial court affirmed the National

Institute on Aging's decision to redact certain documents. On appeal, the court of appeals affirmed the trial court's decision and indicated that the deliberative process exemption is based on the commonsense assumption that officials will not communicate candidly among themselves if each remark is likely to make its way onto the front page of a local newspaper. In protecting these documents from disclosure, the exemption enhances the quality of agency give-and-take by promoting open discourse.[20]

How Do You File a FOIA Request?

If you want documents under FOIA, you submit a written request to the Freedom of Information Office at the agency that you believe maintains the files. Each agency has its own FOIA rules, which are relatively uniform across agencies. Also, each agency reserves the right to charge you for the search and for copying. However, most academics who seek information for academic purposes qualify for a total waiver of search and copying fees.[21] Once you have submitted your request, an agency must advise you within twenty business days whether it will honor your request.[22] Simply acknowledging within the twenty-day window that it has received your request and is processing it is not sufficient. There are little tricks that agencies use to circumvent the twenty-day requirement. Most of these gambits are authorized by FOIA, but are very narrow. For example, an agency can extend the twenty-day period by (1) notifying the requester that the records sought are located in field facilities or elsewhere and the agency needs additional time to gather and examine them or (2) noting that the agency needs to search, collect, and examine such a voluminous amount of records that additional time is necessary.[23]

Some agencies are notoriously slow when it comes to handling FOIA requests. The combined FOIA backlog for government agencies in 2002 was 140,000 requests. HHS has a median processing time of about one hundred days, having received more than 100,000 requests in 2002.[24] The delay, however, is enviable when compared with FDA, which acts on FOIA requests as if the calendar had not been invented. FDA can take years to provide documents—if you're lucky and the agency has not lost your request. Sometimes the materials come so long after they have been requested that the requester may forget having even made the request. In 2008, the *Washington Post* reported that the National Security Archive's fourth annual Rosemary Award—a tongue-in-cheek award—went to the Treasury Department, which gave "'a new meaning to the notion of subprime performance' in its handling of Freedom of Information Act requests."[25] According to the *Post*, one Archive request has been pending for twenty-one years.

Putting aside agency competence, how does a FOIA request play out? There are four ways that an agency can respond to your request after it acknowledges receipt. First, it can promptly provide the documents requested. This will occur, for example, if you are requesting documents that have already been requested by and provided to someone else. Second, the agency can ask you to clarify or focus your request to make it easier for the agency to search their records and respond to your request. Third, the agency can refuse to provide all of the information that you requested. If it declines to provide you with documents or redacts portions of documents that it provides to you, it must tell you why it has declined to satisfy your request. It does this either by citing to one or more of the nine FOIA exemptions or by indicating that it does not possess the documents. Some redactions can be so sweeping that the agency might as well provide you with a ream of blank copying paper.[26] And fourth, it can do nothing, like FDA and some other agencies. It is the third and fourth possibilities that frequently lead to district court.

If an agency refuses to provide information, as was the case with Dr. Kurzon, you have a right to appeal that initial decision. However, if you want to get into federal court, you must work your way through the agency's internal appeals mechanism. In short, you must "exhaust" your administrative remedies before seeking judicial relief. At HHS, for instance, you must file your appeal within thirty days after you are told that the agency is withholding certain records.[27] The appeal must be decided within twenty business days. If the agency fails to do so, you may proceed directly to court.[28] If you lose your administrative appeal, you can also proceed to court.

If an agency fails to respond meaningfully at all to your initial FOIA request within the twenty-day statutory period, you can proceed to court without further ado.[29] This is called constructive exhaustion. As a practical matter, most requesters wait until they receive some form of substantive response and then appeal administratively before going to court. Some requesters, though, who really need the documents quickly and cannot afford to wait one or two years for the agency to process their request have sued immediately after the agency failed to indicate within the twenty-day period whether it would honor the request.[30]

If you prevail, you are entitled to reasonable attorneys' fees, and recent changes in the law may make it somewhat easier to qualify for fees.[31] However, no matter what the standard, getting attorneys' fees out of the federal government is never easy or quick. For example, on March 9, 2000, a group called the Campaign for Responsible Transplantation (CRT) submitted a FOIA request to FDA for all records concerning clinical trials that involved xenotransplantation. After instituting suit and obtaining some of the documents during the course of the litigation, CRT sought attorneys' fees from the government. It took the group three years of litigation before a court decided that CRT was entitled to fees.[32]

How Do You Stop a FOIA Request?

What happens when the shoe is on the other foot and someone requests a copy of documents you have submitted to NIH, NSF, or another funding agency which you consider to contain either private information or trade secrets? Agencies have a limited obligation to let you know that your documents have been requested.[33] You can increase the likelihood that an agency will contact you in the event that your documents are requested by "legending" (i.e., stamping) your commercial confidential or trade secret pages with the appropriate phrase ("commercial confidential" or "trade secret" or both). If the agency contacts you and you object to the release, but the agency disagrees with your analysis and indicates that it will release your documents, you can sue the agency to enjoin it from releasing your documents. These suits are called reverse FOIA suits, and they frequently occur when one company has sought and is about to receive documents that had been submitted to the government by one of its competitors. To prevent the release, the competitor sues the agency; the company that sought the release will normally intervene in the suit on the government's side.

THE PRIVACY ACT OF 1974

The Privacy Act is not what you think it is. When originally enacted it was designed primarily to prevent one federal agency from sharing your personal information (e.g., medical, financial, educational) with other agencies or from publicly releasing that information.[34] It also gave citizens the right to request to see their own files and to demand corrections to those files. In practice, the Privacy Act is narrow and easily circumvented by agencies. The act applies only to records about individuals, and not universities or companies; it applies only to records that the government would be prohibited from disclosing under FOIA. And most of the act applies only to records that are stored in what is called a "system of records." A system of records is collection of information about individuals maintained by an agency where the records are retrieved by the names (or other identifiers) of the individuals. If an agency maintains files that contain information about thousands of individuals, and the information can be retrieved by the names of the individuals, that is not enough to trigger the Privacy Act. If, for example, the agency retrieves the records by the name of a university, the records are not maintained in a system of records because the actual retrieval is not done by the name of an individual.[35] The agency, as a matter of practice, must retrieve the records by individuals' names (or other unique identifiers).

One of the more critical features of the Privacy Act is the requirement that you be given access to your records and be provided an opportunity to correct any errors. In reality, this aspect of the act has not worked well. Agencies are reluctant to acknowledge that their records may be incorrect and will go to great lengths to avoid disclosing records, the first step on the road to correcting errors. One of the more fascinating cases, alluded to earlier (see chapter 2), involved Dr. Bernard Fisher, the head of the National Surgical Adjuvant Breast and Bowel Project (NSABP), funded for decades under a large cooperative agreement from the National Cancer Institute. Fisher and his team discovered during a routine data audit that the principal investigator (Dr. Roger Poisson at St. Luc Hospital in Montreal), at one of the more than one hundred clinical sites involved in one set of studies, had intentionally accrued patients who did not meet the study's eligibility criteria. Fisher reported the incident to NCI. A dispute then arose between NSABP statisticians and some of the administrators at NCI over how to handle data from that site. In the end, Fisher and his team included the data from that site in their published articles because (1) the study was designed to accommodate a specific number of errors from each site, (2) the error rate at the site in question was below that threshold, (3) accrual occurred before randomization, and (4) discarding such data would be inconsistent with accepted statistical practices. As an aside, the data from St. Luc had no impact on the study's overall results or conclusions. When the matter became public and subject to a congressional hearing, Sam Broder, the director of NCI, ordered that Fisher's publications abstracted in Medline, the NIH database of biomedical literature, incorporating any data from St. Luc, be annotated with the phrase "scientific misconduct-data to be reanalyzed." Many of Fisher's articles that contained no data from St. Luc's were also inappropriately annotated with the "scientific misconduct" flag.

Shortly thereafter, on March 16, 1995, Fisher filed suit in district court in Washington, D.C., seeking to enjoin the annotation on the grounds that Medline was a system of records and that by inserting the phrase "scientific misconduct," Broder had violated the Privacy Act. Fisher further argued that the annotation was inaccurate.[36] The court entered a stipulated preliminary injunction ordering the removal of the annotations and the addition of a new annotation—"[prior annotation incorrect]." Fisher also sought damages under the Privacy Act, which permitted someone who has been injured as a result of an intentional violation to obtain damages of $1,000 per violation. The following year the court held that Fisher's articles in Medline were not about Fisher but about the results of his studies, and, as such, the Privacy Act did not apply. Since the information in Medline was not about individuals, they did not constitute "records" within the meaning of the act.[37] The "scientific misconduct" annotation, though, was never reinserted, and Sam Broder quietly resigned following extraordinary criticism over his lack of leadership and common sense in handling the St. Luc affair.

At about the same time that Fisher's case was being decided, another Privacy Act case, this one against the National Science Foundation (NSF) and the Department of Commerce, was before the appeals court. The case involved a grant application submitted in April 1993 by a company, Dynamic In Situ Geotechnical Testing, Inc. (Dynamic); the principal investigators (PIs) were Dynamic's co-owners, Wanda and Robert Henke. The Henkes' NSF proposal, along with thirty-six others, was reviewed by a twelve-member peer review panel of experts from universities and other government agencies; four members of that panel prepared a written assessment of the Henkes' proposal which was shared with the full panel. The full panel recommended against funding the proposal, and it was not funded. Wanda Henke then filed a Privacy Act request seeking the names of the four peer reviewers who prepared the written comments as well as the names of the other eight panel members. NSF conceded that the proposals and the corresponding review materials were retrieved by the names of the PIs and, therefore, the entire packet constituted a system of records within the meaning of the Privacy Act.[38]

NSF disclosed the names of the twelve panel members but refused to identify which four authored the written comments. The Privacy Act has exemptions that permit an agency to decline to provide information. The Privacy Act exemptions differ from the FOIA exemptions. One of the Privacy Act exemptions "protects the identity of confidential sources who provide agencies with information regarding the suitability or qualifications of applicants for 'Federal contracts.'"[39] NSF argued that this exemption applied to the four primary reviewers because it had promised them anonymity and the Privacy Act protects such promises. NSF further argued, and the court agreed, that because PIs and reviewers often switch hats, reviewers may be disinclined out of fear of reprisal to provide blunt and objective reviews if their identities would be revealed. The Henkes argued that the exemption applied only to "Federal contracts" and not grants. The court brushed this argument aside and concluded that NSF was permitted under the Privacy Act exemption to withhold the information, even though it related to the Henkes and was a record within a system of records.

The Privacy Act is complicated, intertwined with other statutes, and, given the way it has been interpreted by the courts, provides significantly less protections than originally envisioned. It requires agencies to do a lot of paperwork, including publishing their systems of records in the *Federal*

Register, but there is little evidence that it is effective in slowing the movement or release of information, other than medical records and the like, and medical records are now covered by HIPAA (see chapter 4). The same is not the case with the Federal Advisory Committee Act and the Government in the Sunshine Act, discussed below, both of which have significantly affected the way in which the government conducts its business. Both are particularly relevant to the research community because peer review groups come within FACA's broad umbrella.

THE FEDERAL ADVISORY COMMITTEE ACT AND GOVERNMENT IN THE SUNSHINE ACT

It is perhaps in the nature of government that national leaders more readily seek advice from their friends and confidants than from those who serve in official capacities. President Andrew Jackson took this practice to an extreme: During the first two years of his administration, he never held an official cabinet meeting. Instead, he relied on the advice of a small group of trusted advisors and old friends, only a few whom held official positions in the government. Jackson's critics dubbed this group of influential advisors the Kitchen Cabinet, a term that has survived. Jackson's practice was quickly emulated and over time amplified by those in the departments and agencies. By the 1960s, the number and influence of these advisory groups—some of which were statutorily recognized—had grown to such a degree that many questioned the wisdom of permitting largely unregulated, and in some cases anonymous, groups of nonofficials to formulate government policy, frequently behind closed doors. Congress responded by enacting in 1972 the Federal Advisory Committee Act (FACA).[40]

FACA sought to bring a semblance of order, openness, and balance to the process of creating, staffing, and operating advisory committees. FACA is particularly significant to the research community, owing to the number and influence of advisory committees that operate under the aegis of NIH, NSF, and FDA.[41] Under the Public Health Service Act, for instance, NIH is not permitted to award a research grant or cooperative agreement in excess of $50,000 unless the prospective award has been first approved by two advisory committees: the initial review group and the advisory council for the institute that would be funding the research (see chapter 2).[42] Fundamental research policy affecting recombinant technology has been and likely will continue to be developed by the Recombinant DNA Advisory Committee (RAC) (see chapter 4).[43] Correspondingly, FDA is required by statute to empanel advisory committees to develop performance standards for certain existing medical devices and to recommend whether new drugs and devices ought to be approved by FDA for marketing.[44] Not surprisingly, more than half of the more than one thousand federal advisory committees are employed by science funding or science regulating agencies.

FACA imposes significant constraints on advisory committees, quite aside from regularly requiring each to justify why it should continue in existence. Thus, FACA requires that each advisory committee file a charter with the General Services Administration.[45] Advisory committee meetings must be announced in advance through publication of a notice in the *Federal Register*[46] and in most cases must be open to the public.[47] Moreover, whether a meeting is to be open to the public or not, the committee is required to maintain detailed minutes of the meeting,[48] and those minutes and related documents must be made available on request subject to the Freedom of Information Act. Most significantly, FACA requires "the membership of the advisory committee to be fairly balanced in terms of points of view. . . ."[49]

What Is an Advisory Committee?

But these rules apply only to a group that is deemed to be an "advisory committee." What precisely is an "advisory committee"? A federal advisory committee is any group of individuals, at least one of whom is not a full-time officer or employee of the federal government, which is established by "statute or reorganization plan, or . . . utilized by the President, or . . . utilized by one or more agencies, in the interest of obtaining advice or recommendations. . . ."[50] This definition is not very helpful because it does not provide any useful way to ordain the boundary between what is and what is not an advisory committee. There was much confusion in the land. Some courts, especially in the earlier years, applied the statute quite literally and held that any group of individuals, even ad hoc groups, convened to provide advice to an agency head constituted an advisory committee.[51] Others adopted a less literal and more agency-friendly approach and declined to extend FACA to include informal or ad hoc group meetings.[52]

Things really did not start to sort themselves out until the Supreme Court decided *Public Citizen v. United States Department of Justice*,[53] a case having nothing to do with science, but which, in a footnote, was to create real heartache for the National Academy of Sciences. At issue in *Public Citizen* was whether the American Bar Association's Standing Committee on the Judiciary (ABA Committee) constituted an advisory committee. The ABA Committee was not established by the government. However, the Department of Justice regularly sought its advice about whether potential nominees for federal judgeships were qualified. The ABA Committee was such an integral part of the judicial selection process that the Department of Justice required putative nominees to complete the ABA Committee's questionnaire and to submit it to both the committee and the assistant attorney general for legal policy. Based on a variety of factors, including the aspirant's answers in the questionnaire, the ABA Committee rated the would-be nominee as "exceptionally well qualified," "well qualified," "qualified," or "unqualified."[54]

The conservative Washington Legal Foundation, later joined by the liberal Public Citizen, instituted suit against the Department of Justice, arguing that the ABA Committee, to the extent that it provided advice to the Department of Justice, was an advisory committee and, as such, its membership had be balanced and its meetings open to the public. The Court, speaking through Justice William J. Brennan, first concluded that the ABA Committee, although not established by the Department of Justice or the president, nevertheless furnished "advice or recommendations" to the president via the Justice Department. According to the Court, "whether the ABA Committee constitutes an 'advisory committee' for purpose of FACA therefore depends upon whether it is 'utilized' by the President or the Justice Department. . . ."[55]

The Court recognized that if FACA were read literally, it would apply to the ABA Committee. The Court then went on to conclude that such a reading would intrude into the president's affairs and would "compel an odd result." Even after reviewing the legislative history, the Court noted that "it seems to us a close question whether FACA should be construed to apply to the ABA Committee. . . ."[56] Ultimately, the Court concluded that if one were to give FACA a literal and expansive reading and to give the term "utilized" its common meaning, then FACA would impermissibly interfere with executive branch decision making in an area of special constitutional significance—the appointments process. According to the Court, Congress, in choosing to anchor the definition of advisory committee on whether it was "established or utilized" by an agency, never "intended to go much beyond" the narrower phrase "established or organized," which appeared in the original version of the legislation.[57] Since the ABA Committee was neither established nor organized by the government, it was, in the Court's view, not an advisory committee.[58] Although *Public Citizen* failed to provide the lower courts with cogent guidelines for differentiating between committees subject to FACA and those that are not, the message that it sent to the lower courts was clear: Agencies are to be given significant leeway when it comes to getting advice from outside groups. Indeed, in the years following *Public Citizen*, the agencies have been remarkably successful in beating back challenges to the unregulated use of ad hoc committees.[59]

The National Academy of Sciences was an unintended victim of the Court's decision in *Public Citizen*. Buried in the opinion was a gratuitous comment that "'a Government-formed advisory committee' such as the groups organized by the National Academy of Sciences and its affiliates . . . would be subject to FACA's requirements."[60] Not missing a beat, the Animal

Legal Defense Fund, Inc., instituted suit against HHS and the United States Department of Agriculture (USDA) over the way in which they developed their *Guide on the Use and Care of Laboratory Animals* (Guide) (see chapter 8). HHS and USDA contracted with the National Academy of Sciences (NAS) to help develop and regularly revise the Guide. NIH and USDA incorporate the Guide into their regulations; those receiving NIH funding are required to follow the principles set out in the Guide. The activities of the National Academy's committee charged with revising the Guide were funded by a grant from NIH. While the National Academy committee held public hearings, its deliberations were closed to the public. In their suit, the Animal Legal Defense Fund and others claimed that the National Academy committee was really an advisory committee under FACA and, therefore, the NAS should be required to hold its deliberations in public and also to provide either minutes or a transcript of its meetings.[61]

One would have thought that in light of *Public Citizen* there would be little doubt but that a committee established by a government contractor or grantee would not constitute an advisory committee. However, the appeals court felt constrained by the language buried in *Public Citizen* to hold that the National Academy committee was an advisory committee. The court, in holding that the Guide Committee was governed by FACA, emphasized that the National Academy was a quasi-public organization, permeated by the federal government.[62] Congress quickly responded by amending FACA to expressly exclude the NAS from the reach of FACA.[63]

Unless a committee is recognized by regulation, statute, or its charter as falling within FACA, it is unlikely that a court will burden that committee with FACA's requirements. The NIH study sections and advisory councils are expressly denoted in the Public Health Service Act as advisory committees, so there is no doubt there—they are all governed by FACA.

What Do Advisory Committees Have to Do?

The most significant features of FACA are the requirements that advisory committee membership be balanced to reflect a diversity of views and that advisory committees meet in public and make their records available in accordance with FOIA. The open meeting provision actually compels agencies to disclose materials that the agency could have otherwise declined to disclose under FOIA.

BALANCED MEMBERSHIP

FACA requires that committee membership be fairly balanced. When an advisory committee is charged with making recommendations concerning controversial issues, it is not uncommon for one group to voice concern that the committee make-up is biased (i.e., does not have enough members whose views coincide with their views). For example, one group sued the Secretary of Agriculture and others claiming that the Secretary's National Advisory Committee on Microbiological Criteria for Foods was not fairly balanced because none of its members were consumer advocates.[64] The twenty-member committee consisted of government scientists or scientists from the private sector or universities with the requisite expertise in microbiology and public health. In turning back the challenge, the court emphasized that FACA's balanced membership requirement does not mean that those who were not appointed can sue. The plaintiffs claimed that the committee was dominated by those from the food industry, although only six members of the committee were from that sector. In the end, two of the three judges held that the committee was appropriately balanced, and the third held that a court had no business even looking to see whether a committee was balanced.

Despite the fact that courts have been reluctant to enter the appointments fray, agencies generally go out of their way to ensure a semblance of balance on each committee. Sometimes they succeed and sometimes they do not. Sitting on an advisory committee can be time consuming, and, as result, agencies often find that it is difficult to find qualified individuals who are willing to serve.

OPEN MEETINGS

FACA, which was enacted in 1972, would be largely irrelevant were it not for another piece of open government legislation passed in 1976, the Government in the Sunshine Act (Sunshine Act).[65] The Sunshine Act requires collegial bodies, namely federal commissions and committees that render decision by majority vote, to hold their meetings and deliberations in public and to make all documents used during the meetings available to the public. The Sunshine Act originally applied to such entities as the Federal Communications Commission, the Federal Election Commission, the Securities and Exchange Commission, and the various other independent federal agencies. FACA was amended to expressly incorporate the Sunshine Act so that part-time advisory committees are obligated to follow the same open-government rules as full-time independent agencies. Thus, each "advisory committee meeting shall be open to the public"[66] unless the president or agency head "determines that such portion of such meeting may be closed to the public in accordance with" the Sunshine Act.[67] The Sunshine Act authorizes an agency head to close a meeting for a variety of reasons, including that an open meeting would "disclose trade secrets and commercial or financial information obtained from a person and privileged or confidential" or "disclose information of a personal nature where disclosure would constitute a clearly unwarranted invasion of personal privacy."[68]

It should be noted that once a meeting is opened to the public, one can make a strong argument that documents used during that meeting, even though they might be predecisional, must be released. Normally, under FOIA agencies are permitted to refuse to disclose so-called predecisional or deliberative process memoranda. The Sunshine Act, though, has no parallel exemption. "The Sunshine Act was designed to open the predecisional process of multi-member agencies to the public."[69] Thus, an advisory committee is not free to close its meeting merely because it will be engaged in predecisional activities. If the contents of predecisional documents from the agency are discussed by the advisory committee at a public hearing, then that could well constitute "public disclosure," thereby forcing the agency to disclose the documents. Once documents have been publicly disclosed, it becomes difficult for the government to decline to release the documents again.[70] Merely because documents are provided to the advisory committee by the agency, however, does not automatically mean that those documents have been "publicly disclosed" within the meaning of FOIA.[71]

As noted above, peer-review committees that consider grant applications are normally closed to the public, invoking an exception in the Sunshine Act that permits closed meetings when an open meeting would "constitute an unwarranted invasion of personal privacy."[72] Funding agencies argue that in assessing a grant application they examine and discuss the qualifications of the PI and that to do so candidly in public would constitute an invasion of privacy. Before a meeting can be closed, though, the agency must publish a notice in the *Federal Register* so indicating, and noting that the general counsel or his or her designee has agreed that the meeting can be closed. What, however, would occur if an applicant were to waive his or her privacy rights? Logically, it would become difficult for the agency to close that portion of the meeting dedicated to a discussion of that researcher's grant application. This actually occurred once in the mid-1990s. The study section still refused to open the meeting, arguing that it would discuss multiple investigators simultaneously and those other investigators had not waived their privacy rights.

THE SHELBY AMENDMENT, THE INFORMATION QUALITY ACT, AND OTHER DATA-SHARING INITIATIVES

As noted above, according to the Supreme Court a researcher's raw data developed with federal grant funds are not subject to FOIA. This seemingly innocuous holding frustrated those who wished to question the scientific validity or reliability of various published studies that were used as the rationale for agency rules, especially in the environmental area. This reached a peak in 1997, with the Environmental Protection Agency's clean air standards, which relied heavily on the results of a federally funded study by researchers from the Harvard School of Public Health suggesting that fine particles played a significant role in negatively affecting health. Harvard, however, declined to provide the raw data to industry scientists. Many in industry believed that if an agency relied on published scientific studies, then they should have access to the raw data, especially if the studies were federally funded. It is difficult to argue with the proposition that government decision making ought to be fully transparent: Those who are to be burdened by a rule, the argument went, ought to be able to see the raw data underlying the studies that supposedly supported the rule.

Two years later, Congress responded with the so-called Shelby Amendment, after its author, Senator Richard C. Shelby (R-AL). The Shelby Amendment was an amendment to an appropriations bill, and requires

> the Director of OMB [to] amend[]. . . . Circular A-110 [Uniform Administrative Requirements for Grants and Agreements with Institutions of Higher Education, Hospitals, and Other Nonprofit Organizations] to require Federal awarding agencies to ensure that all data produced under an award will be made available to the public through procedures established under the Freedom of Information Act.[73]

The research and university communities intensely lobbied the Office of Management and Budget (OMB), and that lobbying paid off. The rule OMB eventually developed, with Shelby's acquiescence, was far narrower than Shelby's original amendment and was aimed at addressing the specific problem that led to the amendment, namely federally funded studies used to support a final rule. As such, the Shelby Rule and ultimately FOIA apply to a request

> for research data[,] relating to published research findings produced under an award that[,] were used by the Federal Government in developing an agency action that has the force and effect of law[.][74]

In such a case, "the Federal awarding agency shall request, and the recipient shall provide, within a reasonable time, the research data so that they can be made available to the public through the procedures established under the FOIA."[75]

Thus, if someone files a FOIA request for your data, the funding agency is obligated to request, and you are obligated to turn over, your raw data relating to published research findings that were used in developing a regulation. The OMB rule is so narrow that it is unlikely to come into play with any frequency. Indeed, since its enactment, I am aware of only one published case in which documents were sought under the Shelby Amendment. That case focused on one study in a series of studies, called the DASH studies or Dietary Approaches to Stop Hypertension, aimed at assessing the health effects of various dietary components (e.g., fat, fruit). The study of interest, funded by the National Heart, Lung, and Blood Institute (NHLBI), examined the effects of dietary salt on high blood pressure. The study authors, in two articles, one published in the *New England Journal of Medicine* and the second in the *Annals of Internal Medicine*, concluded that dietary salt intake correlated with blood pressure (e.g., the higher the salt intake, the higher the blood pressure). Based on these studies, NHLBI issued various press and policy statements about the relationship between dietary salt intake and high blood pressure.

The Salt Institute and other groups expressed concerns about the accuracy of the NHLBI statements and the adequacy of the study design and data analyses. They filed a FOIA request under the Shelby Amendment for the raw data underlying the two publications, along with a request to correct the record under the Information Quality Act. NHLBI denied both requests, and the plaintiffs sued under the Shelby Amendment and under the Information Quality Act, which is discussed below.

The problem that the plaintiffs faced was that the OMB regulation applied only if the study were published and used as the basis of rulemaking. Here, the study at issue was not used in rulemaking. Plaintiffs therefore argued that the OMB regulation was narrower and inconsistent with the Shelby Amendment, and therefore void. The court refused to entertain the argument and dismissed the case on procedural grounds.[76]

The Information Quality Act of 2000 (IQA) differs from the Shelby Amendment in a number of respects.[77] The IQA was designed to ensure that agencies base their policy decisions on sound science and sound data. To that end, OMB was instructed to develop guidelines to ensure that agencies use sound data, and furthermore, that agencies provide interested parties an opportunity to correct information that does not comply with OMB's quality guidelines. The OMB Guidelines require federal agencies (1) to "adopt specific standards of quality that are appropriate for the various categories of information they disseminate," (2) to "develop a process for reviewing the quality . . . of information before it is disseminated," (3) to "establish administrative mechanisms allowing affected persons to seek and obtain, where appropriate, timely correction of information maintained and disseminated by the agency that does not comply with OMB or agency guidelines," and (4) to provide OMB with reports regarding the agencies' information quality guidelines and any information quality complaints they receive.[78] The real problem with the IQA, according to those who advocated its adoption, was that it was not carefully "wordsmithed" and permitted the agencies too much flexibility. In the IQA, Congress directed OMB to develop guidelines for ensuring that national policies are based on "good" scientific data, and the OMB guidelines in turn require the agencies to develop their own guidelines. Each time the responsibility shifted, the legal obligations became more attenuated. HHS, for example, imposed few concrete responsibilities or even deadlines on its own compliance with the IQA.[79]

The plaintiffs in the *Salt* case, in addition to requesting the raw data underlying the two publications, also sought to have the NHLBI correct certain statements that it made in reliance on the DASH study. As with the request for raw data, the court side-stepped the plaintiffs' IQA claim on procedural grounds, holding that there is no private right of action under the IQA because the agencies have no real obligations under the IQA. This too was upheld on appeal.[80]

To date, no one has successfully been able to sue under either the Shelby Amendment or the IQA.[81] Many scientists view the functional demise of the IQA or the Shelby Amendment as a major victory against the forces of evil—industry fat cats who are interested only in corporate coffers and not in sound science. In fact, the battle over access to raw data and whether agencies should be required to use the "best available science" has little to do with political philosophy and more to do with the philosophy of science, which teaches open access. The move toward requiring access to raw data and specimens is therefore not new. Funding agencies had been discussing these issues long before Senator Shelby was even in Congress. NIH and NSF both require their principal investigators to share data and specimens. For example, NSF's Award and Administration Code, Chapter VI (2008), provides as follows:

> Investigators are expected to share with other researchers, at no more than incremental cost and within a reasonable time, the primary data, samples, physical collections and other supporting materials created or gathered in the course of work under NSF grants. Grantees are expected to encourage and facilitate such sharing.[82]

NIH's policy is similar.[83] To my knowledge, neither agency has ever compelled a researcher to share his or her data or specimens with another scientist. Many journals also require, as a condition of publication, that researchers share their reagents and specimens with other qualified scientists who may request them. For example, the journal *Genes & Development* notifies authors that "it is understood that researchers who submit papers to this journal are prepared to make available to qualified academic researchers materials needed to duplicate their research results (probes, plasmids, clones, sequences, and the like)."[84] It is open to question whether an author who denies another's request to share reagents could be compelled by a court through a suit filed by the requester to turn over those materials. It is doubtful whether a journal would have the resources to enforce the pledge through litigation. Peer pressure and the threat of "black-balling" may be the only way to promote compliance with these journal-imposed sharing requirements. As for now, reagent sharing occurs, but not with the frequency that many had hoped to see.

If this trend—preach one thing but do another—continues, Congress will, in a fit of frustration, write a law not as easily evaded as the Shelby Amendment and IQA, one that all scientists will find counterproductive. Indeed, we are beginning to see some interesting developments of this sort in various areas. On December 31, 2007, the president signed into law the OPEN Government Act of 2007.[85] The law was primarily designed to extend FOIA to cover defense contractors. However, it was inartfully drafted, and it is possible that it could be interpreted as extending FOIA to cover raw data collected by grantees. Second, amendments to the Food, Drug, and Cosmetic Act require drug manufacturers to register certain clinical trials with NIH for posting on ClinicalTrials.gov.[86] Also, the International Committee of Medical Journal Editors requires clinical trial registration with NIH as a condition for publishing research results from a trial.[87] The purpose of this registration is to preclude a drug sponsor from publishing only positive results and not sharing with the public negative results or results that may not be as dramatic as those submitted for publication.[88]

Finally, the Consolidated Appropriations Act, 2008, which was signed into law on December 26, 2007, contains a provision that requires investigators funded by NIH to submit to PubMed at the National Library of Medicine electronic copies of their manuscripts of articles accepted for publication in peer-reviewed journals.[89] PubMed would then be required to make the articles publicly available within twelve months of publication. The problem with the legislation is that it collides with the copyright interests of scientific journals. In essence, the provision would limit the length of a copyright of an article to one year, as opposed to the current term, which is the length of the author's life plus 70 years.[90] (See chapter 7.) The NIH director is supposed to implement the legislation "in a manner consistent with copyright law." In response to this, some journals will make affected papers free online on their Web sites after twelve months and will even help authors submit their papers to PubMed.[91] Others may not be as willing to acquiesce. NIH is now implementing this legislation by requiring NIH-funded researchers to "submit [to the National Library of Medicine's PubMed Central] the final version of their peer-reviewed articles . . . upon acceptance for publication." This policy was effective April 7, 2008.[92]

Access to raw data, reagents, and even grant applications has increased in the past decade, but not in response to FOIA or any of the other open-government reforms. FOIA, the Privacy Act, FACA, and the Sunshine Act have proven to be better shields against disclosure than as a means of promoting disclosure. The Shelby Amendment and the IQA provide marginal rights with no remedy. In short, the law has been remarkably unsuccessful at promoting sharing among scientists and between scientists and the government: Most prefer to hold some things in secret, and that is not so surprising.

CASE STUDY AND PROBLEM

The Case of the Careless FOIA Officer

Clyde Movkin is graduate student in physical chemistry at Millard Fillmore Technical University. To earn extra money, Movkin works four to six hours each week for Data Grabb, a small company that does nothing but Freedom of Information Act requests for grant applications, procurement contract applications, and just about anything involving science, procurement, and the government. Movkin's job is to track requests, log in new arrivals, and summarize them briefly so that Data Grabb can then advertise them on its Web site. Most of what Movkin receives is heavily redacted. One afternoon, while working at Data Grabb, he opens a thick envelope from the Department of Defense. It contains a proposal for a cooperative agreement having to do with chemistry. It was submitted by Wilbur-Orville, a subsidiary of Wright-Orville Industries, a major defense contractor. Movkin starts reading it because he understands the material and, amazingly, nothing has been redacted. In fact, items labeled as "Trade Secret" and containing formulae and processes are set out in full. He quickly flips to the budget section and notices that all of the applicant's cost and pricing information and how it will budget things out have not been redacted. The cost and pricing information probably could be used by Wright-Orville's competitors to ascertain Wright-Orville's costs and bidding strategy on major defense systems, such as a new jet fighter that Wright-Orville will be bidding on shortly.

Movkin recognizes that someone in the Defense Department's FOIA office has really screwed up; this information should never have been released. Movkin runs into his boss's office and shows Vincent Grabb, the owner of the company, the proposal; Movkin indicates that they should return it to the Defense Department. Grabb agrees and says he will take care of it. However, rather than returning the materials, Grabb divides the proposal into two parts—one containing the trade secret information and the other containing the budget—and puts a summary of each up on his Web site, offering copies of the budget for $100,000 and copies of the chemistry for $10,000.

The CEO of Wright-Orville Industries sees Grabb's offer and, after screaming for ten minutes, calls the FBI. Has Grabb done something illegal? Does Wright-Orville Industries have a claim against the government? Against Grabb?

NOTES

1. From 1789 to 1795, the Senate met largely behind closed doors and continued to do so for many years thereafter when discussing treaties and defense-related matters.

2. *See, e.g.*, House Rule V, cl. 1, House Rule XI, cl. 4.

3. Press-Enterprise Co. v. Superior Court, 478 U.S. 1, 7 (1986).

4. 5 U.S.C. § 552.

5. Forsham v. Harris, 445 U.S. 169, 180–81 (1980).

6. *Id.* at 182.

7. The recent amendments to FOIA eliminate any doubt that information held for the government by a government contractor is subject to FOIA. *See* OPEN Government Act of 2007, § 9, Pub. L. No. 110-175, 121 Stat. 2524. It is possible that these amendments could be interpreted as applying to grantees as well as contractors, thereby effectively overturning *Forsham*.

8. The nine exemptions are as follows: (1) national defense information; (2) internal personnel rules and practices of an agency; (3) material exempted from disclosure by a statute; (4) trade secrets and commercial or financial information that is privileged or confidential; (5) deliberative process memoranda; (6) personnel files, medical files, and similar records, the disclosure of which would constitute a clearly unwarranted invasion of personal privacy; (7) certain law enforcement information; (8) certain information concerning financial institutions; and (9) geological and geophysical information and data, including maps, concerning wells. *See* 5 U.S.C. § 552(b).

9. *See* Uniform Trade Secrets Act § 1(4) (1985).

10. *See* 18 U.S.C. § 1905.

11. Wash. Research Project v. Dep't of Health, Educ. & Welfare, 504 F.2d 238, 244–45 (D.C. Cir. 1974) (footnotes omitted).

12. *See* Physicians Comm. for Responsible Med. v. NIH, 326 F. Supp. 2d 19 (D.D.C. 2004).

13. Public Citizen Health Research Group v. FDA, 704 F.2d 1280, 1288 (D.C. Cir. 1983) (footnote omitted).

14. Kurzon v. Dep't of Health & Human Servs., 649 F.2d 65, 69 (1st Cir. 1981) (footnote omitted).

15. Kurzon v. Dep't of Health & Human Servs., No. 00-395-JD (D.N.H. July 17, 2001).

16. Wash. Post Co. v. U.S. Dep't of Health & Human Servs., 690 F.2d 252 (D.C. Cir. 1982).

17. I represented the various university and research organizations in *McCutchen*.

18. McCutchen v. U.S. Dep't of Health & Human Servs., 30 F.3d 183, 188 (D.C. Cir. 1994).

19. Although the outcome of the case turned on the law enforcement exemption to FOIA, the court focused on the privacy interests of the accused scientists; therefore, the court's decision is relevant to the privacy exemption.

20. *See* Casad v. U.S. Dep't of Health & Human Servs., 301 F.3d 1247 (10th Cir. 2002).

21. *See, e.g.*, 45 C.F.R. § 5.45.

22. *See* 5 U.S.C. § 552(a)(6)(A)(i).

23. *See id.* § 552(a)(6)(B)(iii)(I) & (II).

24. *See* U.S. General Accounting Office, Update on Freedom of Information Implementation Status 79 (Feb. 2004), *available at* http://www.gao.gov/new.items/d04257.pdf.

25. Al Kamen, *The Loop*, Wash. Post, Mar. 19, 2008, at A13.

26. In 2008, for instance, various news organizations sought a copy of the contract between the State Department and Blackwater Security Consulting, the company that employed security guards who were indicted for killing innocent civilians in Iraq. *See* Press Release, U.S. Dep't of Justice, Five Blackwater Employees Indicted on Manslaughter and Weapons Charges for Fatal Nisur Square Shooting in Iraq (Dec. 8, 2008), *available at* http://www.usdoj.gov/opa/pr/2008/December/08-nsd-1068.html. The contract is 323 pages long, but about 169 of those pages are blank or mostly blank, having been redacted for security reasons. *See* Robert O'Harrow Jr., *Blackwater Contracts, Short on Detail*, Wash. Post, Dec. 8, 2007, at D1; *see also* http://www.washingtonpost.com/governmentinc (displaying the redacted pages).

27. *See* 45 C.F.R. § 5.34.

28. *See id.* § 5.35(b).

29. *See* 5 U.S.C. § 552(a)(6)(C).

30. *See* Taylor v. Appleton, 30 F.3d 1365 (11th Cir. 1994).

31. *See* Openness Promotes Effectiveness in Our National Government Act of 2007, § 4, Pub. L. No. 110-175, 121 Stat. 2523, 2525.

32. *See* Campaign for Responsible Transplantation v. U.S. Food & Drug Admin., No. 00-2849 (D.D.C. Jan. 22, 2009).

33. *See* 45 C.F.R. § 5.65.

34. *See* Pub. L. No. 93-579, § 3, 88 Stat. 1897 (1974) (codified at 5 U.S.C. § 552a).

35. *See* McCready v. Nicholson, 465 F.3d 1, 13 (D.C. Cir. 2006).

36. My colleagues and I at a large Washington, D.C., law firm represented Fisher in this litigation.

37. *See* Fisher v. NIH, 934 F. Supp. 464 (D.D.C. 1996).

38. *See* Henke v. Dep't of Commerce, 83 F.3d 1445, 1448 n.1 (D.C. Cir. 1996).

39. *Id.* at 1449.

40. Pub. L. No. 92-463, 86 Stat. 770 (1972) (codified at 5 U.S.C. appendix §§ 1–14).

41. During FY 2008, HHS operated 271 advisory committees at a cost of about $170 million, or about $53,000 per meeting. Two years earlier, the cost had been about $63,000 per meeting. During that same period, NSF operated 55 advisory committees at a cost of about $41 million, or about $30,600 per meeting. Most HHS committees advised NIH or FDA. *See* http://fido.gov/facadatabase/acr.asp (last visited May 2, 2009). Government-wide there are approximately 1,000 advisory committees. *Id.*

42. *See* 42 U.S.C. § 284(b)(2)(B).

43. *See* Guidelines for Research Involving Recombinant DNA Molecules, 59 Fed. Reg. 34,496 (July 5, 1994). These Guidelines set out the responsibilities of the Recombinant DNA Advisory Committee (RAC) and the director of NIH with regard to recombinant research; they apply to all research, whether federally funded or not, that is sponsored by an entity that receives any support for recombinant DNA research from NIH. In 1996, the NIH director sought to discontinue the RAC and to shift all responsibility for experiments involving human gene transfer to FDA and to shift the remaining responsibility to a new advisory committee that

would operate under the aegis of the Office of Recombinant DNA Activities. *See* 61 Fed. Reg. 35,774 (July 8, 1996). Following public comment, NIH abandoned its proposal to eliminate the RAC. *See* 61 Fed. Reg. 59,726 (Nov. 22, 1996).

44. *See, e.g.*, Federal Food, Drug, and Cosmetic Act § 515(g), 21 U.S.C. § 360e(g).

45. *See* FACA § 9(c).

46. *See id.* § 10(a)(2).

47. *See id.* § 10(a)(1).

48. *See id.* § 10(c).

49. *Id.* § 5(b)(2).

50. *Id.* § 3(2).

51. For example, in National Nutritional Foods Ass'n v. Califano, 603 F.2d 327 (2d Cir. 1979), the court held that FACA applied to meetings between FDA officials and certain physicians who specialized in nutrition, even though the meetings were informal. The court found compelling the fact that the meetings took place at the agency's initiative, that summaries of the meetings were prepared by agency personnel and included recommendations on a regulatory agenda, and that FDA's subsequent notice of proposed rulemaking borrowed heavily from the group's recommendations. Correspondingly, in Food Chemical News Inc. v. Davis, 378 F. Supp. 1048 (D.D.C. 1974), the plaintiff newsletter sought access to two meetings convened by the Bureau of Alcohol, Tobacco and Firearms (ATF) concerning amendments to the agency's regulations on labeling for distilled spirits. One meeting was between consumer groups and ATF officials, and the other was between industry representatives and ATF officials. The court, in holding that FACA applied, noted that an agency is not shielded from FACA by holding a one-time meeting or by calling that meeting "informal." The court found dispositive that the meetings were convened to gather views with an eye toward modifying a regulation. *See id.* at 1051–52.

52. In one of the most influential early cases, Judge Gesell acknowledged that FACA could be read to include ad hoc groups, but declined to adopt such a reading, noting that "Congress was concerned with advisory committees formally organized. . . ." Nader v. Baroody, 396 F. Supp. 1231, 1232 (D.D.C. 1975). In *Baroody*, Ralph Nader unsuccessfully sought an invitation to one of the biweekly meetings held at the White House between the president's assistant for public liaison and various interest groups. The groups varied from meeting to meeting, and the meetings were not aimed at developing specific policies. The court, in concluding that FACA did not apply, opined that the meetings were not concerned with garnering recommendations or advice, but instead "merely wisely provided a mechanism and sounding board to test the pulse of the country. . . ." *Id.* at 1234.

53. 491 U.S. 440 (1989).

54. The ABA Committee no longer rates nominees as "exceptionally well qualified."

55. Public Citizen, 491 U.S. at 451.

56. *Id.* at 465.

57. *Id.* at 462.

58. Chief Justice Rehnquist and Justices O'Connor and Kennedy had difficulty with the majority's expansive use of the legislative history and instead would have held FACA unconstitutional as applied. Justice Scalia did not participate in the case.

59. *See* Wash. Legal Found. v. U.S. Sentencing Comm'n, 17 F.3d 1446 (D.C. Cir. 1994) (holding that the advisory group established by Sentencing Commission was not subject to FACA because the Sentencing Commission is not an agency and further, the group's advice was not being "utilized" by the Department of Justice); Food Chem. News v. Young, 900 F.2d 328 (D.C. Cir. 1990) (holding that an expert panel convened by a government contractor pursuant to its

contract with FDA was not an advisory committee); Grigsby Brandford & Co. v. United States, 869 F. Supp. 984, 1001–2 (D.D.C. 1994) (noting that "one-time," "ad hoc," and "unstructured" meetings do not trigger FACA and that advice on specific issues is a key factor); Northwest Forest Resource Council v. Espy, 846 F. Supp. 1009, 1010 (D.D.C. 1994) (holding that FACA does not apply to every meeting between an agency and advice giver).

60. Public Citizen, 491 U.S. at 462–63.

61. *See* Animal Legal Def. Fund, Inc. v. Shalala, 104 F.3d 424 (D.C. Cir. 1997).

62. *See id.* at 429.

63. *See* Pub. L. No. 105-153, § 2(a), 111 Stat. 2689 (1997) (redefining an "advisory committee" to expressly exclude the National Academy of Sciences and the National Academy of Public Administration).

64. *See* Public Citizen v. Nat'l Advisory Comm., 886 F.2d 419 (D.C. Cir. 1989).

65. Pub. L. No. 94-409, 90 Stat. 1241 (1976) (codified at 5 U.S.C. § 552b).

66. FACA § 10(a)(1).

67. *Id.* § 10(d).

68. Notwithstanding the admonition that advisory committee meetings should be open to the public whenever possible, most are closed in whole or in part. During FY 2007, the General Services Administration reported that advisory committees held 6,815 meetings; of those, only 2,037, or less than one-third, were open to the public. HHS statistics are far worse; more than 90 percent of HHS advisory committee meetings were closed to the public either in whole or in part. During FY 2006, HHS advisory committees convened 3,413 meetings, of which 3,059 were closed either in whole or in part. In other words, only about 10 percent of HHS's advisory committee meetings were totally open to the public. At NSF, only about 2.6 percent (32 out 1,210 meetings) of its advisory committee meetings were fully open to the public. Over the years, the percentage of closed meetings has gradually increased. Most closures by HHS and NSF involve advisory committees evaluating grant proposals, and closures in those cases are usually justified by the agency on trade secret or privacy grounds. *See* 5 U.S.C. § 552b(c).

69. Common Cause v. Nuclear Regulatory Comm'n, 674 F.2d 921, 929 (D.C. Cir. 1982); Public Citizen v. Nat'l Econ. Comm'n, 703 F. Supp. 113, 117 (D.D.C. 1989).

70. Most courts, outside the national security area, have held that once materials have been publicly released by the agency, it becomes difficult for the agency to successfully argue that an exemption should apply. *See* Worthington Compressors, Inc. v. Costle, 662 F.2d 45, 51 (D.C. Cir. 1981) ("[i]f the information is freely or cheaply available from other sources . . . , it can hardly be called confidential"). This is especially the case when exemption 5 is involved. *See* Army Times Publ'g Co. v. Dep't of the Air Force, 998 F.2d 1067 (D.C. Cir. 1993).

71. In a pre–Government in the Sunshine Act case, the court in Aviation Consumer Action Project v. Washburn, 535 F.2d 101 (D.C. Cir. 1976), held that FACA did not provide any greater access to information than FOIA. Thus, if an advisory committee were provided confidential information and did not refer to that information or otherwise use it during a public meeting, it is likely that the agency would prevail if the document satisfied an otherwise valid exemption. The Government in the Sunshine Act restricted the use of exemption 5 only because it made it more difficult for advisory committees to close their meetings to the public and thereby increased the likelihood that information subject to a FOIA exemption would be disclosed during the meeting.

72. 5 U.S.C. § 552b(c)(7)(C).

73. FY 1999 Omnibus Appropriations Act, Pub. L. No. 105-277.

74. ___.36(d)(1), OMB Circular A-110.

75. *Id.*

76. *See* Salt Inst. v. Thompson, No. 04-359 (E.D. Va. Nov. 15, 2004), *aff'd*, 440 F.3d 156 (4th Cir. 2006).

77. *See* Information Quality Act of 2001, Pub. L. No. 106-554, § 515, 114 Stat. 2763A-153–54 (2000).

78. 67 Fed. Reg. 8452, 8458–59 (Feb. 22, 2002).

79. *See* U.S. Dep't Health & Human Services, Guidelines for Ensuring the Integrity of Information Disseminated by HHS Agencies, http://aspe.hhs.gov/infoquality/ (last visited Aug. 3, 2009).

80. *See* Salt Institute v. Leavitt, 440 F.3d 156 (4th Cir. 2006).

81. *See* Single Stick, Inc. v. Johanns, No. 06-1077 (D.D.C. Mar. 10, 2009), *appeal docketed*, No. 09-5099 (Mar. 27, 2009); Americans for Safe Access v. U.S. Dep't of Health & Human Servs., No. C 07-01049 (N.D. Cal. Nov. 20, 2007) (holding that IQA creates no legally cognizable rights); Stephen M. Johnson, *Junking the "Junk Science" Law: Reforming the Information Quality Act*, 58 ADMIN. L. REV. 37 (2006).

82. NATIONAL SCIENCE FOUNDATION, AWARD AND ADMINISTRATIVE GUIDE ch. 6 (Jan. 2008), *available at* http://www.nsf.gov/pubs/policydocs/pappguide/nsf08_1/aag_6.jsp#VID4.

83. *See* NATIONAL INSTITUTES OF HEALTH, OFFICE OF EXTRAMURAL RESEARCH, NIH DATA SHARING POLICY AND IMPLEMENTATION GUIDANCE (Mar. 5, 2003), *available at* http://grants.nih.gov/grants/policy/data_sharing/data_sharing_guidance.htm#time/.

84. http://genesdev.cshlp.org/content/4/3/local/back-matter.pdf

85. *See* Pub. L. No. 110-175, 121 Stat. 2524 (2007); *see supra* note 7.

86. *See* Food and Drug Administration Modernization Act of 1997, § 113, Pub. L. No. 105-115, 111 Stat. 2296.

87. *See* 351 N. ENG. J. MED. 1250 (2004).

88. A manufacturer is required to submit all results to FDA, but FDA is precluded by law from sharing those results with the public.

89. *See* Consolidated Appropriations Act, 2008, Div. G, § 218, Pub. L. No. 110-161, 121 Stat. 1844. This policy was made permanent by the Omnibus Appropriations Act, 2009, Div. F, § 217, Pub. L. No. 111-8, 123 Stat. 524, which provides as follows:

> The Director of the National Institutes of Health ("NIH") shall require in the current fiscal year and thereafter that all investigators funded by the NIH submit or have submitted for them to the National Library of Medicine's PubMed Central an electronic version of their final, peer-reviewed manuscripts upon acceptance for publication, to be made publicly available no later than 12 months after the official date of publication: Provided, that the NIH shall implement the public access policy in a manner consistent with copyright law.

90. *See* Copyright Term Extension Act of 1998 (also known as the Sonny Bono Copyright Term Extension Act), Pub. L. No. 105-298, 112 Stat. 2827; 17 U.S.C. § 302(a).

91. *See* NIH Notice Number: NOT-OD-08-119 (Mar. 19, 2009), *available at* http://grants.nih.gov/grants/guide/notice-files/NOT-OD-08-119.html (implementing § 217).

92. *See* U.S. Dep't Health & Human Services, Nat'l Institutes of Health Public Access, http://publicaccess.nih.gov (last visited Aug. 3, 2009).

CHAPTER 7

Intellectual Property: Who Owns Your Inventions, Words, Data, and Specimens?

Law and science both categorize objects and concepts. In science, clumping likes together makes it easier to see relationships and develop testable hypotheses. In law, clumping likes together makes it easier to apply similar rules and standards to everything in a single category, and so it is with property. As far back as the Babylonian Talmud, the law divided property into two broad categories, real property and personal property. Different rules applied to the different types of property.[1]

This dichotomy carried over into early English law. Real property is land and everything that is attached to it. For example, if you build a house on a parcel of property, the house becomes real property. If you place wall-to-wall carpeting in your house, that too becomes part of the real property because it is affixed to the house, which in turn is affixed to the land. Personal property is everything else—furniture, paintings, clothes, household goods, animals, and the like. An oriental carpet, as opposed to wall-to-wall carpeting, would be personal property (also called personalty) because it is not affixed to the house.

These two categories, real and personal, worked well for years until societies became financially more sophisticated and developed derivative "property," i.e., property that in itself had no value, but rather derived its worth by representing other property that did have value—paper money, shares of common stock, and bonds. Because you cannot touch the "value" of paper money, bonds, or common stock, these property types became known as intangible property. Intellectual property is a form of intangible property. This chapter examines intellectual property—patents and copyrights and the laws that govern ownership, including the Bayh-Dole Act, and the Federal Technology Transfer Act of 1986 (FTTA). It also examines tangible property in the form of biological specimens the value of which derives primarily from the information that can be extracted from those specimens.

PATENTS

What Is a Patent?

PATENT AS A QUID PRO QUO

Monopolies are bad, at least that is what we are frequently taught in school. Because a monopoly is the concentration of market power in the hands of a few, it usually leads to higher prices, fewer choices, and inferior quality. Take the case of two school masters who operated a grammar school in Gloucester, England, and did quite well, until a third school master started another school in the same town. The original two school masters found that they were forced to lower their tuitions by 70 percent to compete with the interloper. They sued the upstart, arguing that they were injured by having to lower their prices. In deciding in favor of the new school and hence against monopolies, the court reasoned that teaching is a "virtuous and charitable thing" and that courts have no legal basis to restrict someone from doing something lawful and beneficial. The interesting thing about the case was that it was decided in 1410.[2]

The early English notion that monopolies should not be enforced by the courts, unless there is a special reason to do so, carried over to the Americas. It is one thing, though, for a court to refuse to enforce a monopoly and quite another for a court to award damages caused by monopolistic practices. To correct this deficiency, Congress in 1890 passed the Sherman Antitrust Act and thereafter the Clayton Antitrust Act of 1914. Together the two laws sought to prevent monopolies and monopolistic practices: price fixing, predatory pricing, improper concentrations of market power, and discriminatory pricing.

Not all monopolies are illegal; some are expressly sanctioned by law, such as public utilities. However, monopoly utilities are not free to charge whatever the market will bear; as the quid pro quo ("something for something") for operating as a legal monopoly, states regulate the prices that utilities can charge for electricity, natural gas, and water. Utilities are not the only type of government-sanctioned monopoly. There is one form of monopoly expressly authorized by the Constitution: "The Congress shall have Power . . . To promote the Progress of Science and useful Arts, by securing for limited Times to Authors and Inventors the exclusive Right to their respective Writings and Discoveries."[3] This provision, sometimes referred to as the patent clause, authorizes Congress to create a system of patents, much like the one in England.

A patent is a government-sanctioned, time-limited monopoly and, like a public utility, is part of a quid pro quo. For a limited term, the government gives the patentee a license allowing him or her to keep others from making, selling, or using the patented invention. In exchange, the patentee has to disclose how to make his or her invention so that when the patent expires, the invention is in the public domain and others can make, sell, or use the product and start competing against the original patent holder. Some inventors have decided that they would prefer not to divulge how to make their inventions. Rather than having a monopoly for a limited term (e.g., twenty years from the date a patent application is filed), an inventor who forgoes patent protection has a monopoly for as long he or she can keep his or her know-how a secret. The most famous example of a company that has opted not to patent its invention is Coca-Cola. The formula for its syrup is a closely guarded secret and has been the subject of books, urban legends, and many lawsuits. As with any trade secret, it is perfectly legal for someone to reverse engineer the product and market it. It is perfectly illegal, though, for someone to steal the secret.

PATENT PROTECTS INVENTIONS

The United States recognizes three types of patents—utility patent, design patent, and plant patent. Here we will focus on utility patents. A utility patent can be obtained for a product or thing, a process for making the thing, or a use for the thing. For example, suppose that a drug company, Pandemic Pharmaceuticals, Inc. (PPI), invents a new molecule, called NM, to treat the "dry scritos" and also invents a way of producing the molecule. PPI would be entitled to three patents: one for the composition (i.e., for NM), one for the method for producing NM, and a third for its use. Fast-forward nineteen years. Just before NM's patent is to expire, PPI discovers that NM can also be used to treat "festering phlort" (FP), another bad disease. Assuming that this new use (to treat FP) is not obvious, PPI can apply for another patent for that use, and it would be entitled to keep others from using NM to treat FP for the term of the new FP patent. Once PPI's original patents expire, it will not be able to prevent others from making NM or from selling it to treat DS, but the new patent will enable PPI to prevent others from selling NM to treat FP for the twenty-year life of that patent.

Conception: The Conceiver Is the Inventor

Patent protection applies to inventions, and an invention is a "concept" that has been "reduced to practice." A concept has to be more than a vague idea. For example, years ago I conceived of a free-hanging sky hook, a novel device that hangs from nowhere. Imagine how useful it would be for hanging pictures without puncturing your walls. But since the idea lacked detail (e.g., how it would work), it would not qualify as a patentable invention. It also appears to be inconsistent with various laws of mechanics and thermodynamics, but that does not affect whether it is a "concept" for patent purposes. By contrast, a helicopter (a real mechanical skyhook of sorts) could be patented if the concept were relatively complete in the inventor's mind.

The patent battle over AZT (3′-azido-3′-deoxythymidine), the first effective therapy against the human immunodeficiency virus (HIV), illustrates what one has to prove to show that he or she conceived an invention for patent purposes. Our story begins in mid-1984 in North Carolina where researchers at Burroughs Wellcome (BW) began screening various compounds to see which ones had antiviral effects against two murine retroviruses, Friend leukemia virus (FLV) and the Harvey sarcoma virus (HaSV), as surrogates for HIV. In late October 1984, BW scientists began screening AZT and obtained positive results. On the basis of those results, BW decided to file a patent application on AZT as an AIDS therapy. After reaching that decision, but before completing and filing the application, BW on February 4, 1985, sent a sample of AZT to Dr. Sam Broder at the National Cancer Institute (NCI) for screening against HIV in human cells. Broder and his colleagues at NCI used live HIV and had developed a way of testing a compound's effectiveness against HIV in humans using a unique line of T cell clones (the ATH8 cell line). The NCI researchers volunteered to screen compounds from private pharmaceutical companies, and BW took NCI up on its offer.

Two days later, BW completed its draft patent application, which described AZT and the process for making it; the application also listed various dosages for treating HIV. Two weeks later, on February 20, 1985, Broder phoned BW to report that AZT impeded HIV replication. BW filed its patent application on March 16, 1985. The application named only BW employees as the inventors.[4]

Thereafter, NCI, working with BW, conducted a series of placebo-controlled clinical trials that were halted prematurely because AZT was proving to be too effective to ethically continue the study. AZT was quickly approved by the Food and Drug Administration (FDA) in 1987, and BW began marketing the drug. However, the government's relations with BW soon soured. Public Health Service (PHS) officials and AIDS patients believed that BW was charging too much for the drug, especially given that BW's research investment, according to PHS, had been relatively modest: Most of the AZT-specific research, from the assaying to the clinical studies, had been conducted by and at the National Institutes of Health (NIH). In response to intense congressional pressure, BW rolled back the price for AZT, but PHS officials viewed the reduction as insufficient. In their minds, the only effective way to reduce the price was by interjecting competition into the market. Enter Barr Laboratories Inc., the Pomona, N.Y., generic drug manufacturer.

Operating under the assumption that Broder and his colleagues should have been included as inventors on the AZT patent applications, the government licensed to Barr whatever interests it might have had in those patents with the understanding that Barr would promptly seek FDA approval to manufacture and market a generic version of AZT. The government believed that conception could not have been complete until NCI had demonstrated that was AZT was active against HIV; BW believed that its limited in-house screening using HIV surrogates was sufficient.

On March 19, 1991, Barr filed an application with FDA seeking approval of its generic AZT. Shortly thereafter, BW sued Barr and NovaPharm, another generic manufacturer, alleging that they were infringing BW's AZT patent. At issue in the lawsuit was whether Broder and Hiroaka Mitsuya, the other NCI scientist, were co-inventors of AZT. If they were co-inventors, then NCI would have become a co-owner with BW on the AZT use patent. Each co-owner has the right to license its interest in the invention separately from the other inventor or inventors.

BW argued that before it received any results back from NCI, it had already prepared its patent application based solely on the results of its murine screening. Barr maintained that murine screening was not predictive of how a compound would work against HIV in humans and therefore, without the NCI results, there was no conception. The trial court, in *Burroughs Wellcome v. Barr Laboratories, Inc.*, rejected this argument. It concluded that BW did not have to prove or even know that AZT would be effective in treating humans having AIDs. According to the court, "[f]or conception to be complete, the law does not require an idea to be proven to actually work."[5] The court of appeals agreed, holding that Broder and Mitsuya helped reduce the invention to practice (e.g., showed that it was workable), but they did not participate in developing the idea. The individual researchers at BW developed the idea, and those individuals were the sole inventors.[6]

Reduction to Practice

Even though the inventor is the individual who conceives the invention, an invention is not complete, at least for patent purposes, until it has been "reduced to practice." In *Burroughs Wellcome*, Barr unsuccessfully argued that under the facts of that case, conception could not occur without some reduction to practice (this is called simultaneous conception and reduction to practice). Barr's legal point was that you could not sufficiently flesh out the concept without doing some relevant experimentation and that BW's surrogate experimentation was not sufficiently relevant.[7]

An invention can be reduced to practice either actually or constructively. Actual reduction to practice requires the inventor to make a sample of the invention and to then show that it works as intended. Constructive reduction to practice is normally much easier to accomplish. The inventor merely has to describe the invention in sufficient detail so that others who are schooled in the discipline can make or use the invention and then file that description in the patent application. The filing of the application marks the time that the invention was constructively reduced to practice. However, the application still has to be sufficiently detailed to enable one skilled in that area to construct and use the invention.

Determining Priority and the One-Year Rule

As you may have gathered from the AZT episode, timing is important in patent law. In most countries, the first person to file the patent application has priority. His or her invention will prevail against those that were filed on a later date, all else being equal. The United States is different. The inventor with priority is the one who conceived the invention first, irrespective of when the application was filed, provided that the person also diligently sought to reduce the invention to practice. As noted above, filing the patent application constitutes "constructive reduction to practice." Suppose Ford conceives the automobile on February 1, 1890, and diligently reduces it to practice three years later. Swiftmobile conceives the automobile on March 1, 1890, but reduces it to practice in 1891, two years before Ford. Ford has priority over Swiftmobile. The date of invention is the date of conception. But how do you prove the date of conception? The best way is by following the procedures used in industry where all researchers record their findings in bound notebooks with prenumbered pages, much like the old laboratory notebooks. At the end of each day, the researcher signs each page and a coworker witnesses that signature. Relatively few university laboratories operate with such formality. In one actual case, involving a breakthrough invention, the researchers recorded their findings on paper towels, message slips, and any piece of paper that they could find other than a laboratory notebook. The papers were then dumped into an envelope for safekeeping and eventually handed over to a patent attorney.

The date of conception is also important in determining whether you have preserved your patent rights. In most other nations, if you market your invention or publish your invention before you file a patent application, you lose your right to a patent. In the United States, there is a one-year grace period, meaning that if you publish an article about your invention or sell it as a product, you have one year from that date to file your patent application. This one-year grace period can lead to interesting legal issues.

Take a famous invention by a Texas Instruments employee for a socket for integrated circuits. The employee conceived the invention and prepared sketches of his concept sometime before March 17, 1981. On April 8, 1981, Texas Instruments (TI) entered into a contract to sell 30,100 of the new sockets; none had yet been built. Indeed, the inventor had not even made a prototype for testing. On April 19, 1982, the inventor filed his patent application for the socket.

When another company sought to make and sell the same type of socket, TI sued it for patent infringement. The other company defended by arguing that the TI patent was invalid because more than one year had passed between the first commercial sale (April 8, 1981) and the date the patent application was filed (April 19, 1982). TI responded by arguing that it had not reduced the invention to practice until many months after the initial sales contract had been executed and the one-year clock should not have begun running until it had reduced its invention to practice, even if a premature sales contract had been executed. The court of appeals disagreed with TI, as did the Supreme Court. Both courts held that at the time of the initial sale (April 8, 1981), the inventor had done enough to qualify for a patent had he filed an application.[8] By waiting more than one year from that initial sale to file, he lost his patent rights.

REQUIREMENTS FOR PATENTABILITY

Not all inventions can be patented. Our patent laws attempt to reconcile our deep-seated antipathy to monopolies with our need to encourage progress. Striking the right balance between these two interests is a delicate undertaking. If it is too difficult to obtain a patent, investment in new technologies might be discouraged. If it is too easy to obtain a patent, it may impede technological and scientific advances by making it difficult to conduct basic research without infringing someone's patent. Mindful of the need to maintain balance, the patent laws impose four requirements to qualify for patent protection. An invention (1) must encompass patentable subject matter and must be (2) novel, (3) not obvious, and (4) useful.

Patentable Subject Matter

Three decades ago, it was a well-established principle of patent law that intellectual concepts, pure ideas, natural phenomena, mathematical algorithms, and laws and products of nature were not patentable subject matter.[9] What would have happened had Newton patented the laws of gravity, Galileo and Kepler the laws of planetary motion, and Maxwell the four equations describing electricity and magnetism? Other researchers would have had to pay them royalties just to use their equations. These restrictions on patentable subject matter were designed to ensure that one researcher could not foreclose others from building on the laws of nature. But recently, the courts have been more willing to permit inventors to patent algorithms and living things.

In 1980, the Supreme Court in *Diamond v. Chakrabarty* expanded the horizons of patent protection when it held that a genetically engineered bacterium that did not occur in nature could be patented.[10] The bacterium in question was designed to digest oil following tanker spills. The Court's ruling in *Chakrabarty* has had profound implications for the biosciences. It stimulated the rapid development of a multibillion industry centered on transgenic organisms, which in turn has affected developments in pharmaceuticals, agriculture, and pest control. Within less than a decade after *Chakrabarty*, one congressional subcommittee estimated that the decision had "opened wide the door for intellectual property protection of biotechnology. A $4 billion industry has arisen in the wake of the decision [as of 1990]."[11] Fifteen years later (as of December 31, 2005) that $4 billion industry had mushroomed with a market capitalization of over $400 billion.[12]

The philosophy underlying the Court's decision in *Chakrabarty*, namely that patent laws should extend to "anything under the sun that is made by man,"[13] affected inventions beyond the biosciences. Mathematical algorithms, previously thought to be beyond patent protection, were suddenly being granted patent protection provided that the algorithm satisfied the other criteria.[14] This expansion occurred at the same time as the computer revolution.

But what about natural relationships that form the basis of a machine or process? Should they too be granted protection, or are they too much akin to a law of nature? In the 1980s, three university researchers found that there was correlation between high levels of homocysteine in the blood and deficiencies of two essential vitamins, folate (folic acid) and cobalamin (vitamin B_{12}). They also found a more accurate way of measuring homocysteine in the blood. They patented a method to detect the vitamin deficiency by measuring homocysteine (based on their discovery of the correlation between the two) and they patented a method to measure homocysteine. The patented technology was ultimately licensed to Metabolite Laboratories, which in turn licensed it to Laboratory Corporation, a commercial clinical laboratory. After a while, other diagnostic companies, such as Abbott Laboratories, developed different tests that could be used to detect homocysteine in the blood. Laboratory Corporation switched and began using the Abbott test, and each time it did, it refused to pay Metabolite royalties. Metabolite, in turn, sued. It claimed that it was due royalties even when the Abbott test was used because it owned the patent on a method based on the "correlation" as well as on one way of testing. Each time a physician receives back a homocysteine test result, he or she automatically correlates those results with the presence or absence of a vitamin deficiency. Thus, the physician became a patent infringer, and Laboratory Corporation was guilty, according to Metabolite, of inducing physicians to infringe its patent. The jury ruled in Metabolite's favor, and the court of appeals upheld that judgment. The court of appeals held that in certain circumstances a correlation can be patented. At first, the Supreme Court agreed to hear the case, but later changed its mind and decided not to hear the case, thus leaving the court of appeals' decision intact.[15]

An Invention Has to Be Novel and Not Obvious

The *sine qua non* of a patent is that the invention is novel, namely something for which there is no evidence in the literature that someone has thought of it before. Even if it is new, it cannot be obvious. Both novelty and obviousness are judged against what is in the literature or otherwise in the public domain, called the "prior art." Novelty in theory is easy to assess: Has someone else come up the same idea before you have? Obviousness is more subtle. For example, suppose that Mr. Schwinn invents and sells a two-wheeled, pedal-propelled vehicle that he names "bicycle." Clown comes along twenty years later and attempts to patent a tricycle. No one had ever thought of this before. The bicycle and tricycle are not identical, and by that measure the tricycle is novel. But is it an obvious leap from the bicycle? If it is obvious, it cannot be patented.

How many times have you caught yourself, after looking at a simple invention that has made someone a fortune, asking "why didn't I think of that, it's so obvious." Many of the great inventions are obvious, but only after the fact: barbed wire (U.S. Pat. No. 157,124, issued November 24, 1874, to Joseph Glidden), zippers (U.S. Pat. No. 1,060,378, issued April 29, 1913, to Gideon Sundback), and intermittent windshield wipers (U.S. Pat. No. 3,351,836, issued November 7, 1967, to Robert Kearns). These three inventions were the subjects of intense and lengthy patent disputes and litigation.

Let's look at the gizmo in a car that permits the driver to electrically adjust the height of the accelerator pedal. Before the 1990s, the accelerator pedal in the typical automobile was attached mechanically to the fuel injection system through cables. When you depressed the accelerator pedal in your car, that act mechanically led to an increase in the flow of gasoline. Starting in the late 1970s, devices had been developed that permitted a driver to electrically adjust the height of the accelerator pedal. The device was a boon, especially to those of us who are vertically challenged.

Starting in the late 1990s, automobile manufacturers began using electronic throttles that controlled gasoline flow using electronic devices instead of cables. The adjustable pedal, however, did not work with these electronic throttles. When the electronically controlled throttles entered the market, those manufacturing adjustable pedals for the auto companies quickly responded by combining various off-the-shelf products. The result was a bulky and expensive system.

In the late 1980s, Teleflex, an automobile parts company, developed electrically adjustable pedal controls that could be used with electronic throttles. Teleflex combined various existing technologies and devices in an artful manner to create its invention, for which it was granted patent protection. When one of its competitors, KSR, began selling jury-rigged adjustable controls, Teleflex sued for patent infringement. KSR argued that Teleflex's invention was obvious because all of the pieces already existed. If the invention were obvious, then the patent would be invalid and the infringement suit had to be dismissed. The court of appeals held that merely because someone combines existing devices and knowledge does not mean that the resulting product is obvious.[16] If that were the case, there would be no such thing as a new molecule. According to the court, things can be combined in novel ways, and it is ultimately up to the jury in a patent infringement suit to decide that question. The court of appeals provided some useful guidance in how one should evaluate whether an invention is obvious.

The Supreme Court, however, disagreed.[17] It reversed the court of appeals and held that based on the record before it, Teleflex's invention was obvious. It noted that if one combines two known technologies in a straightforward way, then that is obvious. As an example, it pointed to its earlier decision in *Anderson's-Black Rock, Inc. v. Pavement Salvage Co.*,[18] a case involving the radiant-heat paving machine, which constantly heated macadam so that it could be uniformly spread over a street. According to the Court in *Anderson's-Black*, the device merely combined two existing inventions—a radiant-heat burner and a paving machine. The device "did not create some new synergy: The radiant-heat burner functioned just as a burner was expected to function, and the paving machine did the same. The two in combination did no more than they would in separate, sequential operations."[19] Following the Supreme Court's decision in *Teleflex*, the contours of what is obvious and what is not are difficult to articulate in a meaningful way. However, one of the key factors still remains, namely whether someone schooled in the subject matter would ordinarily be able to put everything together and come up with the invention. If the answer is "no," then the invention is not obvious, even though its constituents are known to all in the field.

An Invention Has to Be Useful

To be eligible for a patent, an invention has to be useful. This does not mean that it has to work well. But it does mean that it has to do something or accomplish some purpose that has some immediate benefit to the public. The usefulness or utility requirement has taken on added significance with respect to the controversy surrounding the patentability of complementary DNA (cDNA). DNA, through a series of steps, is copied to create messenger RNA (mRNA), which ultimately codes for the manufacture of amino acids in the ribosomes. The mature mRNA is not a base-to-base copy of the DNA. Rather this mRNA carries only the essential bases, leaving out bases not necessary for making amino acids. The cDNA is the mirror image of the mature mRNA, except with the DNA base thymine in place of the RNA base uracil, and cDNA allows one to look at a region of DNA and see which bases are vital to protein manufacture and which are not. cDNA is chemically more stable than mRNA and can be used as a template for making proteins, or as a diagnostic probe.

Originally, J. Craig Venter, then at NIH, isolated large numbers of cDNAs and sought to patent them on behalf of the government. A major policy debate ensued within NIH concerning whether the government—or anyone else for that matter—ought to be patenting molecules that are more related to knowledge than they are to utility. To protect the government's rights in the inventions until the policy issues could be resolved, HHS, as the assignee, filed patent applications in June 1991 and February 1992 for expressed sequence tags (ESTs) from more than 2,700 fragments of human brain cDNA representing more than 300 genes of unknown function. An EST is a short cDNA sequence (150–400 base pairs) that corresponds to the coding sequence of an expressed gene.[20] The Patent and Trademark Office (PTO) found that Venter's inventions lacked utility because the sequences were useful merely as a means for making discoveries. However, many, including the head of the NIH Office of Technology Transfer, believed that the PTO's decision was incorrect because ESTs were useful in the laboratory as probes to identify genes along the chromosome. Under the PTO's logic, Hans Lippershey, Hans Janssen, and his son, Zacharias Janssen, could not have patented their microscopes because a microscope could be used only for making discoveries.[21] Another set of researchers, Dane K. Fisher and Raghunath Lalgudi of Monsanto Co., sought to patent ESTs. When the PTO turned down their applications, they challenged the PTO's action in court, but lost. The court of appeals found that a patent could not be issued without a real use, and the inventors could point to none.[22] The cDNA patent applications generated significant controversy within the scientific community over the morality or appropriateness of patenting genes.[23]

How Do You Obtain a Patent?

Suppose you have invented something and you believe that it may qualify for patent protection. In the United States, if you are employed at a university, government laboratory, or research facility, there are likely forms, sometimes called an invention report form, that you would need to complete and that would be forwarded to someone with relevant patent law experience. That person, or more likely a committee, would decide whether your invention is patentable and shows sufficient economic promise to justify the patent-related expenses (depending on the nature of the invention, $10,000–$25,000, plus filing fees).

If your employer decides that your invention should be patented, what happens next? Probably, the first thing is that you would be asked to assign your invention over to your employer.[24] As a general rule, most employers require, as condition of employment, that an employee assign over to it, if the company so wishes, any intellectual property that the employee develops in the course and scope of his or her employment. Most universities also have this type of policy. For example, Stanford University's policy provides as follows:

> All potentially patentable inventions conceived or first reduced to practice in whole or in part by members of the faculty or staff (including student employees) of the University in the course of their University responsibilities or with more than incidental use of University resources, shall be disclosed on a timely basis to the University. Title to such inventions shall be assigned to the University, regardless of the source of funding, if any.[25]

Notice how the policy does not deal with inventions that a faculty member conceived or reduced to practice on his or her own time while consulting for a company. Some private companies, however, have expansive policies that require employees to assign to the company all intellectual property irrespective of whether it is invented on company time or not.

Remember, at the start of this chapter I indicated that a patent was an intangible property right. Therefore, interests in a patent can be sold in a variety of ways, including by assignment. An "assignment" is a complete sale of the intellectual property. The assignor retains no interest in the property, although he or she may have royalty rights under his or her employment agreement or by operation of federal law or both.

After you have filed an invention disclosure report, you will probably be asked to meet with a patent attorney who will be responsible for preparing the application. The attorney likely represents your employer and not you. The application will set out, among other things, the prior art; demonstrate that the invention is novel, nonobvious, and useful; and explain how to make and use the invention. Finally, the application sets out the claims. The claims are the most important part of the application; they are analogous to the property lines on real estate. The broader the claims, the broader your patent. For example, let's compare three hypothetical bicycle claims. The first is a claim for "any vehicle with two or more wheels." The second is a claim for "a vehicle with two or more wheels which is human-powered." The third is a claim for "a pedal-propelled vehicle with two wheels." The first patent, if it were to issue, would cover virtually every vehicle irrespective of how it is powered. It would arguably cover most, but not all, aircraft. The second claim is far narrower and would cover a scooter, a bicycle, a tricycle, and, yes, even a baby carriage, to name a few. It would not cover a wheelbarrow, though, which has one wheel. The final claim is narrow and would cover only the traditional bicycle. A two-wheeled scooter would not be covered because it is not pedal propelled. Typically, a patent lawyer tries to get the broadest possible claim through the patent office.

Once the patent application is filed with the PTO, it will be assigned to a patent examiner who has technical expertise in the area of your patent. For example, inventions involving chemistry will be assigned to an examiner who was trained as a chemist or chemical engineer. The examiner can take a variety of actions, including issuing the entire patent as requested, declining to issue the patent, or issuing some, but not all, of the claims. Usually, there is a back-and-forth between the examiner and the patent attorney. The process normally takes about thirty months from start to finish and costs about $10,000 to $25,000 in attorneys' fees with additional filing fees. These are averages with significant standard deviations.

In some cases, an examiner declines to issue the patent or refuses to issue the patent for specific claims that the inventor believes ought to be issued. If that occurs, an inventor can appeal using the administrative appeals process within the patent office; if this appeal fails, an inventor can challenge the PTO's "office action" (i.e., decision) in the United States Court of Appeals for the Federal Circuit, which sits in Washington, D.C. Appeals inside the PTO or to the courts increase the time it takes to get (or not get) a patent.

One of the most famous cases dragged on for nearly thirty years in the patent office and then in the courts. It was one of those rare cases where the delay transformed an invention with little immediate value into one worth billions. This story begins at the physics department at Columbia University in the 1950s. In the early 1950s, Columbia University physicist Charles H. Townes built a quantum mechanics-based device that could produce coherent, single-frequency radiation in the microwave spectrum. (Waves of the same frequency are coherent when they propagate in lock-step.) Townes called the device the "maser," an acronym for "microwave amplification by stimulated emission of radiation."[26] Townes's application followed quickly on the heels of theoretical work done by two Russian physicists, Nikolay Basov and Alexander Prokhorov. The three would win the 1964 Nobel Prize for that work.

Townes, in addition to his faculty responsibilities at Columbia, also worked as a consultant at Bell Telephone Laboratories (Bell Labs). He and

Arthur Schawlow, a colleague at Bell Labs, were interested in extending the maser to the optical spectrum. On July 30, 1958, they filed a patent application for a low-power optical maser to be used in communication.[27] About five months later, the two published a seminal paper, "Infrared and Optical Masers," which set out the basic idea for the laser ("light amplification by stimulated emission of radiation").[28]

In 1957, Gordon Gould, a graduate student at Columbia University, also came up with the idea for the laser. Gould discussed his work with Townes, but there is no evidence that either purloined ideas from the other. Gould filed his patent application on April 6, 1959. The Schawlow-Townes patent for the laser issued on March 22, 1960, less than two years after it was filed and about one year after Gould had filed his patent application.

The Schawlow-Townes patent generated relatively few royalties. At the time the patent issued, no one had been able to build a laser. When the first working laser was built on May 16, 1960, the honor went to none of the three East Coasters, but rather to physicist Theodore "Ted" Maiman at the Hughes Research Laboratory in Malibu, California. Shortly thereafter, Maiman started a company called Korad to develop and manufacture lasers, and soon after that Korad was purchased by Union Carbide and operated as a wholly owned subsidiary.

In the early years, there were only two markets for lasers, the military and research laboratories, and neither market was large. In the 1960s, the U.S. Army purchased solid, pulsing lasers (usually of yttrium aluminum garnet—YAG—or pink ruby) as rangefinders that were mounted in tanks. The early YAG and pink ruby lasers were expensive but finicky and unreliable. Researchers used low-powered, noble gas lasers (helium-neon) as convenient sources of coherent, monochromatic light.

Despite the relatively small commercial market, Gould battled Schawlow and Townes over inventorship, first in the patent office and later in the courts. Gould claimed that he had invented the laser before Schawlow and Townes, even though their patent application was filed first. Initially, the PTO and the then–Court of Customs and Patent Appeals declared Schawlow and Townes to be the original inventors. (Note that in 1982, the Court of Customs and Patent Appeals was reorganized and renamed as the United States Court of Appeals for the Federal Circuit.) Gould, however, persisted and ultimately, in 1977, he obtained a fundamental patent on the laser and in 1989 successfully enforced the patent in court.[29] In the nearly twenty years from the time Gould first conceived the invention to the time his patent issued, the market for lasers exploded. After successfully enforcing his patent in 1989, Gould became a near-billionaire overnight. Today, it is difficult to find a U.S. home without at least one laser (e.g., CD or DVD players); lasers, of course, are also used widely in industry.

What Can You Do with a Patent Once You Have It?

THE EFFECT OF A PATENT

Although a patent grants a monopoly, it does not bestow on the owner or inventor the right to make, use, or market the invention. Rather, a patent grants the inventor a negative right: the ability, in most cases, to exclude others from making, using, or selling the invention for the life of the patent, which is twenty years from the date the patent application is filed. (Prior to June 8, 1995, the life of a patent was the greater of seventeen years from the date that the patent issued or twenty years from the date the application was filed.)

For example, suppose that Schwinn invents a bicycle and receives a patent on "a vehicle having at least two wheels." Then, Benz invents a car and receives a patent on "a motorized vehicle having at least four wheels." Schwinn may prevent anyone, including Benz, from making, using, or selling a vehicle with two or more wheels. Although Benz owns the car patent, since a car is also "a vehicle with two or more wheels," Schwinn can prevent Benz from making, using, or selling a car. This does not mean, however, that Schwinn can manufacture a car. He cannot. Benz's patent excludes anyone, including Schwinn, from making, using, or selling "a motorized vehicle having at least four wheels."

Therefore, if Schwinn and Benz do not cooperate, no one can manufacture cars. Schwinn must license Benz under the bicycle patent for Benz to be able to manufacture a car. Similarly, Benz must license Schwinn under the car patent for Schwinn to be able to manufacture a car. This cooperation is known as cross-licensing, and in our example, the bicycle patent would be viewed as the "pioneer," "dominant," or "blocking" patent, and the car patent as the "subservient" or "improvement" patent. Schwinn may decide that he does not want to expand into the car business and instead of getting a license back from Benz, he may require Benz to pay him royalties on each car that Benz sells.

THE NARROW EXPERIMENTAL USE EXEMPTION

One of the more vexing public policy, and ultimately legal, questions has been whether a patented invention can be used in research without the permission of the patent holder. A patent permits its owner to preclude others from making, selling, or using the patented invention. Courts were understandably reluctant to prevent scientists from using the inventions of others for noncommercial research purposes. As a result, there emerged a very narrow "experimental use" exception to the patent laws that allows a researcher under very limited conditions to use another's invention "'for amusement, to satisfy idle curiosity, or for strictly philosophical inquiry.'"[30] The line that separates permissible use from nonpermissible use is fuzzy.

One of the more interesting cases involved a public spat between Duke University and one of its leading physicists, John Madey. Ten years earlier, Duke had recruited Madey away from Stanford, where he had run a free-electron laser laboratory. (A free-electron laser differs dramatically from a typical laser; a free-electron laser uses relativistic electrons, i.e., electrons whose speed has been increased to near that of light where relativistic effects manifest.) At Duke, he reestablished his laboratory where, by all accounts, it prospered in terms of funding and scientific achievements. Eventually, though, Duke came to believe that in spite of the laboratory's success, Madey was an ineffective manager; eventually Duke relieved him of his position as head of the laboratory, and, shortly thereafter, Madey resigned from Duke and took a position at the University of Hawaii. While at Duke, Madey invented a new type of free-electron laser. Duke had the laser patented, but rather than requiring Madey to assign his rights in the laser patent to it, Duke permitted Madey to retain ownership of the patent. When Madey left Duke, the university continued to use the laser in its research efforts.

Madey sued Duke for patent infringement. Duke responded by arguing that it was a research institution and not a commercial enterprise, and therefore, under the "experimental use" exemption, it could lawfully use the invention for research without the inventor's permission and without paying royalties. The court of appeals held that Duke and its research program might not have satisfied the narrow "experimental use" exemption. According to the court, "major research universities, such as Duke, often sanction and fund projects with arguably no commercial application whatsoever. However, these projects unmistakably further the institution's legitimate business objectives, including educating and enlightening students and faculty participating in these projects. These projects also serve, for example, to increase the status of the institution and lure lucrative research grants, students and faculty."[31] In short, most activities conducted by research laboratories at universities would likely not qualify for the narrow experimental use exemption. The court went on to conclude that

> regardless of whether a particular institution or entity is engaged in an endeavor for commercial gain, so long as the act is in furtherance of the alleged infringer's legitimate business and is not solely for amusement, to satisfy idle curiosity, or for strictly philosophical inquiry, the act does not qualify for the very narrow and strictly limited experimental use defense. Moreover, the profit or non-profit status of the user is not determinative.[32]

The FDA Exemption

The narrow experimental use exemption is different from the so-called FDA exemption. The experimental use exemption is a creature of the courts; the FDA exemption is spelled out in statute:

> It shall not be an act of infringement to make, use, or sell a patented invention . . . solely for uses reasonably related to . . . the development and submission of information under a . . . Federal law which regulates the manufacture, use, or sale of drugs.[33]

Under that exemption, a company is permitted to use a patented drug to collect data on that drug as a prelude to seeking regulatory approval for that drug "under a . . . Federal law which regulates the manufacture, use, or sale of drugs."[34]

Although the FDA exemption is narrow and speaks only in terms of drug testing, is it broad enough to encompass clinical testing of medical devices? In *Eli Lilly & Co. v. Medtronic, Inc.*,[35] the Supreme Court was asked to resolve that question. Eli Lilly & Co. owned the patent on an implantable cardioverter defibrillator (ICD), a medical device used in treating certain types of arrhythmias. Medtronic, Inc., a Lilly competitor, sought to test an ICD of its own to develop clinical data for eventual submission to FDA. Lilly claimed that Medtronic's device infringed its patent and that Medtronic's act of "testing" was a commercial use of its product; therefore, Medtronic had infringed Lilly's patent. Medtronic argued that it was collecting data so that it could submit information under a "Federal law" that regulates drugs. The law that regulates drugs is the Food, Drug, and Cosmetic Act. But that law also regulates medical devices and, therefore, Medtronic's activity was exempt. The Court agreed, holding that the phrase "under a . . . Federal law which regulates the manufacture, use, or sale of drugs" incorporates the entire Food, Drug, and Cosmetic Act, including the device provisions.

ASSIGNING, LICENSING, AND USING A PATENT

Patents are property, and like any property they can be sold, leased, or used directly by the owner. As mentioned above, the sale of a patent is called an assignment, and the lease of a patent is called a license. There are two general forms of licenses, exclusive and nonexclusive. An exclusive license is normally used where significant investment is required before the product can actually be marketed. Thus, if you develop a new drug that appears in the laboratory to have action against a specific bacterium or virus, it is unlikely that a pharmaceutical company would be willing to license the invention other than exclusively because of the tremendous investment necessary to bring a new drug to market. The investor would expect to be the only seller of the product in a given geographic market. Drug companies maintain that they can justify their significant investments only if they have monopoly power for a set time. Exclusive licenses come in various varieties, including an exclusive license worldwide, an exclusive license restricted to a geographic area (e.g., the United States, Europe), or an exclusive license to a restricted market (e.g., research laboratories, consumer products).

A nonexclusive license is normally used when the product is fully developed and can be easily marketed by a licensee with minimal special investment. For example, the 1980 Boyer-Cohen gene splicing patent (held by Stanford University, but with royalties split between the University of California and Stanford), Kary Mullis's 1987 polymerase chain reaction patent (held by Cetus Corporation, which sold the patent to Roche Molecular Systems for $300 million), and the Gould laser patents are examples of patents that have been successfully licensed nonexclusively.

Assignments and licenses are memorialized in lengthy written agreements that spell out precisely what is required of each party and what each party expects to get. Added complexities arise when there is more than one inventor. In such a case, one inventor can transfer his or her interest in a patent, but it takes all inventors to assign the full patent. For example, suppose that Alan and Sue invent the gizmo, a new form of widget. Each owns an undivided interest in the entire invention. Alan can assign his interest to Dave, but as long as Sue retains her interest, Dave may still end up having to compete for sales against Sue. Both Dave and Sue can manufacture, use, and market the gizmo. Dave can gain full control of the patent only if Sue assigns her interest to Dave.

When licensing any patent, one normally gets paid through royalties (i.e., a percentage of net revenues from sales). However, what if an inventor exclusively licenses his or her drug patent to a pharmaceutical company and the company fails to commercialize the product? That is, it does not invest in clinical trials or anything else necessary to bring the drug to market. If the licensor gets paid exclusively through royalties, he or she would receive nothing. Clearly, the licensor would want to make certain that this would not occur. Thus, a license agreement would normally require the licensee to pay not only royalties on sales, but also (1) a signing fee, (2) an annual maintenance fee, and (3) benchmark fees. The agreement would also provide the licensor with "march-in" or revocation rights should the licensee not diligently pursue commercialization. To complicate matters further, in many instances the licensor may want to receive common stock in the company licensing the product. This shifts some of the risk back onto the licensor. When Stanford University licensed its interest in the Google algorithms back to Google, it accepted an ownership interest in the new company.

A signing fee, as its name implies, is a fee that the licensee pays up front for the benefit of receiving the license, usually an exclusive license of some sort. The fees vary with the size of the licensor, the financial promise of the licensed product, and the amount that the licensee must invest to bring the product to market. An annual maintenance fee is much like a base rental fee paid each year on the anniversary of the license agreement's execution. Benchmark fees are paid when the licensee reaches certain preset goals. Where a drug is involved, for instance, the licensee may be required to pay $100,000 when FDA clears the product for phase 2 clinical trials, $200,000 when the product is cleared for phase 3 trials, $500,000 when the licensee submits a New Drug Application (NDA) to FDA, and $1 million when the NDA is approved by FDA. A well-crafted agreement permits the licensor to revoke the license (or "march in" and reclaim the license) if the licensee has not attained certain benchmarks by certain designated dates. Assignment and license agreements are complex and should be drafted and reviewed by experienced and competent counsel.

There is a third possibility. You can attempt to commercialize and market your invention yourself. Patents have formed the basis of thousands of small technology companies. These companies operate on the theory that the farther along you can move an invention, the greater the premium you will obtain when you assign or license the product. If you retain your patent, you are shouldering the risks of development and commercialization, and, in many areas, this pays off handsomely should your invention prove successful. For instance, you will earn significantly more if you transfer a patent on a drug that has already successfully completed phase 2 clinical testing than on a new drug that is still being tested at the bench. However, to reap this reward you will likely need investors to help fund your research and the early clinical trial phases. These early investors will normally demand both a significant interest in the company and significant control over both the science and the company's operation.

How Can You Protect Your Patent?

The United States patent laws provide two ways of protecting inventions: (1) interference and (2) infringement. An interference is a proceeding in the PTO to determine which of two overlapping patents (or patent applications, or a patent application and an issued patent) has priority (i.e., which one was conceived first). An interference is an administrative proceeding conducted entirely in the PTO before a hearing board; the losing party can seek review in the United States Court of Appeals for the Federal Circuit. An infringement action, on the other hand, is a lawsuit in a federal district court with a jury usually deciding the outcome.

Suppose that WonCo has just filed a patent application for an improved widget, one that is capable of holding six colors. InfringeCo, WonCo's chief competitor in the widget market, has already been issued a patent on a six-color widget. WonCo is convinced, after rifling through InfringeCo's dumpsters, that InfringeCo's chief scientist Allen Grebdren had conceived the invention three months after WonCo's chief scientist Ella Nerdberg. As noted earlier, under U.S. patent law the first to conceive the invention, not the first to file, is awarded the patent. In this setting, WonCo can seek to provoke an interference in the patent office. To do so, it would have to demonstrate to the patent examiner that there is a pending patent application and that that application has one or more claims that intersect with claims on either another pending application or an issued patent. If the examiner believes that there appears to be interference, he or she refers the matter to Board of Patent Appeals and Interferences (Board), which has the authority to declare and resolve an interference. The Board is an internal appeals body that sits in panels of three. Assuming that it declares an interference, it will then convene a formal hearing to decide whether Nerdberg or Grebdren was the first to conceive the invention. The members of the Board have both patent and technical expertise. While the primary purpose of the hearing is to determine who conceived the invention first, the Board can examine certain other issues as well, including whether an invention is disclosed by prior art or whether the description of the invention is inadequate.

An interference tends to be significantly less expensive than traditional litigation because there are no live witnesses; the issues are resolved on the basis of documents and declarations. A decision of the Board can be appealed to the Court of Appeals for the Federal Circuit.

In contrast, a patent infringement suit is usually an expensive and lengthy process that starts out in a U.S. district court. Suppose that WonCo decides not to provoke an interference. Instead, it begins manufacturing and distributing its widget before the PTO acts on its patent application. InfringeCo could sue WonCo for patent infringement. Usually when this occurs, the defendant argues that the plaintiff's patent is invalid, and therefore the defendant is not infringing a valid patent. WonCo, for instance, would argue that its scientist was the first to conceive the six-color widget and as such, the InfringeCo patent should never have been issued and is not valid. If WonCo prevails, then the InfringeCo patent is declared invalid and WonCo can continue marketing whether it gets a patent or not. If WonCo were to lose, however, it would likely be enjoined from continuing to market the product, and it would be compelled to pay damages to InfringeCo.[36] It might even be required to pay punitive damages if the court finds that its infringement was "willful." Again, the losing party can appeal to the United States Court of Appeals for the Federal Circuit.

Special Infringement Rules for Drugs

During the late 1970s and early 1980s, generic drug manufacturers and innovator manufacturers (i.e., large pharmaceutical companies) waged bitter battles against each other and FDA over certain systemic issues. For years, FDA, without express statutory authorization, had permitted generic copies of patented drugs to be marketed after the patent had expired on the innovator product. A generic manufacturer was permitted to rely on the clinical data generated by the innovator; the generic manufacturer was required to show that its product was similar (i.e., bioequivalent) to the innovator's product. Without having to conduct costly clinical trials, generic manufacturers could sell their products at significantly lower prices than the innovator could. The large pharmaceutical companies argued that FDA was acting outside the law; that each company's safety and efficacy data were trade secrets; and that by permitting generic companies to use the data, FDA was violating the Trade Secrets Act. They also argued that the generic manufacturers were infringing the innovators' patents by using the generic versions in laboratory tests to establish bioequivalence before the innovators' patents had expired. FDA responded by noting that it was not revealing any data to the generic manufacturers. Instead, it was merely permitting the generic manufacturers to reference those data without ever seeing them. To overcome these arguments from the innovator manufacturers and to eliminate the legal uncertainty surrounding generic drugs, Congressman Henry Waxman in the early 1980s drafted legislation that would expressly authorize generic drugs and would permit generic manufacturers to test their products in the laboratory without infringing the innovator's patent. Naturally, Waxman's legislation was opposed by the large pharmaceutical companies.

During this same time, the large pharmaceutical companies were also concerned that the useful lifespan of a drug's patent was effectively shortened by inordinate delays in FDA's drug approval process. For example, if a manufacturer received a patent on a drug in 1970 but it took the manufacturer twelve years to complete its clinical trials and obtain FDA approval, the useful life of the patent was only five years (i.e., seventeen-year life minus twelve years). The large pharmaceutical companies therefore sought legislation that would extend their patents to take into account inordinate delays by FDA. Senator Orrin Hatch in the early 1980s proposed legislation that would extend drug patents under certain circumstances. This legislation was opposed by the generic manufacturers. In an unusual show of bipartisanship, Hatch and Waxman melded their two bills together and, in so doing, gained the support of both the generic industry and large pharmaceutical companies. The Drug Price Competition and Patent Term Restoration Act of 1984, also known as the Hatch-Waxman Amendments,[37] authorizes generic drugs by allowing the generic maker to submit an Abbreviated New Drug Application (ANDA) for a specific drug as soon as the patent on that drug expires. The generic applicant is required to submit data demonstrating that its product is bioequivalent to the brand-name drug and to submit labeling and warnings that are identical to those of the brand-name drug. It can collect the data it needs before the patent on the innovator drug expires. Hatch-Waxman also permitted pharmaceutical companies to petition FDA and PTO to have the patents on their brand-name drugs extended beyond the normal term to offset certain delays in the FDA approval process.

Hatch-Waxman contained one rather fascinating provision. If a generic manufacturer believed that the patent on a brand-name drug was invalid, it could file an ANDA for that drug before that drug's patent expired, provided it certified that it believed that the patent was invalid.[38] The brand-name manufacturer is given forty-five days in which to institute a patent infringement suit against the generic applicant. Once suit is filed, FDA is precluded from approving the generic application for thirty months or until the infringement is resolved in the generic's favor, whichever comes first. Most new drugs with significant sales have been the subject of Hatch-Waxman litigation, including Celebrex,[39] Plavix,[40] and Zoloft.[41]

TECHNOLOGY TRANSFER

A generation ago, most scientists at universities and government laboratories cared little about intellectual property. After all, if a researcher developed a valuable invention under a federal grant, the odds were that the federal government would claim the invention, so why bother even finding out whether something was patentable? Universities were reluctant to patent inventions developed by its faculty under government grants because, if the government claimed the invention, the university was not likely to recover even the cost of patenting. As a result, many inventions developed at universities and at federal laboratories, which could have been patented were never patented, and so ended up in the public domain, enriching manufacturers who could make use of the inventions without the inconvenience of having to pay royalties. This was consistent with the conventional wisdom of the time that the public was better served if scientific advances made with government funding were kept in the public domain or available nonexclusively. Conventional wisdom began changing, though, in the late

1970s. The law began changing in 1980 with enactment of Bayh-Dole and with the Federal Technology Transfer Act of 1986, which gave universities and government laboratories, respectively, a significant financial incentive to patent worthy inventions; it also gave the employees a financial interest in their own inventions by guaranteeing to them a percentage of the royalties. Suddenly, everyone was interested in getting their inventions patented and licensed. This early interest was fueled by serendipity as much as the new legislation. Some of the first inventions to come out of the universities and government in the late 1970s and early 1980s were blockbusters—paying tens and hundreds of millions in royalties to the universities and their faculty inventors. Universities, sensing a hidden treasure trove, rushed to set up offices of "technology transfer" as a way of ensuring that there would be some organizational mechanism to capture and license the next Boyer-Cohen (gene splicing patent owned by Stanford, but with royalties being shared with the University of California), Gallo et al. (HIV test kit patent, owned by the Department of Health and Human Services), or Google (certain patents for search engine algorithms owned by Stanford) inventions.

Today, almost every research university has a technology transfer office, which tries to figure out which inventions ought to be patented and then tries to license those inventions to the private sector. It is still unclear whether the royalties collected by these offices pay for the cost of maintaining the offices. However, it is clear that technology transfer in general has been beneficial to the economy, creating jobs and products, and to the nation by stimulating innovations that improve health and communication.

The Bayh-Dole Patent and Trademark Amendments of 1980

Prior to 1980, U.S. companies that wanted to use federal research and development funds to develop new products were, in the words of one House committee, "[confronted with] a bewildering array of 26 different sets of agency regulations governing their rights to use such research."[42] Moreover, the general rule was that innovations developed with federal funds belonged to the federal government. In short, federal law, either by design or accident, thwarted the efficient transfer of technology. This came at a time when many in Congress voiced open concern that America's position in various world markets was coming under intense foreign competition. There was a perception that many foreign competitors were taking better advantage of American research and development than were domestic companies, and that such a trend only exacerbated our competitive decline.

To partially cure these defects, and to better align the patent laws with Congress's policy objectives, Congress enacted the Bayh-Dole Patent and Trademark Amendments of 1980.[43] Under Bayh-Dole, the patent rights in substantially all inventions developed with federal funds by small businesses and nonprofit entities, including universities, would vest, subject to certain conditions, in the contractor or grantee, as the case may be. A contractor or grantee is required to inform the government within sixty days of the existence of a patentable invention and is then required to prosecute that patent. The government retains certain so-called march-in rights should the contractor or grantee fail to fulfill those obligations. Furthermore, the government is granted by operation of law a paid-up nonexclusive license to use the invention anywhere in the world for government purposes.

While Bayh-Dole marked a major turning point in the government's efforts to transfer technology, its impact was limited for a number of reasons. First, Bayh-Dole did not apply to those entities in the best position to commercialize new technology, namely large profit-making entities. Second, the law precluded small businesses and nonprofit entities from granting exclusive licenses to large profit-making entities, unless those licenses were for less than five years or, in certain instances, eight years. Third, because Bayh-Dole applied only to work performed under a "funding agreement," the status of innovations developed under a collaborative agreement with a federal laboratory remained uncertain. And finally, Bayh-Dole did not apply to inventions developed by a contractor pursuant to a so-called GOCO (i.e., government-owned, contractor-operated facility) arrangement.

Some of the limitations of Bayh-Dole were remedied at the highest level. On April 10, 1987, President Reagan issued Executive Order No. 12,591 which expressly required agency heads to

> promote the commercialization . . . of patentable results of federally funded research by granting to all contractors, *regardless of size*, the title to patents made in whole or in part with Federal funds, in exchange for royalty-free use by or on behalf of the government.[44]

Thus, the benefits of Bayh-Dole were extended to large profit-making entities.

Bayh-Dole also requires that universities and other recipients of federal funds have in place a policy under which income from an invention developed with federal funds would be shared with both the laboratory and the faculty inventor. Bayh-Dole, though, does not specify how much a university has to pay its faculty inventors. As a result, university policies are remarkably diverse. Some universities pay faculty inventors half of the income that the university receives (after subtracting patenting and administrative expenses) without any limit, while other universities provide a much more modest share. For example, the University of Maryland, after deducting its expenses for patenting and administering the inventions, pays the inventors the first $5,000 and then shares the subsequent revenues 50–50 with its faculty inventors.[45] The University of California System has had various policies starting as early as 1963. Under the current policy, the university subtracts and retains 15 percent from the income it receives and then divides the net remaining, with 35 percent going to the inventors, 15 percent to the inventors' laboratories or campus, and 50 percent to the general pool at the inventor's campus or laboratory.[46] Harvard's current system is similar, with the inventors receiving 35 percent for personal use, 15 percent for research use by the inventors, and the remaining divided according to a formula among the department, school, and president's office.[47] Yale has a three-tiered sliding scale: Faculty inventors receive 50 percent of the first $100,000 net royalties, 40 percent of the next $100,000, and 30 percent of everything in excess of $200,000 net royalties.[48] Most institutions figure the faculty share based on the "net royalty," which is a term defined in a university's policy and normally means gross royalties minus various expenses incurred by the university to prosecute and maintain the patent and the licensing agreements. Variation in the way the term "net royalty" is defined can dramatically affect a faculty member's share. In short, there is more to watch than merely the percentages. (See also discussions of royalty income and conflict of interest reporting in chapter 5.) Interestingly, commercial entities tend not to be as generous with inventor-employees as universities and other academic-type institutions. For example, Kary Mullis indicates that all he received from Cetus for inventing PCR was a $10,000 bonus. He received about fifty times that as a Nobel Prize winner in 1993.

Gaining Access to Federal Technology

While Bayh-Dole provided incentives for researchers at private universities to patent their inventions and incentives for various institutions to work together, no such incentives existed at the federal laboratories. To correct this, Congress enacted the Federal Technology Transfer Act (FTTA) of 1986.[49] FTTA was designed to ensure the exchange of commercially valuable information between federal laboratories and the private sector. Accordingly, FTTA creates a system whereby private businesses can in various ways obtain access to technology developed in federal laboratories or at federal expense.

FTTA authorizes the director of any government laboratory, including NIH, to (1) enter into cooperative research and development agreements (CRADAs) with private entities, irrespective of size, including profit-making organizations; and (2) negotiate licensing agreements for government-owned inventions or other intellectual property developed at the federal laboratory.[50] To provide federally employed scientists with an incentive either to participate in CRADAs or to help license innovations that they may have invented, FTTA, as amended, requires federal laboratories to pay federal inventors at least 15 percent of the royalties received by the federal government up to $150,000 per year per inventor. The remainder of the royalties is to be distributed directly to the laboratory and to its parent agency to cover patent expenses and administrative fees.

COOPERATIVE RESEARCH AND DEVELOPMENT AGREEMENTS

The linchpin of FTTA is authority to enter into CRADAs. A CRADA is nothing more than an agreement between a federal laboratory and one or more private parties to jointly develop certain technology or to jointly undertake certain basic or applied research. Under the agreement, each party agrees to contribute personnel, facilities, and equipment toward a common goal. While the private party is free to contribute funds directly, FTTA precludes the federal government from doing so. Significantly, as a matter of law, a CRADA is not considered a procurement contract or a grant, and thus the strictures of the Federal Acquisition Rules and the rules for awarding grants do not apply.

In exchange for participating with the federal laboratory in the joint endeavor, the private party may be granted a license or assignment, or an option to it, in any invention made in whole or in part by a federal employee under the CRADA. The government, however, retains at the very least a nonexclusive, nontransferable, irrevocable paid-up license to use the invention throughout the world for government purposes. Most agencies have adopted policies under which collaborating companies acquire an option at the outset to negotiate for exclusive patent rights. Under these policies, after the invention has been developed, the parties must reach agreement on the type of rights to be licensed (e.g., exclusive versus nonexclusive) and on the royalty to be paid to the federal government. Normally, the scope of the license and royalties to be paid are a function of the relative contributions that the parties have made to the invention in question and the capital costs associated with commercialization. As a rule, most federal agencies do not assign patent rights.

A CRADA can provide a private company with significant advantages at modest cost, including the following: (1) the company can leverage its research and development budget; (2) government scientists can assist in developing a marketable product at no direct cost to the company; (3) the company can obtain a CRADA without going through the government procurement process; (4) the company scientists can share in federal royalties up to $150,000 per year; and (5) the government grants to the company an exclusive license to market the product worldwide at a negotiated royalty.

Inasmuch as a CRADA is neither a procurement contract nor a cooperative agreement, the federal laboratory retains extraordinary discretion in deciding whether a CRADA is appropriate, and if so, which private company should be given the opportunity to negotiate for the CRADA. Not surprisingly, most CRADAs have as their genesis a close, long-standing working relationship between key government scientists and private scientists. Frequently, the private scientists have collaborated previously with the government's scientists on related work or had worked in the federal laboratory. In such cases, the federal scientists might suggest to their privately employed counterparts that a CRADA would provide an appropriate vehicle for further collaborative efforts. Once the respective scientists have communicated their mutual interest in a CRADA to the appropriate administrative personnel, usually some from the office of technology transfer, that office will work with the government scientist and private scientist(s) and legal representatives of the private entity in an effort to craft a mutually satisfactory CRADA.

At NIH, the Office of Technology Transfer has developed a form CRADA; the form is sufficiently flexible to enable the parties to craft provisions to meet the unique needs of the proposed endeavor. Normally, before any CRADA is negotiated, a number of critical issues will have to be resolved. For example, the parties will have to reach agreement on the "scope of work" to be covered by the CRADA, the resources that each party is to commit to the project, and the limitations, if any, on the publication of research results.

In addition, the federal agency that operates the laboratory might attempt to extract from a potential CRADA partner a promise that any product developed under the CRADA and licensed exclusively to the CRADA partner will be sold at a reasonable price. This "pricing restriction" has been the subject of significant controversy within the Department of Health and Human Services, and no clear consensus has been reached. Those favoring a pricing restriction argue that because the federal government ends up paying for most pharmaceutical products either through Medicare or Medicaid, it is unconscionable for the government to be paying inflated prices for products that it helped develop. These critics point to the problems associated with AZT, which was developed at the taxpayers' expense at NCI. Those who object to pricing restrictions argue that NIH is not a regulatory agency, and that imposing such a restriction is contrary to NIH's primary mission and expertise.

Although, as noted above, most CRADAs are "sole source," where the subject of the proposed research is truly significant, as might be the case with an AIDS vaccine, or where a number of qualified companies have expressed an interest in obtaining a CRADA for the same type of research with the same laboratory, a federal agency might publish a notice in the *Federal Register* giving the private companies an opportunity to compete for the CRADA in much the same way that entities would compete for a procurement contract. Similarly, if the development and commercialization potential is of more immediate importance than the basic research aspects, NIH may post a *Federal Register* notice seeking a collaborator that has both the requisite scientific expertise and commercialization capability.

LICENSING INVENTIONS OUTSIDE THE CRADA

The second way in which private businesses can gain access to federal technology is through patent licensing. Under FTTA and Bayh-Dole, federal laboratories can license patents owned by the laboratory to private businesses.[51] FTTA simplified the licensing process by authorizing the head of the laboratory to negotiate the licensing arrangements directly with the private party. Typically, such licensing arrangements consist of initial fees plus running royalties based on sales. For example, the AIDS antibody test kit, which was developed at NCI, was initially licensed to five companies at a royalty rate of 5 percent of gross revenues attributable to the sale of the test kit, plus an up-front fee.

Under the relevant rules, a government laboratory or agency is required to publish in the *Federal Register* announcements of its patents that are available for licensing.[52] An applicant for a license must submit a plan for the development and marketing of the to-be licensed invention.[53]

One of the key issues associated with licensing from a federal laboratory is the ability of a private party to obtain an exclusive license. For those seeking an exclusive license to use or commercialize government inventions, the licensing agency must make formal findings that (1) the federal and public interests are best served by exclusive licensing, (2) expeditious practical application of the invention is unlikely to occur under a nonexclusive license, (3) exclusive licensing is a reasonable and necessary incentive to attract investment of risk capital, (4) exclusive licensing will not tend substantially to lessen competition or result in undue market concentration, and (5) the proposed terms and scope of exclusivity are no greater than reasonably necessary.

An agency does not have to make those five findings when the invention is to be exclusively licensed to an agency's CRADA partner and the invention was developed under the CRADA. Under FTTA, a private CRADA collaborator may obtain an exclusive license upon request, subject to submitting a satisfactory commercial development plan.

Bayh-Dole and FTTA provide significant opportunities to both the private and public sectors. The private parties gain access to advanced government research, to the scientists responsible for that research, and to equipment that may not be available outside a government laboratory. The government actors gain access to applied scientists who know how to commercialize inventions, raise capital, and run companies. The CRADA process, however, has various perils, including a bureaucracy that can be remarkably inflexible and short-sighted and whose short-term goals may be inconsistent with good policy or the public interest. For example, if offices of technology transfer rely for their funding or for their evaluations on licensing revenue, they become more inclined to aim for the large up-front signing fee and annual maintenance fees than they are to aim for larger royalty payments. This can have perverse consequences. First, it drains needed seed money away from early development, when it is needed most. And second, if an invention proves to be successful, the government will end up receiving relatively small payments, having opted to risk less by receiving more up front.

OWNERSHIP OF SPECIMENS AND MATERIAL TRANSFER AGREEMENTS

This chapter deals with intellectual property, which by definition is intangible. Some intellectual property is so closely tied to its physical manifestation that the rules of personal property may govern. Specimens, reagents, cell lines, and the like have generated significant interest. Their worth lies not in the value of their physical being but in the information that they contain or can be used to obtain.

Who, for example, owns specimens that you have collected, cell lines that you have made, or reagents that you have concocted? I am not speaking of ownership in the patent sense but in the physical sense, the one that would guide the desires of a seven-year-old in a candy store or toy shop. Questions of ownership can and have turned ugly. A dispute between Washington University in St. Louis and a former faculty member illustrates this point. John Catalona, an academic urologist, had been on the Washington University (WU) faculty from 1976 to 2003. Although a surgeon, Catalona was interested in exploring the genetic basis of prostate cancer. In 1983, he began collecting tissue and blood specimens that he had removed during surgery for prostate cancer. When his colleagues began doing the same, largely at his recommendation, he established the Genito-Urinary Biorepository, housed at WU. Specimens from the Biorepository were used for research. WU funds the Biorepository either directly through university funds or through private and federal grants obtained by WU as the grantee. A significant portion of the Biorepository's funding was raised by Catalona either as donations to WU or through the grants.

Patients could donate their specimens to the Biorepository, but only after completing an informed consent form, which contained standard language mandated by the Common Rule that "your participation is voluntary and you may choose not to participate in this research study or withdraw your consent at any time."[54] A brochure accompanying the consent form stated that participants' tissue samples

> (1) will be used by "[WU] Medical Center researchers," (2) "may be shared with other authorized researchers doing research in similar fields at [WU] and other research centers," and (3) "may be used for studies currently in progress or studies conducted 10 or 20 years from now.

By 2003, the Biorepository had become the largest such depository in the world, housing "(1) approximately 3,500 prostate tissue samples taken from patients of Dr. Catalona and other WU physicians within the Division[,] (2) about 100,000 blood or serum samples donated by over 28,000 men, 75% of whom were not patients of any WU physician, but rather were volunteers recruited through the media[,] and (3) DNA samples provided by approximately 4,400 men, which included patients of different WU physicians and relatives of those patients."[55]

In early 2003, Catalona accepted a faculty position at Northwestern University, where he intended to continue his study of the genetics of prostate cancer. After accepting the position, he sent a letter to his patients, their relatives, and others who had donated specimens to the Biorepository notifying them of this change in affiliation and asking them to transfer their specimens from the Biorepository at WU to another repository at Northwestern in Chicago. About 6,000 individuals executed the transfer form and returned it to Catalona.

WU, in order to establish that it owned the specimens notwithstanding the transfer documents signed by the subjects, filed suit against Catalona in federal court. The federal court ruled that WU was the rightful owner of the specimens and the transfer forms executed by the subjects had no effect on ownership or possession because the subjects had transferred their complete interest in the specimens to WU; therefore, there was nothing for the subjects to transfer.[56] The court of appeals affirmed. It analyzed the issue of ownership in the same way and using the same factors that a nineteenth-century court would use to analyze ownership of any piece of personal property (e.g., sofa, dining room table, jewelry, paintings, clothing). It did note, though, that the way it was treating the specimens is consistent with WU's policy concerning intellectual property, namely that "all intellectual property (including . . . tangible research property) shall be owned by the University if significant University resources were used or if it is created pursuant to a research project funded through corporate, federal or other external sponsors administered by the University."[57] The court, though, did not address the apparent inconsistency between the language in the informed consent form, which gave subjects the right to withdraw from the study at any time, and their apparent donation of their specimens to WU. Catalona sought Supreme Court review, but the Court declined to take up the case.

Given that most universities and other employers claim ownership of specimens developed by their employees as part of their employment, can university employees transfer specimens to researchers at other universities? Most institutions and companies have developed relatively sophisticated policies governing the transfer of proprietary specimens, reagents, cell lines, and the like, and most make use of what is called a Material Transfer Agreement, or MTA. An MTA is a contract. Among other things, it acknowledges that company A is providing a certain reagent or specimen to company B for certain types of research, and that company A owns the intellectual property associated with the reagent or specimen. Many commercial MTAs indicate that the reagent is confidential and is not to be shared with anyone other than specific employees at company B. Finally, company A may seek to extract as a condition of providing the specimen an agreement that any patent developed by company B's researchers while using the specimen must be assigned over to company A. Most sophisticated recipients refuse to agree to a blanket assignment and may negotiate something far more sensible, such as providing company A with the right to negotiate with company B for a royalty on sales of any product that company B develops using the specimen.

Many academic institutions have different MTA rules and standards depending on whether the recipient or provider of the specimen is a nonprofit institution or a profit-making entity. Virtually all institutions have special rules when the specimens are from humans and have not been fully de-identified. If a specimen can be linked to a patient or subject, both the Common Rule and HIPAA (see chapter 4) likely apply, and IRB approval for the transfer would likely be required. Furthermore, transfer may not be permitted if the original informed consent executed by the patient did not authorize the transfer of the specimen to another institution or entity.

Because MTAs can have significant effects on property rights, most universities and other institutions do not permit researchers to execute an MTA for either materials coming to the university or for materials leaving the university. What happens, though, if a researcher, eager to receive a specimen, executes an MTA with a pharmaceutical company in violation of university policy, and in that MTA, the researcher promises to assign to the company any invention that he or she develops with the specimen? Is the MTA binding? This would normally depend on the facts and circumstances of each case.

Two general rules pertain. Only a person with authority can bind an institution, but a person who does not have actual authority can bind an institution if that person has "apparent authority." Apparent authority means that the institution or university has done something that would lead one reasonably to believe that a given person is actually authorized to execute the document, even though he or she might not be.[58] In one famous case, a visiting researcher at NCI sought an HIV blood specimen from Institut Pasteur in France. When the specimen arrived, the researcher signed the MTA, which conditioned acceptance of the samples by NCI on not using the specimens for commercial purposes. When the government obtained a patent on the HIV test kit, Pasteur claimed that the government breached its contract (i.e., breached the MTA) because the invention must have been developed using the specimens it sent; furthermore, the MTA precluded the recipient from using them for commercial purposes.[59] Pasteur had little chance of prevailing for breach of the MTA because, as far the government was concerned, there is no such thing as apparent authority. A similar rule, though, does not apply in the private sector.

COPYRIGHTS

What Is a Copyright?

Patents protect substance over form (i.e., things that spring from concepts). Copyrights, by contrast, do not. They protect form over substance. A copyright, a form of intellectual property, grants to authors of literary, musical, dramatic, and artistic works the exclusive right to control the reproduction, adaptation, and distribution of the copyrighted works. Copyright, though, provides academics with only minimal protection. Scholarship involves the creation of ideas and the discovery of principles. Yet the federal copyright laws protect neither. The Copyright Act of 1976 draws a sharp distinction between "ideas" and "expression," protecting the latter but not the former. The Supreme Court has consistently held that ideas and facts, whether scientific, historical, or biographical, "may not be copyrighted and are part of the public domain available to every person."[60] However, the way facts or ideas are compiled or expressed is subject to copyright protection, albeit not as much protection as many believe. The facts themselves are not copyrightable.

The distinction between facts and compilations of facts, although critical, is often difficult to clarify. Take the case of Rural Telephone Service, a midwestern telephone company that also compiled white pages for its telephone customers. Another company, Feist Publications, was not in the telephone business. Its sole business was compiling telephone directories for geographic areas larger than would normally be covered by a telephone company's own white pages. Feist and Rural were competitors, and thus, when Feist sought permission to copy Rural's white pages' listings, Rural refused. Feist nevertheless proceeded to copy Rural's white pages, including the four fictitious listings included to detect unauthorized copying. Rural instituted a copyright infringement action against Feist.

When the case reached the Supreme Court, the Justices noted that a compilation of facts is eligible for protection only if the facts were selected, coordinated, or arranged in an original way. Although the standard of originality is low, the Court concluded that Rural's compilation did not satisfy the minimum threshold for creativity for two reasons. First, there was nothing original about Rural's alphabetical listing of names, telephone numbers, and addresses. Second, Rural did not select the names to be included, but instead included all of its customers.

Although a copyright protects original works, that protection is substantially less than is afforded by a patent. A patent vests in the inventor the right to prevent others from using, manufacturing, or distributing the patented invention without the inventor's consent. The fact that someone may have independently developed the patented invention does not prevent the patent holder from instituting an infringement suit. In contrast, a copyright protects only against copying (or paraphrasing); it does not prevent another person from independently producing the same or similar works. For example, Ansel Adams could copyright his photographs. However, I could stand at precisely the same place in Yosemite and snap a picture of the same waterfall, under the same atmospheric conditions as pertained when Adams snapped his picture and then sell my pictures without violating Adams's copyright. Also, as discussed below, a copyright does not protect against so-called fair use of the protected materials.

Although the scope of copyright protection is far narrower than patent protection, the duration of copyright protection is far greater and the legal formalities far fewer. Normally, the life of a patent is twenty years from the date that the application is filed and it cannot be renewed. In contrast, for works created on or after January 1, 1978, the copyright lasts for the life of the author plus seventy years. For works created before January 1, 1978, the copyright lasts for ninety-five years from the date of publication.[61] Special rules apply if the materials were prepared as "work for hire." Moreover, one can only enjoy patent protection if the Patent and Trademark Office actually issues a patent. Such is not the case with a copyright. Any original work is entitled to protection, under the 1976 Act, even if that work has not been registered with the Library of Congress. Registration with the Library provides certain significant benefits. First, registration is a prerequisite to an infringement suit. Second, an author may recover attorneys' fees and certain statutory damages for infringement, but only if the author registers his or her copyright within three months after publication of the work or prior to the infringement. Otherwise, only an award of actual damages and lost profits are available to the copyright owner. In short, a patent must be issued to the inventor by the PTO; a copyright, in contrast, exists by operation of the common law with or without registration.

What Is Fair Use?

Although the owner of a copyright can prevent others from copying his or her works, there is one important exception. The Copyright Act expressly permits a person or entity to make "fair use" of a copyrighted writing, namely to reproduce and distribute a copyrighted document "for purposes such as criticism, comment, news reporting, teaching (including multiple copies for classroom use), scholarship, or research."[62] In deciding whether a particular use is "fair," courts consider four factors: (1) the purpose of the use (i.e., whether the use is commercial or nonprofit), (2) the nature of the copyrighted work, (3) the amount of copyrighted work to be copied, and (4) the effect of the use on the potential market of the copyrighted work. A number of cases, including one in 1994 that pitted Texaco Inc. and its scientists against the publishers of various scientific journals, illustrate the difficulty that courts have had in deciding whether a use is fair or not.

Texaco Inc., headquartered at the time in Eianison, New York, employed more than four hundred scientists and engineers at various research facilities throughout the United States, and it subscribed to hundreds of scientific and technical journals that were maintained in large libraries. At one of its facilities in Beacon, New York, Texaco scientists frequently photocopied research articles from the library for use in their ongoing research efforts. A group of publishers of scientific journals, including Academic Press and the American Geophysical Union, instituted a copyright infringement suit against Texaco.[63] Texaco, in turn, argued that such photocopying by its scientists was a "fair use" and not copyright infringement.

The analysis of the case by the United States Court of Appeals in New York is instructive. First, the court held that Texaco's use was not purely commercial in that Texaco was not selling the copies, but rather retaining them for future reference. Nonetheless, the court did find that the commercial character of Texaco's operations weighed against the company. It should be noted that when NIH was sued over a similar practice twenty years earlier, the court held that NIH was free of liability because its work was inherently nonprofit.[64]

Second, the court considered the nature of the copyrighted materials that Texaco had copied and concluded that because research articles are heavily factual and dissemination is an essential element of science, the extent of protection is more limited than might otherwise be the case.

The final two factors—the amount of information copied and the impact of the infringement on loss of revenues—both favored the publishers.

Because three of the four "fair use" factors weighed in favor of the publishers, the court concluded that the fair use doctrine was not appropriate in this case.

Although the commercial character of Texaco's operations was certainly a factor weighing against the applicability of the fair use doctrine, the importance of this factor is somewhat less now following the Supreme Court's decision in *Campbell v. Acuff-Rose Music, Inc.*[65] In that case, the Court was called upon to resolve whether a parody by a rap group of the hit song "Pretty Woman" was a "fair use." A lower court had held that the commercial character of the parody automatically undermined fair use. In reversing the lower court's decision, the Supreme Court reemphasized that the commercial character of a copy is only one of the four factors that must be weighed and that that factor should not be given "inflated" significance.

How Much Can You Copy without Violating the Law?

One of the most frequently asked questions is, how much of someone else's work can I quote without running afoul of the copyright laws? This question normally arises in the context of deciding whether a particular use is "fair" or not. In an educational setting, the guidelines are fairly clear. Multiple copies of a complete poem of fewer than 250 words (if printed on not more than two pages) and a complete article of fewer than 2,500 words may normally be reproduced for classroom purposes.[66]

Outside the classroom, though, the lines become murkier. One has to consider the amount of the copyrighted material that is reproduced both relative to the original work and to the new work, as well as its importance to the new work. For example, one could conceivably copy 250 words from a book without running afoul of the copyright laws if those 250 words were incorporated into a 400-page book. However, if the same 250 words were incorporated into a 280-word article, that might constitute a violation of the Copyright Act. Any assessment, though, would largely depend on the other three fair use factors.

What about copying that bridges classroom and commerce? In the late 1980s, Kinko's Graphics Corporation, the large copying chain now owned by Federal Express, sold "course packets" of materials compiled by professors from published materials for students in their classes. Neither the professors nor Kinko's obtained permission from the copyright holders. In 1989, many of the major publishing houses in New York sued Kinko's for copyright infringement.[67] The actual case involved a series of specific course packets. One packet at issue, for example, was compiled by a professor at the New School for Social Research in New York City and consisted of 388 pages of materials copied from 25 copyrighted books. The packet contained a 34-page excerpt copied from a 146-page book and various excerpts that represented from 8 to 15 percent of each copied book. The court found that the so-called education exemption did not apply here because Kinko's was a commercial enterprise and was copying the materials not for educational purposes but for commercial purposes. The court also found that the amount copied was too significant—going to the heart of many of the works—to qualify for fair use: "[T]he passages copied ranged from 14 to 110 pages, representing 5.2% to 25.1% of the works."[68]

Who Owns Your Copyright?

The Copyright Act recognizes a concept of "work for hire," which essentially means that if you create something that would ordinarily be subject to copyright, but do so as part of your job or under a contract with a third party to produce the work, then the work can viewed as "work for hire," and the copyright would be owned by your employer or other party. Indeed, under the work-for-hire doctrine, the employer is considered the author of the works and not the individual.[69] This differs conceptually from the way individuals are treated under the patent laws. Under the patent laws, only individuals can be inventors, but under the Copyright Act corporations and other entities can be authors. There is little practical difference:[70] The employee-inventor is obligated to assign his invention to his employer. In the copyright world, assignment may not be necessary because the employer can copyright the work in its own name.

Copyrighting an article or book is simple. Protecting your article or book from misappropriation, whether innocent or otherwise, can be difficult, costly, and time consuming. There is a silver lining: A roster of copyright infringement plaintiffs reads like a veritable who's who in contemporary world history, encompassing the greatest heroes and vilest villains. Martin Luther King Jr. successfully sued to block publication of his famous "I Have a Dream" speech,[71] the late-night comic Johnny Carson prevented a manufacturer from calling its outhouses "Here's Johnny,"[72] and Adolf Hitler's publisher managed to halt the sale of an unauthorized version of his infamous diatribe *Mein Kampf*.[73] Copyright disputes can provide an unusually rich glimpse into the personality quirks of the rich and famous. For example, a Los Angeles jury decided that the heirs of one of the Three Stooges had turned the heirs of the other two into real stooges. The jury found that the heirs of "Larry" Fine and "Curly Joe" DeRita were robbed of their shares of the syndication rights to the Three Stooges films by the heirs of "Moe" Howard; the jury awarded the two Stooges' heirs $3.6 million.

CASE STUDIES AND PROBLEMS

Case 1: The Case of the Culpable Chemist

Dennis N. Appleton is vice president for research of Pillpher Pharmaceuticals. Pillpher has been working on a drug to treat presbyopia for the past seven years with little to show for the company's $100 million investment. Pillpher's researchers believe that they have identified the chemical structure of the type of drug that is needed but have been unable to consistently synthesize the product, making even laboratory work difficult. The researchers believe and Appleton confirms that if they can synthesize the product in small amounts and prove that it works, they will able to get additional funding from the higher ups in the company which they can use to figure out how to make the product in large commercial batches.

One of Phillpher's young chemists, Ted A. George, while attending a chemical society meeting, learns that Paul Linus, a professor of chemistry at a nearby institution, has figured out how to synthesize small quantities of the chemical that Pillpher needs. George knows, though, that Linus is a real lefty and does not believe in private drug companies or in licensing technology. On one New Year's Eve, rather than attending any parties, George and another Pillpher chemist pay a midnight visit to Linus's laboratory and make off with a small vial of the chemical.

They use the purloined chemical to prove that it works conceptually in a Petri dish to increase elasticity of tissue necessary to reverse presbyopia. Armed with this information, Pillpher increases its budget for researching the chemical. George and his colleagues figure out on their own how to synthesize the product. They have no idea whether their method is the same as Linus's method. Pillpher patents the chemical, its use to treat presbyopia, and a method for synthesizing the chemical. After six years of clinical trials, FDA approves the drug and Pillpher begins marketing the product.

Linus reads over the Pillpher patent applications and immediately recognizes that the synthesis method is identical to the one he developed, but never published, years ago. He gets his university to institute an infringement action against Pillpher with respect to all three patents. Analyze.

Case 2: The Invention of the Mirror Image

Jerry Bond is an assistant professor of organic chemistry at Synthesis University. He earns about $60,000 per year, but spends very little. He is also independently wealthy and contributes his salary back to the university anonymously. Bond decides to look at all approved drugs that are organic compounds and see which ones are chiral molecules (i.e., molecules that are either "left-handed" or "right-handed"). After about three months of research, he discovers that about fifty of the most common drugs are chiral molecules. Bond, for the heck of it, starts synthesizing the other "handedness" of each patented drug. So, for instance, if drug A is right-handed, Bond synthesizes the left-handed version of the drug. Using his own money, he files patent applications for the method of synthesizing each drug; in some cases, where no one has synthesized the chemical before, he seeks to patent the chemical itself. He has no idea whether any of these compounds would be useful for anything. Should the patents be issued?

NOTES

1. Compare *Baba M'tzia*, which deals primarily with ownership and transfer of personal property, and *Baba Basra*, which deals primarily with ownership and transfer of real property.

2. *See* The Schoolmaster Case, 11 Hen. IV, f. 47, pl. 21 (Ct. of Common Pleas, Hilary Term 1410).

3. U.S. CONST. art. I, § 8, cl. 8.

4. There were actually six U.S. patent applications and patents involved, which were filed at various times. The first issued on February 8, 1988. The March 16, 1985 application, referenced in the text, was filed in the United Kingdom.

5. Burroughs Wellcome v. Barr Labs., Inc., 828 F. Supp. 1208, 1212 (E.D.N.C. 1993).

6. *See* Burroughs Wellcome v. Barr Labs., Inc., 40 F.3d 1223, 1231 (Fed. Cir. 1994).

7. *See id.* at 1229–31.

8. *See* Pfaff v. Wells Elecs., Inc., 525 U.S. 55 (1998).

9. Also, one cannot patent, in the traditional sense, anything having to do with nuclear weapons. Instead, if you develop an otherwise patentable invention involving nuclear weaponry, you are obligated to send it to the Department of Energy for consideration. *See* Atomic Energy Act of 1946, §10, Pub. L. No. 79-585, as amended by the Energy Reorganization Act of 1974, Pub. L. No. 93-438 (codified at 42 U.S.C. § 5801 *et seq.*). The Department of Energy will pay you for your invention, but any patent that issues will be in the name of the Department of Energy and kept secret. Imagine a terrorist or home-grown sociopath logging on to the Patent and Trademark Office Web site, pulling up the patent for a nuclear weapon, and reading precisely how to construct such a device. The fact that the patent may not have expired would likely have little deterrent effect.

10. *See* Diamond v. Chakrabarty, 447 U.S. 303 (1980).

11. Robert Charrow, *Intellectual Property Rights: Who Owns What, and When*? 2 J. NIH RES. 88 (1990).

12. *See* Biotechnology Industry Organization, Biotechnology Industry Facts, http://bio.org/speeches/pubs/er/statistics.asp (last visited Aug. 7, 2009).

13. Chakrabarty, 447 U.S. at 309.

14. *See* AT&T Corp. v Excel Communications Inc., 172 F.3d 1352 (Fed. Cir. 1999); State Street Bank & Trust v. Signature Fin. Group Inc., 149 F.3d 1368 (Fed. Cir. 1998).

15. *See* Metabolite Lab. Inc. v. Lab. Corp. of Am. Holdings, 370 F.3d 1354 (Fed. Cir. 2004), *cert. dismissed*, 548 U.S. 124 (2006).

16. *See* Teleflex, Inc. v. KSR Int'l Co., No. 04-1152 (Fed. Cir. Jan. 6, 2005).

17. KSR Int'l Co. v. Teleflex, Inc., 550 U.S. 398 (2007).

18. 396 U. S. 57 (1969).

19. KSR Int'l Co. v. Teleflex, Inc., 550 U.S. at 417.

20. *See* Anna Maria Gillis, *The Patent Question of the Year*, 42 BIOSCIENCE 336 (1992); Christopher J. Harnett, *The Human Genome Project and the Downside of Federal Technology Transfer*, http://www.piercelaw.edu/risk/vol5/spring/harnett.htm (last visited Aug. 7, 2009).

21. Interestingly, Lippershey is the first to have sought a patent on the telescope in 1602; the Dutch patent office rejected the application in 1608. There is a historical dispute over whether the Janssens (father and son, Dutch eyeglass makers) or Lippershey, also a Dutch eyeglass maker, invented the microscope.

22. *See In re* Fisher & Lalgudi, 421 F.3d 1365 (Fed. Cir. 2005).

23. *See* Christopher Anderson, *US Patent Application Stirs Up Gene Hunters*, 353 NATURE 485 (1991); L. Roberts, *Genome Patent Fight Erupts*, 254 SCIENCE 184 (1991).

24. If your employer declines to patent your invention, it may permit you to retain the invention. If that were to occur, you would have to pay the patenting costs, assuming you decided that the invention was worth patenting.

25. STANFORD UNIVERSITY, OFFICE OF THE VICE PROVOST & DEAN OF RESEARCH, RESEARCH POLICY HANDBOOK doc. 5.1 (July 15, 1999), *available at* http://rph.stanford.edu/5-1.html.

26. Albert Einstein first theorized simulated emission of radiation. *See Zur Quantentheorie der Strahlung* (On the Quantum Theory of Radiation), 18 PHYSIKA ZEITSCHRIFT 121 (1917).

27. *See* Maser and Maser Communications Systems, U.S. Patent No. 2,929,222 (filed July 30, 1958).

28. *See* Arthur L. Schawlow & Charles H. Townes, *Infrared and Optical Masers*, 112 PHYSICAL REV. 1940 (1958).

29. *See* Gould v. Control Laser Corp., 866 F.2d 1391 (Fed. Cir. 1989).

30. Madey v. Duke Univ., 307 F.3d 1351, 1362 (Fed. Cir. 2002) (quoting Embrex, Inc. v. Service Eng'g Corp., 216 F.3d 1343, 1349 (Fed. Cir. 2000)).

31. *Id.*

32. *Id.*

33. 35 U.S.C. § 271(e)(1).

34. *See* Merck KGAA v. Integra LifeSciences I, Ltd., 545 U.S. 193 (2005).

35. 496 U.S. 661 (1990).

36. *See* eBay Inc. v. MercExchange, L.L.C., 547 U.S. 388 (2006) (holding that proof of infringement does not automatically entitle the plaintiff to injunctive relief).

37. Pub. L. No. 98-417, 98 Stat. 1585 (1984) (codified at 21 U.S.C. § 355(j), 35 U.S.C. §§ 156, 271, 281).

38. The first generic manufacturer to file an ANDA challenging the validity of an innovator's patent is usually entitled to 180-day marketing exclusivity over the other generic applicants once its ANDA is approved.

39. *See* Pfizer, Inc. v. Teva Pharms. USA, Inc., No. 2007-1271 (Fed. Cir. Mar. 7, 2008).

40. *See* Sanofi-Synthelabo v. Apotex, 470 F.3d 1368 (Fed. Cir. 2006).

41. *See* Teva Pharms. USA, Inc. v. Pfizer, Inc., 395 F.3d 1324 (Fed. Cir. 2005).

42. H.R. Rep. No. 96-1307, pt. 1, at 2 (1980), *reprinted in* 1980 U.S.C.C.A.N. 6460, 6461.

43. *See* Amendments to the Patent and Trademark Laws § 6 (Bayh-Dole Act), Pub. L. No. 96-517, 94 Stat. 3015, 3018 *et seq.* (1980) (codified at 35 U.S.C. §§ 200–211).

44. Exec. Order No. 12,591, § 1(b)(4), 3 C.F.R. 220 (1987) (emphasis supplied).

45. *See* University of Maryland, Baltimore County, 107.0 IV-3.00-Policy on Patents (approved May 31, 1990), *available at* http://www.umbc.edu/otd/Old%20Patent%20Policy.html (last visited Aug. 7, 2009).

46. *See* University of California, Office of the President, Frequently Asked Questions, http://www.ucop.edu/ott/inventorshares/faq.htm (eff. Oct. 1, 1997) (last visited Aug. 7, 2009).

47. *See* Harvard University, Office of Technology Development, Royalty Sharing, http://otd.harvard.edu/resources/policies/royalty (eff. Jan. 1, 2008) (last visited Aug. 7, 2009).

48. See Yale University, Office of Cooperative Research, Patent Policy, http://www.yale.edu/ocr/pfg/policies/patents.html (rev. Feb. 1998) (last visited May 3, 2009).

49. Pub. L. No. 99-502, 100 Stat. 1785 (1986) (amending 15 U.S.C. § 3701 *et seq.*). The Federal Technology Transfer Act forms part of Chapter 63, Technology Innovation, of Title 15, United States Code. Chapter 63 originated with the Stevenson-Wydler Technology Innovation Act of 1980, Pub. L. No. 96-480, 94 Stat. 2311. The Stevenson-Wydler Act consists of broad legislation setting out the United States' policy to stimulate technology, *inter alia*, by promoting the use by private businesses of technology developed at federal laboratories. *See* 15 U.S.C. § 3701.

50. *See* 35 U.S.C. § 207.

51. *See* 15 U.S.C. § 3710a(a)(2), 35 U.S.C. § 207, 37 C.F.R. pt. 404.

52. *See* 37 C.F.R. § 404.7.

53. *See* 35 U.S.C. § 209(a).

54. Washington Univ. v. Catalona, 490 F.3d 667, 671 (8th Cir. 2007), *cert. denied*, ___ U.S.___, 07-525 (U.S. Jan. 22, 2008).

55. *Id.* at 672.

56. *See* Wash. Univ. v. Catalona, 437 F. Supp. 2d 985, 1002 (E.D. Mo. 2006).

57. Catalona, 490 F.3d at 676 n.8.

58. *See* RESTATEMENT (THIRD) OF AGENCY § 2.03 & cmt. c (2006).

59. *See* Institut Pasteur v. United States, 814 F.2d 624 (Fed. Cir. 1987).

60. Feist Pubs. v. Rural Tel. Serv. Co., 499 U.S. 340, 348 (1991).

61. *See* 17 U.S.C. § 304(a) & (b).

62. 17 U.S.C. § 107.

63. *See* Am. Geophysical Union v. Texaco Inc., 60 F.3d 913 (2d Cir. 1994).

64. *See* Williams & Wilkins Co. v. United States, 487 F.2d 1345 (Ct. Cl. 1973).

65. *See* Campbell v. Acuff-Rose Music, Inc., 510 U.S. 569 (1994).

66. *See* Agreement on Guidelines for Classroom Copying in Not-for-Profit Educational Institutions with Respect to Books and Periodicals (Mar. 19, 1976). This agreement was between the Ad Hoc Committee on Copyright Law Revision, Author-Publisher Group, through Authors League of America, and the Association of American Publishers, Inc., and was made part of the House Report accompanying the Copyright Act of 1976; it is reproduced in LIBRARY OF CONGRESS, U.S. COPYRIGHT OFFICE, REPRODUCTION OF COPYRIGHTED WORKS BY EDUCATORS AND LIBRARIANS 7–8 (June 1998), *available at* http://www.copyright.gov/circs/circ21.pdf.

67. *See generally* Basic Books v. Kinko's Graphics Corp., 758 F. Supp. 1522 (S.D.N.Y. 1991).

68. *Id.* at 1533.

69. *See* 17 U.S.C. § 201(b); Shaul v. Cherry Valley-Springfield Cent. Sch. Dist., 363 F.3d 177, 185 (2d Cir. 2004).

70. The term of a copyright in a work made for hire is "95 years from the year of its first publication, or a term of 120 years from the year of its creation, whichever expires first." 17 U.S.C. § 302(c) (for works made on or after January 1, 1978).

71. *See In re* Estate of Martin Luther King v. CBS, Inc., 184 F. Supp. 2d 1353 (N.D. Ga. 2002) (discussing the various copyright infringement suits instituted by the King Estate against various media organizations).

72. *See* Carson v. Here's Johnny Portable Toilets, Inc., 698 F.2d 831 (6th Cir. 1983).

73. *See* Houghton Mifflin Co. v. Stackpole Sons, 113 F.2d 627 (2d Cir. 1940).

CHAPTER 8

What Regulations Govern the Welfare of Laboratory Animals?

To a biologist, a goose is an animal. To the Supreme Court of Tennessee it is not. The conflict in taxonomy all started when a gaggle of hapless geese was run down by a speeding train. The owner of the geese sued the railway under a Tennessee law requiring engineers to take evasive action when "an animal or obstruction" appeared on the track. The court reasoned that if the purpose of the statute were to protect animals from being injured by trains, then the geese would be animals within the meaning of the statute and their owner could make use of the statute in his suit against the railway. If, on the other hand, the purpose of the statute were to protect passengers by preventing collisions with roaming large animals, then the geese would not be animals because they lacked the size necessary to upend a train or otherwise harm passengers. Ultimately, the court ruled that the purpose of the law was to protect passengers and not geese, and therefore, geese were not animals under the law.[1]

Traditionally, both in England and the United States, animals were viewed as personal property, simple chattels, and the owner of a chattel was free to do with his property as he saw fit, including destroying it. In the 1820s, Maine and New York became two of the first states to enact an anticruelty statute.[2] The New York law covered horses, oxen, cattle, and sheep. By 1860 many states had enacted laws like New York's that protected specific species against torture, maiming, and other outrageous conduct.

In the 1860s, though, societal norms were changing both in England and the United States, especially with the publication of *Black Beauty: The Autobiography of a Horse* in 1877 in England and the advent of the humane societies that successfully lobbied the New York legislature to enact tougher, more robust anticruelty laws. Even so, the revised laws in the United States remained relatively modest. Typically, killing an animal was not a criminal offense unless it was done in a cruel manner. And beating a horse was deemed noncriminal if it were done for legitimate training purposes. The so-called training defense still remains viable in various jurisdictions.[3] Most state anticruelty statutes did not extend to laboratory experimentation.

The federal anticruelty statutes were narrower than the state laws and regulated only in specific areas, such as transportation and slaughter. For example, federal law specified how animals were to be transported and the way in which they were to be slaughtered for meat.[4] Animal cruelty laws at both the federal and state levels remained static until the mid-1960s, when Pepper entered the picture.

Pepper was a Dalmatian who disappeared from her owner's yard. Shortly thereafter, her owner, while watching the local news, recognized his missing dog in a picture of an overcrowded truck belonging to an animal dealer. The owner tried to contact the dealer, and when that failed he contacted Congressman Joseph Resnick, a junior member from New York. Resnick too was rebuffed; he was unable to gain entry into the dog farm. When Resnick learned that Pepper had been euthanized as part of a medical experiment in a nearby hospital, he introduced what would become the Laboratory Animal Welfare Act (AWA),[5] the first federal legislation aimed at regulating the use of animals in laboratory experiments.

OVERVIEW OF FEDERAL REGULATIONS ON ANIMALS USED IN RESEARCH

Today, two federal laws regulate those conducting research using animals. The AWA, which was the first such law, applies to all research irrespective of the funding source, but it covers only certain animals—warm-blooded animals except "birds, mice of the genus Mus, and rats of the genus Rattus, bred for use in research." [6] The Health Research Extension Act of 1985[7] applies only to research funded by the Public Health Service (PHS), but it covers all vertebrates, even those expressly excluded from the coverage of the AWA. The AWA is administered by the United States Department of Agriculture (USDA) through its Animal and Plant Health Inspection Service (APHIS), while the Health Research Extension Act is administered by the Department of Health and Human Services (HHS) through the National Institutes of Health (NIH) Office of Laboratory Animal Welfare.

The AWA, which has been amended many times, requires the Secretary of Agriculture to issue regulations for the humane treatment of certain animals in certain settings.[8] As originally enacted, the AWA imposed minimum regulation on research facilities.[9] In 1985, the same year that Congress enacted the Health Research Extension Act, it also amended the AWA by instructing USDA to "promulgate standards to govern the humane handling, care, treatment, and transportation of animals by dealers, research facilities, and exhibitors."[10] The statute contains a special proviso for primates requiring that their physical environment be adequate to "promote the[ir] psychological well-being."[11]

Overall, the AWA contemplates a system for animals that is structurally (although not substantively) similar to the one for human subjects. Under the AWA and its implementing regulations, a laboratory that maintains "covered" animals for research (i.e., warm-blooded animals excluding birds, *Mus musculus*, and *Rattus rattus*, bred for research) must register with APHIS as such, and its mother institution must establish an Institutional Animal Care and Use Committee (IACUC), which is somewhat analogous to the human Institutional Review Board (IRB).[12] Each IACUC must consist of at least three members (NIH requires a minimum of five), one of whom is unaffiliated with the institution and one of whom is a veterinarian with experience or training in maintaining animals intended for research. If the committee has more than three members, no more than three may come from the same organizational unit within the institution. An IACUC is required to review all animal facilities at least once every six months and approve research that makes use of covered animals and any significant changes to that research. In reviewing the research, an IACUC is to be guided by various principles, including, among others, that no covered animal be made to suffer unnecessarily, that the principal investigator (PI) has considered alternatives to the use of covered animals, that the protocol has been designed to minimize pain or suffering, and that the protocol not duplicate a similar experiment unless there is overriding reason to do so.[13] The AWA recognizes that there will be experiments where animals must be sacrificed. In those instances, euthanasia must be accomplished by "a method that produces rapid unconsciousness and subsequent death without evidence of pain or distress, or a method that utilizes anesthesia produced by an agent that causes painless loss of consciousness and subsequent death."[14] The regulations impose significant record-keeping requirements on the facility and the PI. The IACUC has the authority to terminate or suspend that portion of a protocol involving animals when there have been uncorrected violations of the AWA.

In contrast to the AWA, the Extension Act covers *all* vertebrate animals involved in any PHS-funded research.[15] Thus, whereas rats, mice, and birds bred for research are not subject to the AWA, live rats, mice, and birds are subject to the Extension Act if they are involved in PHS-funded research. While the AWA requires animal facilities to register with APHIS, the Extension Act requires institutions to file an assurance with the Secretary of Health and Human Services promising to comply with the AWA and with all relevant guidelines. IACUC approval of the protocol is required before

NIH will release grant funds for the research.[16] There is another significant difference between the AWA and the Extension Act: The AWA can be implemented only by rulemaking by the Secretary of Agriculture, while certain aspects of the Extension Act can be implemented through "guidelines," a process that is more opaque than rulemaking because public participation is not required. The Extension Act also contemplates regulations, but none has ever been issued. PHS has issued guidelines in the form of a policy statement entitled "Public Health Service Policy on Humane Care and Use of Laboratory Animals," and a Guide, published by the National Academy of Sciences (NAS), entitled "Guide for the Care and Use of Laboratory Animals."[17] The NAS Guide has taken on added significance as its guidelines are incorporated into the HHS regulations, which require that institutions funded by NIH follow the most recent version of the Guide in adopting programs involving laboratory animals.[18] Indeed, as discussed in chapter 6, the National Academy was sued under the Federal Advisory Committee Act because of the way in which it developed its Guide. There are also special guides for transporting animals, for chimpanzees, for surgical intervention, and for euthanasia.

The Extension Act has been interpreted by HHS to cover only research funded by PHS; however, it would be impractical for an institution to maintain separate systems for PHS-funded and non-PHS-funded research because APHIS, whose jurisdiction is not funding dependent, and NIH use the same general rules and guidelines. Thus, an institution that wished to avoid regulation on all research other than PHS-funded research would have to maintain two systems—one for PHS-funded research and those species covered by the AWA and another for non-PHS-funded research using species not covered by the AWA. Maintaining such a system, while possible, would be administratively difficult. It could also maintain a third system, if it wished, for research covered by the AWA but not covered by PHS, but this could prove to be an administrative nightmare.

Although the AWA and its regulations are generally consistent with the PHS policy, there are certain differences aside from the species covered, including the minimum number of members of the IACUC. The AWA sets the minimum at three with at least one veterinarian and one member not affiliated with the institution. PHS, however, sets the minimum at five, with at least

- one veterinarian with training or experience in laboratory animal science and medicine, who has direct or delegated authority and responsibility for activities involving animals at the institution;
- one practicing scientist experienced in research with animals;

- one member whose primary concerns are in a nonscientific area (e.g., ethicist, lawyer, member of the clergy); and
- one member who is not affiliated with the institution other than as a member of the IACUC.[19]

To ensure compliance with both the AWA and the Extension Act, institutions usually adopt the most stringent policies of each.

FEDERAL ENFORCEMENT OF ANIMAL WELFARE STANDARDS

USDA and NIH have each sought to enforce the animal welfare standards within their respective spheres. Most enforcement actions against universities are taken by USDA. For example, following routine inspections of animal facilities and records at the University of California at San Francisco, APHIS charged that the university had "failed to use local anesthesia and postoperative analgesics in surgery on a pregnant sheep, failed to ensure that its Institutional Animal Care and Use Committee reviewed the most current procedures used on animals, and failed to maintain a program of adequate veterinary care."[20] UCSF settled the case, without admitting liability, by paying a $92,500 fine.[21] A similar plight befell Johns Hopkins University, which agreed to pay a $25,000 fine to USDA following a series of inspections that led the government to charge that Hopkins had performed painful procedures without proper sedatives or anesthetics, had failed to provide appropriate postoperative care, and had failed to euthanize animals promptly after they had been advised that the animals ought to have been euthanized. The university was also charged with failing to provide appropriate cage space and, for the primates, failing to keep the facility in good repair. The allegations spanned a four-year period starting in 1999.[22]

USDA fines tend to be highly publicized, but are not the only enforcement mechanisms available. The Office of Laboratory Animal Welfare (OLAW) within NIH and university IACUCs can and have meted out sanctions of their own. In 2007, for instance, a neurosurgeon at the Cleveland Clinic diverted a dog approved by the school's IACUC to participate in a research protocol and instead used the dog to demonstrate an aneurysm coil for sales representatives of the coil's manufacturer. The Cleveland Clinic conducted an internal investigation, suspended the neurosurgeon from animal research for two years, and self-reported the incident to USDA, which conducted an inspection of its own and confirmed the Clinic's findings.[23] OLAW and IACUC each has the authority under appropriate circumstances to suspend research, or, in the case of OLAW, to seek to have the grant terminated, a sanction that has never been applied, to my knowledge.

GODZILLA MEETS ARTICLE III OF THE CONSTITUTION—SUING ABOUT ANIMALS

Animal advocacy groups, many of which are opposed to all research involving animals, have sought to get the courts involved. In some instances, they have challenged research practices at specific institutions and in other instances they have challenged the way in which USDA has implemented the AWA. No matter what the case is about, though, it invariably devolves into a dispute over whether the courts have jurisdiction to hear these types of controversies. Article III of the Constitution restricts the jurisdiction of federal courts to "cases" or "controversies." Underlying Article III is the idea that the authority of a federal court is limited to resolving concrete disputes between individuals, persons, and governmental entities, and that federal courts should not be in the business of fashioning social policy. That is the job of the Congress and the president. One of the doctrines that helps courts determine whether a dispute is sufficiently real to trigger its jurisdiction is "standing." A plaintiff cannot maintain an action in federal court unless he or she has "standing." A plaintiff has to show that without court intervention he or she will suffer some imminent "concrete" injury, that the defendant's actions caused that injury, and that a court can somehow prevent that injury.

Cases involving the AWA have tested the outer limits of standing. For example, how is it that a plaintiff suffers injury if the AWA is not interpreted to cover certain species that the plaintiff believes ought to be covered? The fact that you may have an emotional attachment to animals or that you believe that animals ought to be treated with the same dignity as humans does not translate into a concrete injury. Merely because a person disagrees with the government's policy is not enough. If it were, everyone could sue the government over anything at any time. A number of years ago, the Secretary of the Interior and the Secretary of Commerce jointly issued a rule that interpreted the Endangered Species Act (ESA) as applying to governmental activities in the United States and on the high seas. Plaintiffs argued that the ESA imposed no geographic borders and the government should not be able to take action that adversely affects an endangered or threatened species overseas. The plaintiffs argued that if the U.S. government were to fund a dam to be constructed somewhere in Africa, ESA applies, and the government would have to ascertain that no endangered or threatened species would be put at risk by the dam. The Supreme Court ultimately held that none of the plaintiffs had standing to challenge the rule because no plaintiff could show that he or she would be directly injured by governmental action abroad. A plaintiff would have to identify the specific government action involved and the endangered or threatened species involved and then would have to show

how an injury to that species overseas would adversely affect the plaintiff. None of the plaintiffs could meet this burden.[24]

Few actions have generated as much litigation as USDA's attempt to issue rules implementing the special nonhuman primate provisions of the AWA. The statute envisioned government-imposed requirements that would take into account the psychological aspects of confinement.[25] In 1989, the Secretary of Agriculture proposed a series of specific requirements aimed at better ensuring the psychological health of nonhuman primates used in research or placed on exhibit. In 1991, when the final rule issued, the specific requirements were replaced by a flexible policy that allowed the institution to determine how to ensure primate welfare. Shortly thereafter, the Animal Legal Defense Fund instituted suit challenging the new policy; one of plaintiffs claimed that he regularly visited primate preserves in the United States and the way that they were being treated interfered with his ability to enjoy watching the primates in those preserves. The court of appeals in Washington held that that plaintiff had demonstrated a sufficiently concrete injury to satisfy standing requirements and allowed the suit to proceed.[26] The court sent the case back so that the district court could decide whether the USDA regulations were consistent with the AWA. Two years later, the case was back before the appeals court, which this time held that the 1991 regulations were sufficient and the Secretary ought to be afforded significant leeway in issuing these types of regulations.[27] However, that did not end the matter.

Three years later, USDA issued a report concluding that "after five years of experience enforcing [the USDA regulation], regulated entities did not necessarily understand how to develop an environment enhancement plan that would adequately promote the psychological well-being of nonhuman primates."[28] Rather than modify its flexible rule, USDA issued a Draft Policy statement outlining specific actions that a facility could take to improve a nonhuman primate's psychological experience in captivity.[29]

In 2003, apparently frustrated that USDA never had finalized its Draft Policy, the Animal Legal Defense Fund and various individuals again sued USDA, this time in San Francisco, seeking to have the agency finalize its Draft Policy. The National Association for Biomedical Research, a nonprofit organization representing universities and other entities subject to the AWA, intervened in the suit. USDA also announced that it had decided to withdraw its Draft Policy. The trial court dismissed the suit, finding that the agency had no duty to finalize the Draft Policy and that that decision was not reviewable in court.

On appeal, a three-judge panel of the Court of Appeals for the Ninth Circuit first focused on whether any of the plaintiffs even had standing to sue.[30] Two of the three judges held that at least one of the plaintiffs, an individual who had formed a close relationship with one of the chimpanzees housed at a zoo in Nevada, had standing because she regularly visited the chimpanzee and she intended to do so in the future. The way in which the chimpanzee was treated, she argued, would have a direct effect on her ability to continue to interact with the animal. The court also held that the decision to withdraw the Draft Policy was reviewable in court and therefore sent the case back to the trial court to determine whether the agency's withdrawal of the Draft Policy was arbitrary.

The third judge, Chief Judge Alex Kozinski, dissented. He noted that the "majority wastes much toner trying to show that USDA's Draft Policy at issue here would, if adopted, ameliorate the conditions plaintiffs complain about."[31] He went on to note that no matter what the court holds, it has no authority to force USDA to adopt a Draft Policy. Because a court cannot force the agency to adopt a Draft Policy, it can provide no meaningful remedy to the plaintiffs. Without a remedy, reasoned Kozinksi, the plaintiffs have no standing. This did not end the saga. The government sought to have the case heard by an eleven-judge panel of the court of appeals,[32] not just a three-judge panel. The petition for rehearing *en banc*, as it is called, was granted.[33] Two months later, the plaintiffs, apparently having gotten some concessions from USDA, moved to dismiss the case, and that motion was granted.[34] It is unclear from the public record precisely what led the plaintiffs to dismiss their suit.

The two cases against USDA illustrate the zeal and tenacity of animal advocacy organizations. Institutions operating laboratories that use animals in research should be ever vigilant, not only to ensure that they are following the law, but also that animal extremists who might not follow the law are in no position to sabotage research, as has happened in the past. Sabotage and harassment have become so prevalent with respect to certain types of research that institutions have successfully obtained state and federal court injunctions barring these groups from interfering with activities of the institution or its employees. In 2008, for instance, a California court temporarily barred certain animal extremists from harassing and threatening UCLA employees and a UC Santa Cruz researcher. [35] Similarly, a court in Philadelphia prohibited animal extremists from protesting with bullhorns and amplifiers GlaxoSmithKline facilities and those of its business associates.[36]

Ironically, in some instances the conduct of animal activists had been so extreme, such as destroying laboratories, stealing animals, and injuring scientists, that Congress enacted a law to protect scientists and others from these extremists. The Animal Enterprise Protection Act of 1992 (AEPA) [37] makes it a federal crime to intentionally disrupt "a commercial or academic enterprise that uses animals for food or fiber production, agriculture, research, or testing; . . . a zoo, aquarium, circus, rodeo, or lawful competitive animal event; or . . . any fair or similar event intended to advance agricultural arts and sciences."[38] In 2006, a federal court found six extremists and their organization (Stop Huntingdon Animal Cruelty USA Inc.) guilty of conspiring to e-mail and to use other forms of communication to incite others to engage in unlawful acts against Huntingdon Life Sciences of East Millstone, N.J.[39] Three of the individual defendants were sentenced to federal prison; the leader received a six-year sentence.[40] All defendants were required to pay $1 million in restitution. At about the same time, the government indicted eleven members of the Earth Liberation Front and the Animal Liberation Front on charges of arson and property destruction at various research facilities throughout the western states.[41]

It appears that laws with respect to animal welfare have come full circle, and that those most needing protection now appear to be the researchers themselves, who must be protected from physical violence perpetrated by those who claim to love all animals—other than *Homo sapiens.*

CASE STUDIES AND PROBLEMS

Case 1: The Case of the Double-Crossed Dogs

Dalton Krebs, a senior researcher at the Medical Academy of Louisiana (MAL), has received an NIH grant to conduct a study examining how heart muscle responds to various chemical stimuli. Krebs indicated in his proposal that he intends to use twelve dogs, that each would be given a nonlethal dose of sodium thiopental, and that their chests would be opened and their heart muscles studied under varying conditions. At the end, each dog would be sewn back up and allowed to recover. Krebs's proposal is reviewed and approved by the institution's IACUC.

After the research is funded, Krebs alters his protocol without telling anyone, so that now near the end of the experiment each dog is to be given a lethal dose of sodium thiopental, and the heart is to be removed and dissected. Krebs also expands the scope of his experiment from twelve to eighteen dogs. Rather than going through MAL's vivarium, he gets his extra six dogs at the local pound.

After Krebs publishes his results, the head of the IACUC pays him a visit. Does Krebs have problems?

Case 2: The Case of the Seafaring Sharks

Peggy Smith is a young and ambitious biophysicist at the Upper Maine Peninsular Institute of Research. She has been keenly interested in how sea creatures manage to navigate the oceans. She hypothesizes that most such creatures use water temperature, light, and the earth's magnetic fields alone or in combination. Smith is receiving funding from the Defense Advanced Research Projects Agency. She does most of her work about twenty miles out in the ocean. She captures a group of baby sharks and intends to study them in her small laboratory on board her boat by altering the temperature, light, and magnetic fields. She wants to see what has the greatest effect on their movements. She also implants radio-transmitting devices that monitor and transmit the animals' brain waves. When released, the device will remain implanted and can receive information for up to two years. Upper Maine gets no NIH funding, and therefore, it does not follow the Extension Act. What are Smith's obligations under the AWA? Would it matter if she used dolphins rather than sharks?

NOTES

1. *See* Nashville & K.R. Co. v. Davis, 78 S.W. 1050 (Tenn. 1902).

2. *See* N.Y. Rev. Stat. tit. 6, § 26 (1829); *see also* Me. Laws ch. IV, § 7 (1821); Joseph G. Sauder, *Enacting and Enforcing Felony Animal Cruelty Laws to Prevent Violence Against Humans*, 6 Animal L. 1 (2000) (providing an excellent overview of the historical development of anticruelty laws and federal legislation).

3. *See, e.g.*, Ariz. Rev. Stat. Ann. § 13-2910.06; Regalado v. United States, 572 A.2d 416, 420 (D.C. 1990).

4. *See* Act of June 29, 1906, ch. 3594, 34 Stat. 607 (regulating the interstate transport of "cattle, sheep, swine and other animals"); Humane Slaughter Law, 7 U.S.C. § 1901 (imposing humane methods of slaughter on meat processing facilities).

5. Pub. L. No. 89-544, 80 Stat. 350 (1966) (codified and amended at 7 U.S.C. §§ 2131–2159).

6. *See* Farm Security and Rural Investment Act of 2002, § 10301, Pub. L. No. 107-171, 116 Stat. 134, amending section 2(g) of the Laboratory Animal Welfare Act (7 U.S.C. § 2132(g)) by redefining the term "animal" to exclude birds, mice (*Mus*), and rats (*Rattus*) bred for research.

7. Pub. L. No. 99-158, 99 Stat. 820 (1985).

8. The AWA regulations are at 9 C.F.R. pt. 2.

9. *See, e.g.*, 7 U.S.C. § 2143(a) (1984).

10. 7 U.S.C. § 2143(a)(1) (2009).

11. *Id.* § 2143(a)(2)(B).

12. *See* 9 C.F.R. §§ 2.30–2.31.

13. *See id.* § 2.31(d).

14. *Id.* § 1.1 (defining euthanasia and incorporated by reference into 9 C.F.R. § 2.31(d)(xi)).

15. Congress also enacted, as part of the National Institutes of Health Revitalization Act of 1993, § 205, Pub. L. No. 103-43, 107 Stat. 133, a new section 404C to the Public Health Service Act (codified at 42 U.S.C. § 283e) requiring the NIH director to study methods to reduce the use of animals in research.

16. *See* 67 Fed. Reg. 51,290 (Aug. 7, 2002).

17. *See, e.g.*, National Research Council, Guidelines for the Care and Use of Mammals in Neuroscience and Behavioral Research (2003), *avail-*

able at http://grants.nih.gov/grants/olaw/National_Academies_guidelines_for_Use_and_Care .pdf.

18. *See, e.g.*, 42 C.F.R. § 52b.14(d)(5) (construction grants).

19. An unaffiliated member does not become affiliated merely because the institution pays that individual a modest stipend for serving as a member of the Committee. *See* Institutional Animal Care and Use Committee, http://grants.nih.gov/grants/olaw/tutorial/iacuc.htm (last visited Aug. 7, 2009).

20. 4 MED. RES. L. & POL'Y REP. 748 (2005).

21. *See id.*

22. *See id.* at 640.

23. *See* 6 MED. RES. L. & POL'Y REP. 88 (2007).

24. *See* Lujan v. Defenders of Wildlife, 504 U.S. 555 (1992).

25. *See* 7 U.S.C. § 2143(a)(2)(B).

26. *See* Animal Legal Defense Fund, Inc. v. Glickman, 154 F.3d 426 (D.C. Cir. 1998) (en banc).

27. *See* Animal Legal Defense Fund, Inc. v. Glickman, 204 F.3d 229 (D.C. Cir. 2000).

28. 64 Fed. Reg. 38,145, 38,146 (July 15, 1999).

29. *See id.*

30. *See* Animal Legal Defense Fund v. Veneman, 469 F.3d 826 (9th Cir. 2006).

31. *Id.* at 844 (Kozinski, J., dissenting).

32. This is called en banc review. In most circuits, en banc cases are heard by all active judges in that circuit. The Ninth Circuit has so many judges that en banc review involves a panel of eleven judges.

33. *See* Animal Legal Defense Fund v. Veneman, 482 F.3d 1156 (9th Cir. 2007).

34. *See* Animal Legal Defense Fund v. Veneman, 490 F.3d 725 (9th Cir. 2007).

35. *See* Univ. of Cal. v. UCLA Primate Freedom, No. SC097145 (Super. Ct. L.A. Cty. Feb. 25, 2008).

36. *See* SmithKline Beecham Corp. v. Stop Huntingdon Animal Cruelty Inc., No. 06-11-00366 (Pa. Ct. Com. Pl. Nov. 27, 2006); *see also* Novartis v. Stop Huntingdon Animal Cruelty, No. A107538 (Cal. Ct. App. Oct. 12, 2006) (affirming on appeal preliminary injunction barring defendants from harassing Chiron employees and their families).

37. See Pub. L. No. 102-346, 106 Stat. 928 (1992) (codified at 18 U.S.C. § 43).

38. 18 U.S.C. § 43(d)(1).

39. *See* United States v. Stop Huntingdon Animal Cruelty USA Inc., No. 04-373 (D.N.J. Mar. 2, 2006).

40. *See id.* (sentencing Sept. 12, 13, 19, 2006).

41. *See* United States v. Dibee, No. 6:06-CR-60011 (D. Or. Jan. 19, 2006); 5 MED. RES. L. & POL'Y REP. 86 (2006).

APPENDIX A

A Short Guide to the Unusual World of Legal Citations

The legal citation system does not resemble any other system, nor is it based on the sage advice of renowned legal scholars. To the contrary, the citation system was developed and is updated episodically by law students at Yale, Harvard, Pennsylvania, and Columbia and is compiled in *The Bluebook: A Uniform System of Citation*, now in its eighteenth edition. Some of the citation conventions are just weird and appear to be outgrowths of frat parties gone amok; others are inadequate or antiquated. For example, in citing patents, the *Bluebook* uses the date the patent was issued. This was useful when the life of a patent was seventeen years from the date of its issue. But that is no longer the case; the term of a patent is twenty years from the date the application is filed, making the filing date the critical point.

A number of conventions are unique to legal citation, and these the reader will spot immediately. For example, the volume number of any journal or other multivolume set precedes the name of the journal or set. Thus, in citing an article in *Science*, the information would appear in the following order: authors' names, article title (in italics), volume number, journal name (in small capital letters), starting page of article, page where information is to be found (called a spot cite or jump cite), and the year (in parentheses). The precise date of the publication, March 23, is omitted; the year alone is cited. Thus, the journal cite for an article that appears in the March 23, 2007, issue of *Science* would look as follows:

Yaniv Brandvain, Michael S. Barker & Michael J. Wade, *Gene Co-inheritance and Gene Transfer*, 315 SCIENCE 1685 (2007).

I occasionally refer to cases that have been decided by courts and to federal statutes or rules. Cases are compiled in books called case reporters that identify the court. For example, cases decided by the Supreme Court of the United States are compiled in *United States Reports*, published by the Government Printing Office. Cases decided by the U.S. courts of appeals (also known as circuit courts) are found in the *Federal Reporter* (abbreviated as F., F.2d, or F.3d, depending on the series), published by West Publishing Co.[1] Decisions from the federal trial courts (also called district courts) are published in the *Federal Supplement* (abbreviated F. Supp. or F. Supp. 2d), also published by West. Take, for example, the citation for the following case:

Animal Legal Defense Fund v. Veneman, 469 F.3d 826 (9th Cir. 2006).

This citation indicates that the case is found on page 826 of volume 469 of the *Federal Reporter, Third Series*, and was decided by the United States Court of Appeals for the Ninth Circuit (which sits in San Francisco and Pasadena), the circuit most frequently reversed by the Supreme Court. The defendant, Ann Veneman, is being sued in her official capacity as the secretary of agriculture; this bit of information you can glean only from reading the case. Case names, by the way, are either italicized or underlined; I prefer the former. (This rule has an exception: Case names appear in Roman when used in footnotes.) Usually, the plaintiff (the person or entity bringing the suit) appears on the left side of the "v." and the defendant (the person or entity being sued) appears to the right of the "v."

Federal statutes are codified in the United States Code by subject area.[2] The United States Code is divided into fifty titles; each title is devoted to a specific subject or set of related subjects or nearly related subjects.[3] For example, titles 1–5 of the United States Code are devoted to organizing the government, title 21 is largely reserved for the Food and Drug Administration and the Controlled Substances Act; title 42 contains, among other things, the Public Health Service Act, Medicare, and Medicaid. Thus, 42 U.S.C. § 1395 is section 1395 of title 42 and is the first section of Medicare; 21 U.S.C. § 321(h) defines what a drug is; and so on.

Federal regulations are compiled in the Code of Federal Regulations (C.F.R.), which is published by the Government Printing Office and is also arranged by titles. Remember, the volume or title number always comes first. When regulations are first issued, they appear in the *Federal Register*. The *Federal Register* is published daily by the Government Printing Office and contains not only regulations and proposed regulations, but also notices about federal agency meetings, most of which are open to the public. The *Federal Register* contains no Sudoku, no crosswords, no comics, and no sports page; it is probably the single most boring journal in the world.

NOTES

1. The first series carries cases from 1880 to 1924, the second series carries cases from 1925 to 1993, and the third series carries cases from 1993 through the present. Also, not all cases are published. Each court decides which cases it wants published and which it does not.

2. *See* 1 U.S.C. § 204.

3. Not all statutes make it into the United States Code. When Congress passes a law, it indicates whether that new law or portions of it will be codified or not. All statutes enacted by Congress (whether codified or not) can be found in the *Statutes at Large*, published by the Government Printing Office since 1874.

APPENDIX B

Often-Used Abbreviations

AWA—Animal Welfare Act
CDC—Centers for Disease Control and Prevention
C.F.R.—Code of Federal Regulations
CRADA—Cooperative Research and Development Agreement
DAB—Departmental Appeals Board
DARPA—Defense Advanced Research Projects Agency
FACA—Federal Advisory Committee Act
FCA—False Claims Act
FDA—Food and Drug Administration
FOIA—Freedom of Information Act
FTTA—Federal Technology Transfer Act of 1986
HHS—Department of Health and Human Services
HIPAA—Health Insurance Portability and Accountability Act of 1996
IACUC—Institutional Animal Care and Use Committee
IBC—Institutional Biosafety Committee
ICMJE—International Committee of Medical Journal Editors
IDE—Investigational Device Exemption
IND—Investigational New Drug
IQA—Information Quality Act of 2000
IRB—Institutional Review Board
MTA—Material Transfer Agreement
NASA—National Aeronautics and Space Administration
NCI—National Cancer Institute
NHLBI—National Heart, Lung, and Blood Institute
NIAID—National Institute of Allergy and Infectious Diseases
NIH—National Institutes of Health
NOAA—National Oceanic and Atmospheric Administration
NSF—National Science Foundation
OBA—Office of Biotechnology Activities

OHRP—Office for Human Research Protections
OIG—Office of Inspector General
ONR—Office of Naval Research
ORI—Office of Research Integrity
OSI—Office of Scientific Integrity
PHI—Protected Health Information
PHS—Public Health Service
PTO—United States Patent and Trademark Office
RAC—Recombinant DNA Advisory Committee
U.S.C.—United States Code
USDA—United States Department of Agriculture

APPENDIX C

Links to Relevant Laws and Regulations

A. Research Misconduct (Select Departments and Agencies)
  1. Public Health Service—42 C.F.R. pt. 93
     http://www.access.gpo.gov/nara/cfr/waisidx_08/42cfr93_08.html
  2. National Science Foundation—45 C.F.R. pt. 689
     http://www.access.gpo.gov/nara/cfr/waisidx_08/45cfr689_08.html
  3. National Aeronautics and Space Administration—14 C.F.R. pt. 1275
     http://www.access.gpo.gov/nara/cfr/waisidx_09/14cfr1275_09.html
  4. Department of Energy—10 C.F.R. pt. 733 (Allegations) & § 600.31 (Definitions & Assurance)
     http://www.access.gpo.gov/nara/cfr/waisidx_09/10cfr733_09.html
     http://edocket.access.gpo.gov/cfr_2009/janqtr/10cfr600.31.htm

B. Common Rule (Select Departments and Agencies) & FDA Rules
  1. Department of Health and Human Services—45 C.F.R. pt. 46
     http://www.hhs.gov/ohrp/humansubjects/guidance/45cfr46.htm
  2. Food and Drug Administration Rules—21 C.F.R. pts. 50 and 56
     http://www.access.gpo.gov/nara/cfr/waisidx_08/21cfr50_08.html
     http://www.access.gpo.gov/nara/cfr/waisidx_08/21cfr56_08.html
  3. National Science Foundation—45 C.F.R. pt. 690
     http://www.access.gpo.gov/nara/cfr/waisidx_08/45cfr690_08.html
  4. Consumer Product Safety Commission—16 C.F.R. pt. 1028
     http://www.access.gpo.gov/nara/cfr/waisidx_09/16cfr1028_09.html
  5. Department of Agriculture—7 C.F.R. pt. 1c
     http://www.access.gpo.gov/nara/cfr/waisidx_08/7cfr1c_08.html
  6. Department of Commerce—15 C.F.R. pt. 27
     http://www.access.gpo.gov/nara/cfr/waisidx_08/15cfr27_08.html
  7. Department of Defense—32 C.F.R. pt. 219
     http://www.access.gpo.gov/nara/cfr/waisidx_08/32cfr219_08.html
  8. Department of Education—34 C.F.R. pt. 97
     http://www.access.gpo.gov/nara/cfr/waisidx_08/34cfr97_08.html

9. Department of Energy—10 C.F.R. pt. 745
   http://www.access.gpo.gov/nara/cfr/waisidx_08/10cfr745_08.html
10. Department of Housing and Urban Development—24 C.F.R. pt. 60
    Adopts the HHS Common Rule at 45 C.F.R. pt. 46
    http://www.access.gpo.gov/nara/cfr/waisidx_09/24cfr60_09.html
11. Department of Justice—28 C.F.R. pt. 46
    http://www.access.gpo.gov/nara/cfr/waisidx_08/28cfr46_08.html
12. Department of Transportation—49 C.F.R. pt. 11
    http://www.access.gpo.gov/nara/cfr/waisidx_08/49cfr11_08.html
13. Department of Veterans Affairs—38 C.F.R. pt. 16
    http://www.access.gpo.gov/nara/cfr/waisidx_07/38cfr16_07.html
14. Environmental Protection Agency—40 C.F.R. pt. 26
    http://www.access.gpo.gov/nara/cfr/waisidx_08/40cfr26_08.html
15. Agency for International Development—22 C.F.R. pt. 225
    http://www.access.gpo.gov/nara/cfr/waisidx_07/22cfr225_07.html
16. National Aeronautics and Space Administration—14 C.F.R. pt. 1230
    http://www.access.gpo.gov/nara/cfr/waisidx_09/14cfr1230_09.html

C. Conflict of Interest Materials

1. PHS Research Objectivity Regulations—42 C.F.R. § 50.601 *et seq.*
   http://www.access.gpo.gov/nara/cfr/waisidx_08/42cfr50_08.html
2. FDA Financial Disclosure by Clinical Investigators—21 C.F.R. pt. 54
   http://www.access.gpo.gov/nara/cfr/waisidx_08/21cfr54_08.html
3. NSF Research Integrity Policy
   http://www.nsf.gov/pubs/manuals/gpm05_131/gpm5.jsp#510

D. Animal Welfare Act Materials

1. Animal Welfare Act, as amended—7 U.S.C. § 2131 *et seq.*
   http://www.aphis.usda.gov/animal_welfare/downloads/awa/awa.pdf
2. Animal Welfare Act Regulations—9 C.F.R. pt. 2
   http://www.access.gpo.gov/nara/cfr/waisidx_08/9cfr2_08.html

APPENDIX D

Approaches to Problems

This is not an answer key to the problems in the text. Many of the problems have no right or wrong answers. Rather, the materials below highlight some of the issues that each problem raises and discuss those issues. You are invited to disagree with my conclusions, but if you do, provide cogent reasons for doing so.

CHAPTER 2

Case 1: The Case of the Sporting Scientist

This case is modeled after two incidents involving assistant secretaries at the Department of Health and Human Services in the mid-1980s. An assistant secretary is usually appointed by the president with the advice and consent of the Senate. One assistant secretary had a son who played professional football; she developed her speaking schedule based on the city her son would be playing that weekend. Her speeches were invariably given on Fridays. She was asked to resign because a federal employee's travel schedule, which is funded by the government, must be designed to further the mission and interests of the agency, not to accommodate personal desires. Ironically, her replacement also had a son who played football at the college level. The replacement engaged in precisely the same conduct as her predecessor, including asking organizations to invite her to speak in a given city on a specific Friday. She too was asked to resign. When it comes to spending federal funds, there is no reason to treat a principal investigator differently from a federal employee. The focus should always be on the predominant reason why a trip is being scheduled. If the predominant reason underlying the travel is personal (e.g., seeing a son play football) and the meeting or speech merely provides a convenient vehicle for having the grant pay the travel costs, then the travel should not be reimbursed by the grant. The fact that the cost of staying the extra day, which Defarge billed the grant, is offset by flying back on Sunday (at a much lower fare) rather than on Friday is not material. The focus should be the trip itself. The overarching question is would Defarge have scheduled the trip if her son were not playing football in that city on that weekend. If the answer is "no," then the cost of the entire trip should not have been billed to the grant or the university.

The trip to Acapulco raises different but related issues. First, it always looks bad to travel to a resort for a scientific conference, and therefore, grantees and PIs should carefully scrutinize such travel using the *New York Times* test. Would they want to see their government-paid travel to Acapulco appearing on the front page of the *Times*? Second, grantees should assess whether to take a trip based on its overall reasonableness—the cost of the travel versus the benefit. Here, the cost seems great and the benefit minor. Finally, using the federal grant funds to pay for third parties (boyfriend) is usually illegal.

Case 2: The Case of the Dictatorial Director

This problem is not based on real events, but it does raise questions about the types of behavior that may create serious ethical and legal issues. First, it is questionable whether grants or cooperative agreements should be used to fund the types of research discussed in this problem. Cookie-cutter research is more properly funded through procurement contracts, which are subject to greater scrutiny than either grants or cooperative agreements.

Second, favoritism—especially when it is part of a quid pro quo—is inappropriate and possibly criminal.

Third, receiving "kickbacks" from grantees is criminal, even if those kickbacks are merely ploughed back into the agency's account. This is because Congress appropriates funds for specific purposes (e.g., research), and kickbacks to the agency are being used here to circumvent that purpose. Here, the kickbacks are being used to transform funds appropriated for research into funds appropriated for good times. This sort of activity—"reprogramming without congressional authorization"—violates the Antideficiency Act[1] and is criminal. The fact that members of Congress are being wined and dined with these misappropriated funds also adds to the scandal.

CHAPTER 3

Case 1: The Mystery of the Missing Data

This case is designed to highlight various basic issues. First, does sharing data and results with a few colleagues, as part of a lab meeting, constitute "publishing" data for purposes of the misconduct rule? The answer is likely "yes." However, what about sharing data with a friend in the hallway of the lab building? Is that a publication? What about sharing data with a friend while drinking at a bar? Is that a publication?

Second, are the data false or fabricated? Here, there is a factual dispute, with the graduate student maintaining that his data are in fact accurate and that the data point he discarded was a true outlier. Does a researcher have the right to discard outliers? Was 109 a true outlier? Does the fact that he failed to discard the value 12, which is farther from the mean than 109 but which supported his hypothesis, influence your decision? How do you define an "outlier" for purposes of discarding the point? When you discard data, do you have record-keeping obligations?

The third issue relates to intent. The fact that Green tossed away a sheet with all of the data and was less than fully truthful supports the notion that he was acting to deceive. What about the fact that his original data were publishable? This fact is not relevant. For purposes of determining Green's state of mind, what counts is what he thought, and he thought that the data with 109 were not publishable.

In light of these facts, Powers probably has little choice but to report the incident to the appropriate university official as a possible case of scientific misconduct.

Case 2: Revenge of the Disgruntled Postdoc

This case blends the facts of various actual cases. Laboratory romances are not uncommon, but like any workplace affair, they are ill advised and, in many employment settings, against the rules. This case raises at least three basic issues: (1) so-called self-plagiarism, (2) anticipatory publication, and (3) rifling through files of others without asking permission. First, as discussed in the text, each coauthor of a jointly authored paper "owns" the whole paper, so using portions of it without attribution would constitute "self-plagiarism" which is an oxymoron. Many, though, would argue that in such a setting a researcher has an ethical obligation to cite to the paper listing all the authors. This point is certainly open to debate.

Second, it is normally improper to submit papers or proposals setting out results that you anticipate obtaining but have not yet obtained. Here, though, there is significant factual dispute over whether this occurred. The statement in the grant application is as follows: "We have already run some of these experiments on rabbits and they demonstrate that our method is viable." One can argue that when the statement is examined clause by clause it is literally true. Mulligan had run some of the experiments; therefore the first clause is true. Mulligan does not state that those experiments were successful, merely that they demonstrated the method's viability, whatever that means. The second clause is too vague and opinion-based to be either true or false. However, one can argue that the effect of the two clauses when taken together was to convey a false impression that the experiments worked. Should statements be judged for misconduct purposes on the imprecision that they convey or on what they literally state?

Third is the question of whether Frank's conduct is improper. Many would argue that it is a breach of privacy or etiquette, or both, to rifle through files that do not belong to you unless you have permission to do so. The fact that one's search is driven by a motive that is not pure only adds to the intrusion's improper character.

Case 3: The Problem of the Prolific Professors

This is based on an actual case with facts modified somewhat. At issue is whether universities and the government have any business enforcing private publication agreements between faculty and private journals, and whether these agreements, when they foreclose further publication of the data, are inconsistent with sound public policy. Also at issue is whether "multiple publication" is improper.

In my view there is nothing inherently improper with multiple publication, especially where it serves a legitimate purpose—one other than increasing the length of one's c.v. On the other hand, editors of the private journals will argue that they can survive only if the articles that they publish are "exclusive," and further, publishing the same data twice robs other scientists of the opportunity to publish once. The editors are correct that authors who engage in multiple publication may be selfish and egotistical, but that does not make the practice wrong, in theory.

Multiple publication can become improper when one intentionally breaches a contract with a publisher (e.g., one in which you agree not to publish the same data elsewhere for a specified period). However, even here one has to question whether a contract that limits one's ability to republish data (as opposed to reprinting the copyrighted article) is against public policy and ought not to be enforced.

Case 4: The Case of the Understanding Urologist

This case is modeled after the case concerning the National Surgical Adjuvant Breast and Bowel Project (NSABP) at the University of Pittsburgh, which was discussed briefly in chapters 2 and 6. At issue here is how to handle tainted data. For example, when a clinical trial is designed to accommodate an error up to a ceiling, and the error rate for a given center is less than that ceiling, does it matter what caused the errors? If you discard erroneous data not because of the nature of error but rather the motive of researcher, are you threatening the rigor of the statistics and possibly undermining the integrity of the clinical trial? These are tough questions with few clear answers.

CHAPTER 4

Case 1: The Case of the Compromised Collaboration

This is a case that is difficult to read without spotting legal or ethical issues at every turn. Under HIPAA an institution is not permitted to use a patient's protected health information for research purposes without the patient's authorization. Van Husen probably did not violate any federal ethics law by merely saving specimens with associated information. However, does he violate federal law by contacting patients to see if they are willing to participate in a study? He is contacting patients based on their protected health information for nontreatment purposes (e.g., to see if they will participate in a study). The IRB has even approved this. The IRB, though, cannot waive federal law. The question turns on whether Van Husen is considered a treating physician. If he is, then he may be free to call the patients to see if they want to sign the authorization. However, if he is not their treating physician, he could not call them without their approval (which can turn into a catch-22).

Van Husen's problems arise when he decides to collaborate with Crickson, who is at another institution. First, neither the patients nor the IRB granted Van Husen authority to conduct genetic research. So any genetic research is unapproved and unauthorized. Second, no one gave Van Husen any authority to share data. To the contrary, the informed consent form promised that the information would be kept confidential. Third, as to Crickson, he had not received approval from his IRB to receive the data or conduct the research.

Case 2: The Case of the Random Regime

With respect to the Common Rule, we first have to determine whether Vegas is conducting human subjects research. Even though he is treating the patients, the method of treatment is determined by a research protocol, albeit informal, and that is enough to transform Vegas's activities into research involving human subjects. Vegas has violated the Common Rule in a number ways. First, he did not have IRB approval. Second, subjects did not give informed consent. In fact, subjects were unaware that they were participating in a research project. Third, if patients did not read or speak English, the informed consent form would have had to be in the subject's native language or a language spoken by the subject, or the consent form would have had to be read to the subject in a language he or she understood. Fourth, conducting research on minors without the consent of a guardian or parent is improper under the Common Rule and likely illegal under the laws of most states.

Assuming that Vegas had presented the protocol to his IRB, is the protocol with respect to minors something that an IRB could approve? This research probably falls within the second category of risk: research involving greater than minimal risk, but presenting the prospect of direct benefit to the individual subject. This is so because both drugs are approved by FDA: Defenestratia has been approved by the FDA to treat Bollix Disease (BD) while OmniAll has been approved to treat a disease related to BD. The standard treatment—Defenestratia—has side effects, while the experimental treatment does not. Therefore, the subjects could be spared side effects.

The FDA rules present different issues. The FDA rules, which require a researcher to submit an Investigational New Drug application, do not apply if the following conditions pertain:

(i) The investigation is not intended to be reported to FDA as a well-controlled study in support of a new indication for use nor intended to be used to support any other significant change in the labeling for the drug;
(ii) If the drug that is undergoing investigation is lawfully marketed as a prescription drug product, the investigation is not intended to support a significant change in the advertising for the product;
(iii) The investigation does not involve a route of administration or dosage level or use in a patient population or other factor that significantly increases the risks (or decreases the acceptability of the risks) associated with the use of the drug product;
(iv) The investigation is conducted in compliance with the requirements for institutional review set forth in part 56 and with the requirements for informed consent set forth in part 50; and
(v) The investigation is conducted in compliance with the requirements of 21 C.F.R. § 312.7, which preclude the Sponsor of the research from charging for the drug.[2]

The FDA exemption does not apply because Vegas did not receive IRB approval nor did he obtain informed consent. It is also difficult to measure the risk of the protocol with respect to children.

Even though the exemption does not apply, has Vegas still violated the Food, Drug, and Cosmetic Act? One could argue that a physician is free to use a drug "off label" and FDA lacks jurisdiction to regulate how physicians use drugs. Vegas was using OmniAll "off label," namely to treat BD. Under normal circumstances, this would fall outside FDA's jurisdiction. However, Vegas was doing more than merely using the drug as a treatment—he was using it in a clinical trial off label. As such, FDA would arguably have jurisdiction, and Vegas theoretically could be charged with distributing a misbranded and adulterated drug.

The personal injury aspect of this case is interesting. The plaintiff may find it difficult to prove a causal link between the experimental treatment and the injury. Here, although the plaintiff was part of an experiment, he received the standard treatment rather than the experimental treatment. If there had been no experiment, more probably than not he would have received precisely the same drug that he did receive.

Case 3: The Case of the Naïve Nephrologist

This problem is based on two real cases that occurred at neighboring institutions in the Midwest and combines human subjects issues with research integrity issues. First, Newcomb and her IRB have problems because their approved protocol calls for eligibility tests to be administered to patients before an informed consent had been obtained. This is relatively minor when compared with the rest of the events.

Second, is Newcomb responsible for the actions of her subordinates, especially when she was unaware of what one of those subordinates was doing? FDA takes the position that the primary investigator—Newcomb—is responsible for the actions of everyone on the team. If one of the subordinates does something wrong, then that is Newcomb's responsibility. FDA argues that Newcomb agreed to live by this standard when she signed the Form 1572, which stated: "I agree to personally conduct or supervise the described investigation(s)." FDA would then argue that any deviations from the protocol, as occurred here, are the result of substandard supervision.

On the other hand, one could argue that the FDA standard is unrealistic, and, if applied to FDA, its commissioners would be forced to resign on a monthly basis. For example, using the agency's logic, if tainted peanuts are missed by an FDA inspector and end up killing children (as actually occurred), then the commissioner of food and drugs ought to resign since all authority flows from the commissioner. Here, Newcomb arguably should have fired Wagner because her error rate was unacceptable.

Third, FDA would argue that by not taking sonograms at the six-month interval, Newcomb had deviated from the protocol. This is nothing more than a recap of the charge of failure to supervise. It is interesting to note that Newcomb could possibly have checked those sonograms and did not. How was she supposed to check blood pressures and temperatures? Was she supposed to retake them? If she was, then what is the purpose of having subordinates?

CHAPTER 5

Case 1: The Case of the Curious Conflict

This is a cross-over question involving both human subjects and conflict of interest issues. First, are Schute and Ladder engaged in human subjects research requiring IRB approval and informed consent? Drawing blood is not exempt from the normal IRB rules, and using that blood to sequence a genome or a portion of one's genome is research and subject to the Common Rule if it involves "humans." As discussed in chapter 4, there is an arguable debate as to whether "self-experimentation" constitutes human subjects research. What if Schute draws Ladder's blood and vice versa? In my view, that is not the crux of this research; the sequencing creates the risks. Therefore, does Schute sequence his own genome? If he does his own, then this probably is self-experimentation. If Ladder sequences Schute's and vice versa, then it is questionable whether it is still self-experimentation.

There is less debate, however, when Schute and Ladder involve their employees. Once others are involved—even lab employees—the research is no longer self-experimentation. One can argue that greater risks are present when employees are being asked to participate because of the element of coercion, whether spoken or otherwise. Once we determine that the lab employees are subjects, then the full array of safeguards in the Common Rule apply—IRB approval, informed consent, and HIPAA authorization (if CTI is a covered entity). Schute and Ladder ignored all of these requirements, and therefore likely violated HHS's and NSF's Common Rule, assuming the research was funded by HHS and NSF.

What about altering research to accommodate one's financial self-interest? Do Schute and Ladder have a significant financial interest? Probably yes, if their interests in M^3 exceed $10,000 each. If we decide they do have a significant financial interest, then we must determine whether it is reportable. It is reportable if

(i) the significant financial interest would reasonably appear to be affected by the PHS/NSF funded research; *and*
(ii) the interest is in entities whose financial interests would reasonably appear to be affected by the research.

Here, both prongs of the test are met, but not in the way the effect would normally occur. If the experiments move forward and are successful, then Schute and Ladder's financial interests in M^3 would decrease as would the financial well-being of M^3. The Common Rule looks to whether there is a likely effect and not the direction of that effect. We do not know whether Schute or Ladder reported their respective interests to CTI. If they had, then CTI would have to determine whether a "reportable significant financial interest" creates a conflict or potential conflict of interest. A conflict is deemed to exist when the designated official "reasonably determines that the Significant Financial Interest *could directly and significantly* affect the design, conduct, or reporting of the PHS [or NSF] funded research."[3] What if the designated CTI official decided that there was no conflict? Is it proper under the regulations for Schute and Ladder to have changed the course of their research for commercial gain? Morally and scientifically, the answer is "no," but it is less clear whether the conduct is illegal.

Schute and Ladder arguably have violated the scientific misconduct rules of HHS, NSF, and CTI by lying about the data at a lab meeting. As we discussed earlier, lab meetings are likely sufficiently formal to trigger the misconduct rules. By firing Wise, Schute and Ladder also violated the whistleblower laws concerning misconduct, possibly violated the Rehabilitation Act of 1973 that bars discrimination in federally funded projects based on a disability or perceived disability, and possibly violated various state laws as well.

Case 2: The Case of the Painful Panacea

This case involves both financial conflicts and intellectual conflicts. By revealing his financial interests in a competing modality, McCash has discharged his obligations under the federal conflict rules—the research rules and rules governing those who sit on federal advisory committees. McCash, though, has an overriding intellectual and perhaps ego conflict. By failing to reveal his subsequent inconsistent results, has he violated a norm? Should FDA have forced McCash to recuse himself from voting in the meeting? If FDA did that, though, they would be removing from the panel someone with a keen understanding of the underlying biology. Does McCash have a responsibility to report that his prior research was incorrect, or is he free to republish his original letter knowing that it is incorrect? The real issue is whether McCash has violated the False Statements Act, which makes it a crime for anyone to knowingly tell a falsehood with respect to a matter within the jurisdiction of a federal agency. If he just kept his mouth shut and voted, then there would be no violation. However, by trumpeting the letter, has he violated the law? Merely referring to the results in a published article—even knowing that the results are wrong—is likely not a crime. Stating affirmatively, though, that the results are correct knowing they are not is potentially a crime.

Case 3: The Case of the Stock-Swapping Scientist

This study is not federally funded, and therefore neither the HHS nor NSF conflict rules would apply. But NICE U. has rules similar to the ones adopted by HHS and NSF, and those rules apply to all research irrespective of the funding source. The fact that Pabst accepts payment personally, rather than through the university, for conducting research on the university's premises raises separate issues (e.g., using university property for private gain). The drug company is paying him to conduct research, and that payment ($1,000 per patient in Rex stock) constitutes a "significant financial interest." The minute he signs the contract, he has a reasonable expectation of receiving more than $10,000 from Rex over the next twelve months; that constitutes a significant financial interest. It is also reportable because the value of his Rex stock will be affected by the outcome of the research. And if NICE U. is like most institutions, it would not have permitted Pabst to conduct the clinical trial since he is receiving separate remuneration directly from the company. If NICE U. permitted him to participate in the study, the informed consent form would likely have been modified by the IRB to reflect Pabst's financial interests in Rex.

The two sales of his stock also raise questions under the securities laws.

Case 4: The Case of the Consulting Theoretician

Swann certainly has significant outside income. Each item of income or equity easily qualifies as a "significant financial interest." However, are any of those interests "reportable"? None of his research bears any relationship to either his income or equity, and therefore there is nothing to report. It is difficult to see any relationship between losing weight and metabolism, on the one hand, and string theory, on the other hand. He may be violating some ethics code of the television and radio stations or programs, but those codes are private codes and have nothing to do with Swann's research. The viewers of those programs may wish to know that Swann is getting paid by a drug company to give talks to physicians, but Swann has no duty to provide that information to the television stations. After all, they never reveal to us the holdings of their anchors or writers.

CHAPTER 6

The Case of the Careless FOIA Officer

A variant of this actually occurred when a company, much like Data Grabb, offered for a substantial sum cost and pricing information that had been erroneously released to it by a brain-dead FOIA officer.

First, in theory the FOIA officer could be subject to criminal prosecution for violating 18 U.S.C. § 1905, which makes it a crime for a government employee to knowingly reveal trade secret information. The cost and pricing information may not qualify as trade secret information, but the chemical formulae would likely qualify. The fact that they were labeled as a "trade secret" makes the FOIA officer's actions more difficult to defend.

Second, the government itself could be subject to suit under the Federal Tort Claims Act, which permits a private company to sue the U.S. government for negligence—although we do not know enough to assess the viability of such a suit. Also, it may be possible for Wright-Orville to sue the government for "taking" its property without due process. A trade secret is considered property, and by releasing the trade secret publicly, the government potentially destroyed its value.

Third, Orville probably has a claim against Data Grabb for holding property that it knew was a trade secret and commercial confidential information and not returning that property to the government or the owner.

In the real case, the government put pressure on the FOIA company to stop advertising the information and to return it to the government. It should be noted that any competitor that buys the information and uses it to bid against Orville could be disqualified. Obtaining and using private information can form the basis for disqualification for bidding on a procurement contract.

CHAPTER 7

Case 1: The Case of the Culpable Chemist

This case highlights the difference between patent law and science. What Pillpher researchers did was reprehensible and illegal. One can view what they did as stealing and then publicly revealing a trade secret. This essentially eliminates the value of the trade secret. In theory, Linus and the university could sue Pillpher for damages.

But what effect does it have on the validity of the patents? It may have surprisingly little impact on the patents themselves. First, one could argue that the Pillpher researchers had enough—the concept that a specific chemical could retard and reverse presbyopia—to satisfy the "conception" requirement for purposes of getting a patent. They had not been able to reduce their invention to practice by showing that their theory was correct. They used the purloined materials to do this. This, however, would affect at most the validity of the use patent.

Second, the method for producing the chemical, which was the subject of a separate patent, was not stolen from Linus. It was independently developed and was not based on Linus's materials. The fact that Linus developed the method first would not invalidate the Pillpher patent because Linus never published his result. Also, Linus probably waited too long to file and prosecute his own patent. If he filed a patent application on the process for synthesizing the chemical, that would lead to an interference between the two patents. To prevail in an interference, Linus would have had to show, among other things, that he diligently prosecuted the patent. Waiting six years to file an application would likely not come close to satisfying this requirement.

This case is similar in certain respects to the litigation between Genentech, Inc., the South San Francisco pharmaceutical company, and the University of California. In that case, UC claimed that Genentech agents broke into a UC laboratory, purloined laboratory materials, and used those materials as the basis of patents. Genentech denied the allegations, and the matter eventually settled.

Case 2: The Invention of the Mirror Image

One might conclude that the chemicals themselves have to be useful and the putative patentee has to show what that use is. Here, there are two types of patents—a patent for the process of creating a chemical and a patent for the chemical itself. For the process patent, the utility lies in synthesizing the chemical itself. One could argue that it does not matter that the chemical might have no known use; the process, though, does have a use, namely creating the chemical. However logical that view might be, it was squarely rejected by the Supreme Court in *Brenner v. Manson*.[4] There, the Court held that the chemical itself must have some use. This holding necessarily dooms Bond's patent on the chemical itself, unless he can posit some use for that chemical.

The founder of one drug company actually did what is portrayed in the problem, but was able to posit some utility for the chemical produced through each process that he patented, or in some cases, uses that he patented.

CHAPTER 8

Case 1: The Case of the Double-Crossed Dogs

The animal research reviewed and approved by the Institutional Animal Care and Use Committee (IACUC) is not the same animal research that Krebs conducted. The number of dogs increased and their fate changed. Both would have required another review by IACUC. Since Krebs is now sacrificing dogs, the IACUC would want to know whether similar research had been done elsewhere and may have had other questions as well. Also, with respect to animals, the grant application must be relatively specific, specifying the species and number of animals to be involved in the research and the nature of the research. The information in the grant application differs from what Krebs actually did.

Finally, MAL has a vivarium, and usually it is responsible for purchasing all research animals. That policy has likely been incorporated into MAL's assurance to HHS that it will abide by the Extension Act and the Guide. Also, it is likely that in acquiring the additional dogs from the pound, Krebs did not advise the pound that he intended to use the animals in an experiment and then dispatch them. In short, there is a good chance he acquired them under false pretenses. The regulations implementing the AWA prohibit obtaining "live dogs, cats, or other animals by use of false pretenses, misrepresentation, or deception."[5]

Krebs therefore violated the AWA and the Extension Act.

Case 2: The Case of the Seafaring Sharks

It is open to question whether the AWA applies to Smith's research. The AWA applies to a "research facility" that "uses or intends to use live animals in research, tests, or experiments, and that (1) purchases or transports live animals in commerce. . . ."[6] Further, the term "animal" means only "warm-blooded" animals. "The term 'animal' means any live or dead dog, cat, monkey (nonhuman primate mammal), guinea pig, hamster, rabbit, or such other warmblooded animal, as the Secretary may determine is being used, or is intended for use, for research, testing, experimentation, or exhibition purposes, or as a pet. . . ."[7] It is doubtful that the research would come within the AWA for two reasons. First, a shark is not a warm-blooded animal. And second, it is questionable whether Smith's boat would qualify as a "research facility" subject to the AWA because it is questionable whether she is "transporting" the sharks in commerce. All of her activities are on the high seas outside of U.S. waters. One could argue that the boat might be viewed as U.S. territory, but even so, there is a question whether she is engaged in "commerce." The term "commerce" implies interstate commerce or foreign commerce (between the United States and a foreign country). It is open to question whether research on the high seas, not funded by any federal agency, would fall within the U.S. Constitution's commerce clause.

NOTES

1. *See* 31 U.S.C. §§ 1341 and 1517, which impose the substantive limitations on using funds appropriated for one purpose for another; and 31 U.S.C. §§ 1350, 1519, which makes that conduct criminal (an officer or employee who "knowingly and willfully" violates any of the three provisions cited above shall be fined not more than $5,000, imprisoned for not more than two years, or both).

2. 21 C.F.R. § 312.2(a).

3. 42 C.F.R. § 50.605.

4. 383 U.S. 519 (1966).

5. 9 C.F.R. § 2.132(b).

6. 7 U.S.C. § 2132(e).

7. *Id.* § 2132(g).

Index